MERCURY CONTAMINATION: A Human Tragedy
Patricia A. D'Itri and Frank M. D'Itri

POLLUTANTS AND HIGH RISK GROUPS
Edward J. Calabrese

METHODOLOGICAL APPROACHES TO DERIVING ENVIRONMENTAL AND
OCCUPATIONAL HEALTH STANDARDS .
Edward J. Calabrese

NUTRITION AND ENVIRONMENTAL HEALTH—Volume I: The Vitamins
Edward J. Calabrese

NUTRITION AND ENVIRONMENTAL HEALTH—Volume II: Minerals and Macronutrients
Edward J. Calabrese

SULFUR IN THE ENVIRONMENT, Parts I and II
Jerome O. Nriagu, Editor

COPPER IN THE ENVIRONMENT, Parts I and II
Jerome O. Nriagu, Editor

ZINC IN THE ENVIRONMENT, Parts I and II
Jerome O. Nriagu, Editor

CADMIUM IN THE ENVIRONMENT, Parts I and II
Jerome O. Nriagu, Editor

NICKEL IN THE ENVIRONMENT
Jerome O. Nriagu, Editor

ENERGY UTILIZATION AND ENVIRONMENTAL HEALTH
Richard A. Wadden, Editor

FOOD, CLIMATE AND MAN
Margaret R. Biswas and Asit K. Biswas, Editors

CHEMICAL CONCEPTS IN POLLUTANT BEHAVIOR
Ian J. Tinsley

RESOURCE RECOVERY AND RECYCLING
A. F. M. Barton

QUANTITATIVE TOXICOLOGY
V. A. Filov, A. A. Golubev, E. I. Liublina, and N.A. Tolokontsev

ATMOSPHERIC MOTION AND AIR POLLUTION
Richard A. Dobbins

CHEMISTRY AND
ECOTOXICOLOGY
OF POLLUTION

CHEMISTRY AND ECOTOXICOLOGY OF POLLUTION

DES W. CONNELL
GREGORY J. MILLER

School of Australian Environmental Studies
Griffith University

A Wiley-Interscience Publication
JOHN WILEY & SONS
New York • Chichester • Brisbane • Toronto • Singapore

Copyright © 1984 by John Wiley & Sons, Inc.

All rights reserved. Published simultaneously in Canada.

Reproduction or translation of any part of this work
beyond that permitted by Section 107 or 108 of the
1976 United States Copyright Act without the permission
of the copyright owner is unlawful. Requests for
permission or further information should be addressed to
the Permissions Department, John Wiley & Sons, Inc.

Library of Congress Cataloging in Publication Data:

Connell, D. W.
    Chemistry and ecotoxicology of pollution.

    (Environmental science and technology, ISSN 0194-0287)
    "A Wiley-Interscience publication."
    Includes bibliographical references and index.
    1. Pollution–Environmental aspects.   2. Pollution–Toxicology.
    I. Miller, Gregory J.   II. Title.   III. Series.

QH545.AlC65   1984   628.5   83-16794
ISBN 0-471-86249-5

Printed in the United States of America

10 9 8 7 6 5 4 3 2 1

SERIES PREFACE
Environmental Science and Technology

The Environmental Science and Technology Series of Monographs, Textbooks, and Advances is devoted to the study of the quality of the environment and to the technology of its conservation. Environmental science therefore relates to the chemical, physical, and biological changes in the environment through contamination or modification, to the physical nature and biological behavior of air, water, soil, food, and waste as they are affected by man's agricultural, industrial, and social activities, and to the application of science and technology to the control and improvement of environmental quality.

The deterioration of environmental quality, which began when man first collected into villages and utilized fire, has existed as a serious problem under the ever-increasing impacts of exponentially increasing population and of industrializing society. Environmental contamination of air, water, soil, and food has become a threat to the continued existence of many plant and animal communities of the ecosystem and may ultimately threaten the very survival of the human race.

It seems clear that if we are to preserve for future generations some semblance of the biological order of the world of the past and hope to improve on the deteriorating standards of urban public health, environmental science and technology must quickly come to play a dominant role in designing our social and industrial structure for tomorrow. Scientifically rigorous criteria of environmental quality must be developed. Based in part on these criteria, realistic standards must be established and our technological progress must be tailored to meet them. It is obvious that civilization will continue to require increasing amounts of fuel, transportation, industrial chemicals, fertilizers, pesticides, and countless other products; and that it will continue to produce waste products of all descriptions. What is urgently needed is a total systems approach to modern civilization through which the pooled talents of scientists and engineers, in cooperation with social scientists and the medical profession, can be focused on the development of order and equilibrium in the presently disparate segments of the human environment. Most of the skills and tools that are needed are already in existence. We surely have a right to hope a technology that has created such manifold environmental problems is also capable of solving them. It is our hope that this Series in Environmental Science and Technology will not only serve to make this challenge more explicit to the established professionals, but that it also will help to stimulate the student toward the career opportunities in this vital area.

Robert L. Metcalf
Werner Stumm

PREFACE

The control of pollution is now one of the most serious problems in environmental management. Pollution arises from a wide variety of sources and has increased with growth in production of goods, services, and population. It occurs in all countries and affects localized areas, regions, and, in some cases, the entire ecosphere.

Socio-economic and political forces play a major role in the control of pollution. Nevertheless, to allow pollution to be seen in its correct perspective and devise appropriate control measures, we need a clear understanding of its nature and effects. This book aims to provide a basic understanding of the chemical, toxicological, and ecological factors involved when the major classes of pollutants act on natural systems. It does not deal with the direct effects on humans and human health since this is another topic requiring full and detailed consideration.

A chemical and ecotoxicological approach is taken in that the nature and effects of pollutants are considered from their sources and chemical properties, through dynamics in the environment to toxic and other detrimental effects on organisms and ecosystems. A vast number of scientific papers and books are available on many different aspects of pollution. However, only a limited number address the basic scientific principles involved in the various pollution processes. In this book the available information has been collated and evaluated leading to the development of some unifying theories as to the fundamental chemical and ecological nature of pollution processes.

We believe a book with this approach is timely. Although most research has been directed toward short-term management objectives, a substantial body of information is available and it is appropriate that this be assessed to provide directions for future activities. There is little doubt that pollution research, particularly ecotoxicology, will be further developed in future years as government initiatives in controlling chemicals take effect.

In writing this book we were assisted in a variety of ways by many people — in fact, too many to appropriately acknowledge here. However, there are some to whom we owe a special thanks. Brian Gilbert, actively assisted by the University Library, thoroughly searched the literature for reference material, thus providing us with a firm working base. The long and difficult manuscript was patiently and efficiently typed by Judith Davies and Denise Bigwood while Aubrey Chandica, Richard Blundell, and Annette Henderson skillfully prepared the diagrams. Diane Barron was most helpful in obtaining permissions, proofreading, and so on. However, without the willing assistance of the School of Australian Environmental Studies this book could not have been completed. Finally, we have greatly appreciated the active support and encouragement of our wives, Patricia Connell and Gillian Miller.

DES W. CONNELL
GREGORY J. MILLER

Brisbane, Australia
October 1983

CONTENTS

PART TWO.　CHEMICAL BEHAVIOR AND ECOTOXICOLOGY OF POLLUTANTS

CHEMISTRY AND
ECOTOXICOLOGY
OF POLLUTION

1

INTRODUCTION

Our apparently limitless habitat has long been taken for granted with its supposedly vast capacity to absorb wastes. In recent years it has become apparent that our environment has limitations to the waste it can destroy or dilute to insignificance. Atmospheric wastes aggregate around our urban centers while domestic, agricultural, and industrial wastes contaminate our soils and waters.

The response to this situation has been a massive research effort and many waste control measures have been introduced in recent years. Nevertheless, many socio-economic and political factors need to be addressed, as well as scientific and engineering matters, to arrive at an effective solution to pollution problems.

Pollution occurs when substances resulting from human activities are added to the environment, causing a detrimental alteration to its physical, chemical, biological, or aesthetic characteristics. Of course, all nonhuman organisms also produce wastes which are discharged to the environment, but these are generally considered part of the natural system, whether they have detrimental effects or not. Pollution is usually considered to occur as a result of human actions. However, natural processes can result in situations in the natural environment which closely resemble those due to pollutants.

It is important to note that the occurrence of pollution requires a subjective judgement as to whether a detrimental effect has resulted or not. In fact, there can be conflicting opinions on this. For example, when plant nutrients are discharged to waterways, this causes an increase in the number of plants present and often a resultant increase in fish numbers. Thus, a fisherman would regard this action as beneficial and therefore not pollution. On the other hand, domestic water supply authorities may find that the algal content of water would increase and remedial measures are required to provide adequate quality of domestic water. Thus these authorities would consider that pollution had occurred. Since the subjective judgments are made by humans, the "human factors" are critical in pollution management.

The maintenance of water, air, and soil quality is a central concern of environmental management agencies. Quality can only be defined in terms of a perceived use and quality for one purpose may not be suitable for another. This may be simply a difference in the quantities involved or it may be that entirely different factors are considered. For example, fecal coliform contamination is very important for domestic water supply but of little significance for irrigation purposes. Thus overall, the maintenance of environmental quality may be seen in entirely different terms by different environmental management agencies, responsible for the maintenance of quality for different uses.

In the past, water pollution management has concentrated on protection of water for domestic purposes, stock consumption, agriculture, industry, and to a lesser extent fisheries. In recent years, in addition to those uses, environmental management agencies are now concerned with the protection of water quality for the conservation of natural ecosystems, recreation, and aesthetic purposes. These latter purposes pose some new difficulties in management. Recreation and aesthetic purposes require subjective judgment and, thus, standards and criteria are difficult to define. The conservation of natural ecosystems requires a detailed understanding of the long-term effects on natural ecosystems. This understanding is limited at the present time.

Scientific investigation of the effects of pollutants in the natural environment has focused on the effects on individual biological species. Such investigations can only give suggestions as to the likely effects on the complex ecosystems existing in the natural environment. Recently, there has been increasing interest in the effects of pollutants on whole ecosystems, since these are of primary management concern. This has led to the development of the relatively new field of *ecotoxicology* — an infant science involving a new concept and approach to toxicology (see Butler, 1978).

In this book the ecotoxicology of pollutants is considered as a sequence of interactions and effects controlled by their chemical and physical properties, as indicated in Figure 1.1. A pollutant discharged into the environment can be subject to physical dispersal in the atmosphere, water, or soils and sediment depending on its physicochemical properties. At the same time, it can be chemically modified and degraded by abiotic processes or more often by microorganisms present in the environment. Often the degradation products are harmless but on occasion they can, of themselves, have a detrimental impact greater than the original pollutant. In some cases, the environment can be modified by the degradation process rather than by the pollutant itself. For example, in water bodies degradation of organic materials, such as carbohydrates, results in the loss of dissolved oxygen in the water mass due to the increased activity of microorganisms. The organisms present are now exposed to the toxicant in its degraded or transformed state or react to the conditions produced by the presence of the pollutant.

Organisms exhibit a variety of different reactions from very little to sublethal effects such as reduced growth, reproduction, behavioral effects, or ultimately lethality. The complex natural ecosystem of which the organisms are an integral part can react in a variety of ways to the effects on the component organisms. Food chain relationships, energy flows, and so on may be altered.

A somewhat different process occurs with the addition of plant nutrients to a water body. The organism response is stimulation of plant growth leading to a modification of the ecosystems characteristics which can be deleterious to some organisms present. It can also be quite severely damaging to water quality as considered for a variety of usages.

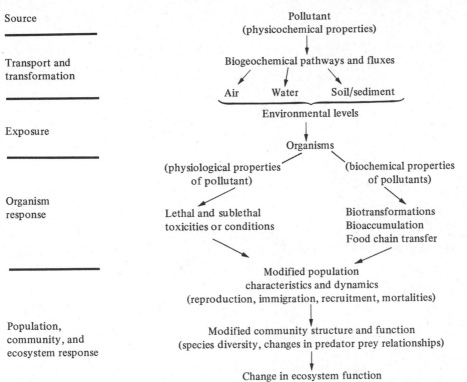

Source

Transport and
transformation

Exposure

Organism
response

Population,
community, and
ecosystem response

Pollutant
(physicochemical properties)

Biogeochemical pathways and fluxes

Air   Water   Soil/sediment

Environmental levels

Organisms

(physiological properties
of pollutant)

(biochemical properties
of pollutants)

Lethal and sublethal
toxicities or conditions

Biotransformations
Bioaccumulation
Food chain transfer

Modified population
characteristics and dynamics
(reproduction, immigration, recruitment, mortalities)

Modified community structure and function
(species diversity, changes in predator prey relationships)

Change in ecosystem function
(respiration to photosynthesis ratio, nutrient cycling rates,
patterns of nutrient flow)

Figure 1.1.   Diagrammatic illustration of the impact of pollutants on the components and functions of natural ecosystems. (Some important characteristics causing, or modified by, change indicated in parentheses.)

The basic sequence in Figure 1.1 has been used as the framework for the organization of this book. This treatment is based on the actual sequence of events in the environment starting with the initial pollutant and following its environmental behavior as a consequence of its chemical and physical properties. Thus Part One considers the principles of how the chemical and physical properties of pollutants influence transport and transformation, toxicity to individuals, and effects on ecosystems. This approach allows pollutants to be placed in classes with similar environmental characteristics and simplifies discussion of the many substances that can be pollutants. Part Two takes the various classes and considers each according to the same sequence (see Figure 1.1). However, in each class the content is varied according to the nature of the pollutant, which may require an emphasis on certain aspects, and the information available. Part Three outlines how knowledge of the chemistry and ecotoxicology of pollution can be utilized in environmental management.

REFERENCES

Butler, G. C. (1978). Principles of Ecotoxicology, SCOPE 12, John Wiley and Sons, New York.

PART ONE

PRINCIPLES GOVERNING THE INTERACTION OF POLLUTANTS WITH NATURAL SYSTEMS

2

CHEMODYNAMICS OF
POLLUTANTS

The behavior and effects of environmental pollutants are related to their dynamics in the four major compartments or phases which constitute the earth's ecosphere, that is, air, water, soil, and biota. Thus, the properties of chemicals and the mechanisms that control their forms and distribution within and between environmental compartments are of importance. Aspects of particular interest are those that control the rates of chemical and energy transfer within environmental compartments and across the major interfaces between compartments.

SOME FUNDAMENTAL PRINCIPLES OF POLLUTANT BEHAVIOR

Environmental Interfaces and Chemical Equilibria

An environmental interface can be described as the junction where two different compartments meet and interact across a common boundary. Spontaneous transfer of chemical and thermal energy occurs across interfaces between compartments or phases until net movement of chemicals and thermal energy ceases at equilibrium. However, chemical species continue to cross the interface but the rates of transfer between the two phases are equal.

If we assume a chemical equilibrium results from reversible transfers between environmental phases, thermodynamic considerations require that the chemical potentials (μ) or fugacities (f) of a component A in both phases of a multicomponent system be equal for constant temperature and pressure (Thibodeaux, 1979). Thus for component A between two phases, air (a) and water (w):

$$f_{Aa} = f_{Aw} \tag{1}$$

In each phase the fugacity of component A can be related to the activity coefficient (γ), mole fractions X or Y, concentration C, phase molar volume V_A, and the reference fugacity f° as

$$f_A = X_A \gamma_A f_A^\circ = C_A V_A \gamma_A f_A^\circ \tag{2}$$

Thus a more useful form of (1) can be derived from (2) as

$$(Y_A \gamma_A f_A^\circ)_a = (X_A \gamma_A f_A^\circ)_w$$

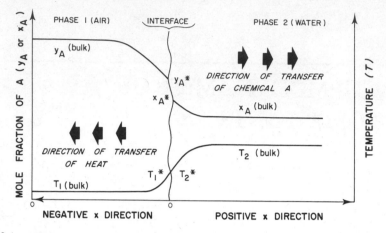

Figure 2.1.   Heat and mass transfer through environmental interfaces. [From Thibodeaux (1979). Reprinted with permission from John Wiley & Sons, Inc.]

The equilibrium distribution of a chemical substance between two immiscible phases occurs as a fixed proportion partition coefficient, which depends on solubility in each phase and is independent of the quantity of solute.

The transfer of heat and mass through environmental interfaces can be further defined in terms of equilibrium and nonequilibrium concepts, as illustrated by Thibodeaux (1979) in Figure 2.1. Profiles of concentration and temperature in the region near the interface represent time-averaged values. But equilibrium, both chemical and thermal, is assumed to exist at the interface itself. Thus the chemical concentration profile is usually discontinuous at the interface since concentrations usually differ at equilibrium. On the other hand, the thermal profile is continuous since thermal equilibrium occurs when the temperatures of two regions at the interface are the same.

The bulk phases, that is, the regions far from the interface, are not at thermal or chemical equilibrium so interface movement of heat and mass spontaneously proceeds. The mass of material associated with the interface regions where temperature and concentration gradients exist is a diminishingly small fraction of the total mass, and, by definition, the equilibrium interface itself is assumed to be a hypothetical physical region two molecules thick, one monomolecular layer in each phase (Thibodeaux, 1979). In natural systems, although not usually truly reversible, it is realistic to assume an equilibrium between compartments in appropriate situations. This provides a basis for the theoretical evaluation of transport and transformation of chemicals and energy in the environment.

Abiotic Transport and Transformation Processes

The transport and transformation of pollutants in the environment are related to: (1) physicochemical properties of the pollutants, (2) transport processes in the

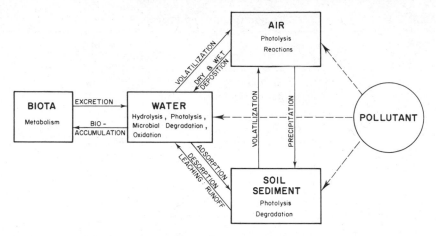

Figure 2.2. Transport and transformation processes for pollutants. [Adapted from Haque et al. (1980). Reprinted with permission from Ann Arbor Science Publishers, Inc.]

environment, and (3) pollutant transformation processes. The introduction of a chemical into the environment will result in intercompartmental transfers to establish equilibrium which is dependent on the physicochemical properties of the substance. For example, in the movement of a chemical across the water–soil interface, properties such as solubility, partition coefficients, and heats of solution are significant factors (Tinsley, 1979).

Within a compartment, that is, air, water, soil, and biota, movement of a chemical is primarily a function of the characteristic transport processes of that compartment. For example, aqueous systems move substances to the extent that water is moving whether or not the chemical is in solution or adsorbed onto a particle. Accordingly, its movement would be defined by the appropriate hydrological parameters (Tinsley, 1979). Finally, there are several transformation processes which function individually or simultaneously to convert a chemical from one form to another or to degrade the chemical. These processes are illustrated in Figure 2.2.

Environmental Fate of Chemicals

There are a number of environmental processes which are important in controlling the fate of chemicals in the aquatic, atmospheric, and terrestrial environments (see Table 2.1). The net concentration of a chemical may be defined in terms of the kinetic processes and equilibrium rates of the major processes acting on the substance at a given location (Mill, 1980). This procedure is based on a description of all major pathways of movement and transformation of a chemical, and its transformation products, including concentration as a function of time and location. Kinetic and equilibrium rates are widely used as quantitative measures for evaluating the contributions of the various individual processes (Mill, 1980). The procedure requires two basic assumptions: (1) The concentrations of chemicals and their

Table 2.1.   Environmental Compartments and Processes

Compartment	Transport	Transformation
Air	Meteorological transport: diffusion and dispersion Precipitation/Fallout	Photolysis Oxidation
Water	Sorption Volatilization Bio-uptake	Photolysis Hydrolysis Oxidation Metabolism/biodegradation
Soil	Sorption and sediments Runoff Volatilization Leaching Bio-uptake	Hydrolysis Oxidation Photolysis Reduction Metabolism/biodegradation

Source: Mill (1980). Reprinted with permission from Ann Arbor Science Publishers, Inc.

loss rates at any location and time are determined by combinations of independent rate or distribution processes. (2) For each process a quantitative relationship exists between chemical and environmental properties.

With the latter assumption, almost all processes in the environment which control rates and concentrations involve interaction of a pollutant with some environmental property. Table 2.2 lists some examples of environmental processes and controlling environmental properties. If chemicals are present in the environment in very low concentrations, the rate of reaction for a specific environmental process depends usually on the first power of concentration, and the kinetic-rate law takes the following form:

$$R_n = \frac{d[A]}{dt} = k_n[A][P]_n$$

where R_n is the rate of process n; k_n the rate constant which characterizes that process; [A] the concentration of chemical A, and [P] the environmental property for the nth process expressed in concentration units compatible with k. The values are characteristic of specific environments. Thus, the net rate of change in concentration of chemical A, that is, the sum of all equilibrium and kinetic processes, can be given as

$$R_T = k_n[A][P]_n \tag{3}$$

where R_T is the net rate of change and k_n the rate constant for the nth process. This expression is simplified by assuming that the environmental property P_n, is constant or $[\Delta P] \ll [P]$, which is usually the case in low-level polluted systems. Equation (3) then becomes a set of simple first-order processes:

$$R_T = \sum k'_n, \quad [A] = k_T[A]$$

Table 2.2.   Environmental Processes and Properties

Process	Property[a]
Physical Transport	
Meteorological transport	Wind velocity
Bio-uptake	Biomass
Sorption	Organic content of soil or sediments, mass loading of aquatic systems
Volatilization	Turbulence, evaporation rate, reaeration coefficients, soil organic content
Runoff	Precipitation rate
Leaching	Adsorption coefficient
Fall out	Particulate concentration, wind velocity
Chemical	
Photolysis	Solar irradiance, transmissivity of water or air
Oxidation	Concentrations of oxidants and retarders
Hydrolysis	pH, Sediment or soil basicity or acidity
Reduction	Oxygen concentration, ferrous ion concentration and complexation state
Biological	
Biotransformation	Microorganism population and acclimation level

Source: Mill (1980). Reprinted with permission from Ann Arbor Science Publishers, Inc.

[a] At constant temperature.

where k_n' is the pseudo-first-order constant for the nth process, and

$$k_T = k_a + k_b + \cdots k_n$$

The assumption of first-order conditions is a practical approach to estimating the contributions of individual rate processes to the total rate of loss.

Estimates of Environmental Persistence

For the single application of a pollutant to a compartment, first-order loss conditions can apply giving the rate of loss as

$$\frac{d[A]}{dt} = -k_T[A]$$

From this the following expression can be derived:

$$\ln \frac{[A_0]}{[A]} = k_T t \tag{4}$$

where $[A_0]$ is the concentration of A at time zero and $[A]$ the concentration after time t. When half of A has been lost $[A_0]/[A] = 2$ and the time period t, a constant, is obtained by substituting 2 in Equation (4). Thus,

$$t_{1/2} = \frac{\ln 2}{k_T} \tag{5}$$

This half-life constant can be used to characterize the environmental persistence of a pollutant in a compartment under specified conditions.

Continuous addition of pollutant to a compartment, for example, point discharge to a stream, results in a different situation. In these circumstances persistence is measured as the steady-state concentration resulting from a balance of input and loss processes:

$$\frac{d[A]}{dt} = 0 = R_i - R_L \quad (\text{thus } R_i = R_L)$$

where R_i is the rate of input and R_L is the total rate of loss by transport and transformations of A.

Since $R_i = \Sigma k_n[A] = k_L[A]$

$$[A] = \frac{R_i}{k_L} \tag{6}$$

where k_L is the rate loss constant for A.

Effect of Sorption

Environmental fate processes also include one or more mechanisms for the reversible sorption of chemicals to sorbates (S) such as sediments, soils, biomass, or airborne particulates.

$$[A] + [S] \underset{k_{-s}}{\overset{k_s}{\rightleftharpoons}} [AS]$$

and

$$K_s = \frac{k_s}{k_{-s}} = \frac{[AS]}{[A][S]}$$

$$[A] + [AS] = [A_T] \quad (\text{mass balance})$$

where $[A]$, $[AS]$, $[S]$, and $[A_T]$ are the concentrations of unadsorbed A, adsorbed A, sorbant, and total A; k_s and k_{-s} are rate constants for the forward and reverse reactions; and K_s represents the partition coefficient.

The loss of A is represented by

$$R_L = \frac{k_L[A_T]}{K_s[S] + 1}$$

Thus the net effect of sorption on the overall rate of loss of A from a compartment is to reduce the process by the factor $1/(K_s[S] + 1)$. Consequently, sorption leads to longer half-lives or higher steady-state (ss) concentrations as derived from Equations (5) and (6).

and

$$t_{1/2} = (K_s[S] + 1) \ln 2/k_T$$

$$[A]_{ss} = \frac{R_i(K_s[S] + 1)}{k_L}$$

The above equations allow the prediction of environmental persistence using appropriate parameters (Mill, 1980).

POLLUTANT BEHAVIOR IN AQUATIC SYSTEMS

Solubility Equilibria

Solubility is the extent to which a chemical substance mixes with a liquid to form a homogeneous system (Thibodeaux, 1979). Water solubility is an intrinsic property of a pollutant and is a determinant of the transport of the substance in the aquatic environment (Haque et al., 1980).

Solubility equilibria for pure chemicals in water can be represented by equating the fugacities in both equilibrium phases as previously given by Equation (2) (Thibodeaux, 1979). Thus for component A,

$$Y_A \gamma_{Aa} f_{Aa}^\circ = X_A \gamma_{Aw} f_{Aw}^\circ \qquad (7)$$

For a pure gas, $Y_A = 1$ and $\gamma_{Aa} = 1$ (ideal gas behavior assumed) and at 1 atm, $f_{Aa}^\circ = 1$, and Equation (7) becomes

$$1 = X_A \gamma_{Aw} f_{Aw}^\circ \qquad (8)$$

where γ_{Aw} is the activity coefficient of chemical A in water, f_{Aw}° is the pure-component fugacity of A in water, and X_A is the mole fraction solubility of A in water.

For a pure liquid, Equation (8) applies also. For this system, since $f_A^\circ \simeq 1$, the solubility of a pure liquid can be represented by

$$X_A = \frac{1}{\gamma_{Aw}}$$

For a pure solid (s) Equation (2) becomes

$$f_{As}^\circ = X_A \gamma_{Aw} f_{Aw}^\circ$$

where f_{As}° is the fugacity of pure-solid chemical A. When a solid chemical is in equilibrium with an aqueous solution of concentration A, f_{As}° is estimated as the vapor pressure of the solid. With components of low solubility it can be further shown that components for a pure chemical A (solid or liquid) in equilibrium with water give

$$P_A^\circ = X_A^* \gamma_{Aw} f_{Aw}^\circ \qquad (9)$$

where P_A° is the vapor pressure of the pure solid or liquid and X_A^* is the solubility in water at equilibrium (see Thibodeaux, 1979).

Table 2.3.   Evaporation Parameters for Various Compounds at 25°C

Compound	Molecular Weight	Solubility (mg/L)	P_A° Vapor Pressure (mm Hg)	$K_{Aw}{}^a$ (m/hr)
Alkanes				
n-Octane	114	0.66	14.1	0.124
2,2,4-Trimethyl pentane	114	2.44	49.3	0.124
Aromatics				
Benzene	78	1780	95.2	0.144
Toluene	92	515	28.4	0.133
o-Xylene	106	175	6.6	0.123
Cumene	120	50	4.6	0.119
Naphthalene	128	33	0.23	0.096
Biphenyl	154	7.48	0.057	0.092
Pesticides				
DDT($C_{14}H_9Cl_5$)	355	0.0012	1×10^{-7}	9.34×10^{-3}
Lindane	291	7.3	9.4×10^{-6}	1.5×10^{-4}
Dieldrin	381	0.25	1×10^{-7}	5.33×10^{-5}
Aldrin	365	0.2	6×10^{-6}	3.72×10^{-3}
Polychlorinated Biphenyls (PCBs)				
Aroclor 1242	258	0.24	4.06×10^{-4}	0.057
Aroclor 1248 ($C_{12}H_6Cl_4$)		5.4×10^{-2}	4.94×10^{-4}	0.072
Aroclor 1254 ($C_{12}H_5Cl_5$)		1.2×10^{-2}	7.71×10^{-5}	0.067
Aroclor 1260 ($C_{12}H_4Cl_6$)		2.7×10^{-3}	4.05×10^{-5}	0.067
Other				
Mercury	201^b	3×10^{-2}	1.3×10^{-3}	0.092

Source: Mackay and Leinonen (1975). Reprinted with permission from American Chemical Society.
[a] K_{Aw}, overall mass transfer coefficient for chemical A from water to air.
[b] Atomic weight.

Phase equilibria relationships have been applied by Thibodeaux (1979) to low-solubility chemicals which are of importance in environmental pollution. Using Equation (2), the partial pressure of chemical A and its concentration in aqueous solution may be related by assuming that the vapor phase fugacity coefficient and the activity coefficient are unity. Hence the following equation can be derived from Equation (7):

$$P_A = X_A \gamma_{Aw} f_{Aw}^\circ \qquad (10)$$

in which X_A, the mole fraction, is typically about 10^{-7}, γ_{Aw}, the activity coefficient, is about 10^7, and f_{Aw}° is the fugacity of the pure liquid or solid A at the same temperature and pressure.

The product $\gamma_{Aw} f^{\circ}_{Aw}$ is equal to P°_{A}/X^{*}_{A}, as derived from Equation (9), and substituting this into Equation (10), the following relationship is obtained:

$$P_A = \frac{P^{\circ}_A}{X^{*}_A} X_A \tag{11}$$

Pure-component vapor pressure P°_A and solubility data, X^{*}_A or ρ^{*}_{Aw} (mg/L), are readily available for many chemicals (e.g., see Table 2.3).

Transport Processes

The distribution of pollutants in aquatic environments is strongly influenced by a number of interactive transport processes, such as volatilization, precipitation from air, leaching, and runoff. Some, such as volatilization, decrease the concentrations in water, whereas others, including precipitation from air, leaching, and runoff, increase the concentrations (Haque et al., 1980). The input processes are considered in following sections on the atmospheric, soil, and sediment environments.

Volatilization from water to the atmosphere has been recognized as a major pathway for pollutants with low solubilities and low polarity. Despite low vapor pressures many pollutants of high molecular weight can volatilize at relatively high rates because of their high activity coefficients in solution (Thibodeaux, 1979; Haque et al., 1980). This is an important factor in the volatilization of pollutants such as pesticides, PCBs, and petroleum hydrocarbons.

Volatilization processes are usually treated as equilibrium systems. However, the loss of a chemical from an environmental compartment involves kinetic processes such as diffusion to and away from the surface (Tinsley, 1979). In general, the volatilization rate of a chemical R_V is assumed to be a first-order process (Liss and Slater, 1974; Mackay and Leinonen, 1975) and can be described by

$$R_V = \frac{dC_w}{dt} = K_{Vw} C_w$$

The rate constant for volatilization from water (K_{Vw}) is given by the relation

$$K_{Vw} = \frac{A'}{v} \left(\frac{1}{K_l} + \frac{RT}{H_C K_g} \right)^{-1} \tag{12}$$

where R_V is the volatilization rate of a chemical A (mol L^{-1} hr^{-1}); C_w concentration of A in water (mol $L^{-1} = M$); K_{Vw} volatilization rate constant (hr^{-1}); A surface area (cm^2); V liquid volume (cm^3); K_l liquid phase mass transport coefficient ($cm\,hr^{-1}$); H_C Henry's Law constant (torr M^{-1}); K_g gas phase mass transport coefficient ($cm\,hr^{-1}$); R gas constant (torr K^{-1} M^{-1}); and T temperature (K).

In both the gas and liquid phases,

$$K_l = \frac{D_l}{\delta_l} \quad \text{and} \quad K_g = \frac{D_g}{\delta_g}$$

where D is the diffusion coefficient and δ the boundary layer thickness.

Mackay and Wolkoff (1973) showed that an estimate of Henry's Law constant, H_C, can be obtained from

$$H_C = \frac{P_{sat}}{C_{sat}}$$

where P_{sat} is the vapor pressure of pure chemical and C_{sat} the solubility of chemical A in mol L^{-1}. Equation (12) can be expressed for volatile compounds if $H_C > 1000$ as

$$K_{Vw} = \frac{A'}{V}\left(\frac{K_l RT + H_C K_g}{RT}\right)$$

where mass transfer is liquid phase limited.

If $H_C \ll 1000$, then Equation (12) becomes

$$K_{Vw} = \frac{A'}{V}\frac{H_C K_g}{RT}$$

where mass transfer is gas phase limited.

For highly volatile chemicals, volatility can be estimated from the following equation:

$$[K_{Vw}^c]_{env} = \left[\frac{K_{Vw}^c}{K_{Vw}^O}\right]_{lab}[K_V^O]_{env} \qquad (13)$$

where K_{Vw}^c is the volatilization rate constant for the chemical (hr^{-1}) and K_{Vw}^O the oxygen reaeration rate constant (hr^{-1}) in the laboratory or the environment.

In Equation (13) the rate constant for volatilization of the chemical is proportional to the rate constant for volatilization or reaeration of oxygen from the same solution under the same environmental conditions (Mill, 1979). The ratio K_{Vw}^c/K_{Vw}^O can be established in the laboratory by measuring simultaneously the concentration of oxygen redissolving in deaerated stirred solution and the concentration of chemical volatilizing from the same solution (Smith et al., 1978).

Estimates of the volatilization of low-solubility contaminants from water may be derived from the relationship below, which is analogous to Equation (11):

$$P_A = C_A\left[\frac{P_A^o}{C_A^*}\right]$$

where P_A is the partial pressure of chemical A in air, P_A^o the vapor pressure of pure A, C_A^* the equilibrium solubility of A in water, and C_A the concentration of A in water.

At time t in a given water sample of unit cross-section and depth d, the concentration of chemical A may be expressed as

$$C_{At} = C_{A0}\,e^{-Kt/d}$$

where C_{A0} is the concentration of $t = 0$, and K is the overall liquid exchange constant and is a function of the exchange constants K_l and K_g and H_C.

$$K = \frac{H_C K_l K_g}{H_C K_g + K_l}$$

Table 2.4   Volatilization of Selected Chemicals from Water

Chemical	Half-Life (hr) for $d = 1$ m
n-Octane	5.55
Benzene	4.81
Toluene	5.18
o-Xylene	5.61
Naphthalene	7.15
Biphenyl	7.52
DDT	73.9
Arochlor 1254	10.3
Mercury	7.53

Source: Mackay and Leinonen (1975). Reprinted with permission from American Chemical Society.

Since this is a first-order process the half-life of chemical A can be given as

$$t_{1/2} = \frac{0.693d}{K} \quad \text{[see Equation (5)]}$$

This equation can be used to estimate volatilization half-lives for low-solubility contaminants (see Table 2.4).

Transformation Processes

Aquatic transformation processes include hydrolysis, photolysis, microbiological degradation, and oxidation. Of these, oxidation is considered to be the least important (Haque et al., 1980).

Hydrolysis

Pollutants can undergo reactions with ubiquitous water molecules in the environment, resulting in the introduction of a hydroxyl group into the chemical structure, in place of a leaving group ($-X$) (Callahan et al., 1979). For example,

Organic chemicals: $RX + H_2O \longrightarrow ROH + HX$ (or H^+, X^-)

$$R-\overset{\overset{\displaystyle O}{\|}}{C}-X + H_2O \longrightarrow R-\overset{\overset{\displaystyle O}{\|}}{C}-OH + HX$$

Metal salts: $MX + H_2O \longrightarrow MOH + H^+ + X^-$

These reactions are catalyzed mainly by hydronium (H_3O^+) and/or hydroxyl ions (OH^-) so that the rate of hydrolysis is given by the equation

$$R = -\frac{dC}{dt} K_h[A]$$

$$= K_A[H^+][A] + K_B[OH^-] + K_N'[H_2O][A] \tag{14}$$

Table 2.5. Hydrolysis Rates at 25°C and pH 7 of Some Halogenated Compounds

Compound	Rate Constants (sec^{-1})			
	K_N	$K_B[OH^-]$	K_h	$t_{1/2}$
CH_3F[a]	7.44×10^{-10}	5.82×10^{-14}	7.44×10^{-10}	30 yr
CH_3Cl[a]	2.37×10^{-8}	6.18×10^{-13}	2.37×10^{-8}	339 days
CH_3Br[a]	4.09×10^{-7}	1.41×10^{-11}	4.09×10^{-7}	20 days
CH_3I[a]	7.28×10^{-8}	6.47×10^{-12}	7.28×10^{-8}	110 days
$CH_3CHClCH_3$[a]	2.12×10^{-7}	–	2.12×10^{-7}	38 days
$CH_3CH_2CH_2Br$[a]	3.86×10^{-6}	–	3.86×10^{-6}	26 days
$CH_3\text{-}\underset{\underset{Cl}{\vert}}{\overset{\overset{CH_3}{\vert}}{C}}\text{-}CH_3$[a]	3.02×10^{-2}	–	3.02×10^{-2}	23 sec
CH_2Cl_2[a]	3.2×10^{-11}	2.3×10^{-15}	3.2×10^{-11}	704 yr
$CHCl_3$[b]	–	6.9×10^{-12}	6.9×10^{-12}	3500 yr
$CHBr_3$[b]	–	3.2×10^{-11}	3.2×10^{-11}	686 yr
CCl_4[b,c]	–	4.8×10^{-7}	4.8×10^{-7}	7000 yr (1 ppm)
$C_6H_5CH_2Cl$	1.28×10^{-5}	–	1.28×10^{-5}	15 hr

Source: Mabey and Mill (1978). Reprinted with permission from John Wiley & Sons, Inc.
[a] $K_h = K_N; K_B \ll K_N$.
[b] $K_h = K_B; K_N \ll K_B$.
[c] Rate second order with respect to $[CCl_4]$, K_h (L mol^{-1} sec^{-1}).

where K_h is the first-order rate constant at a given pH; K_A and K_B are second-order rate constants for the acid and base catalyzed processes, respectively; and K_N is the second-order rate constant for the neutral reaction of a chemical with water, which may be expressed as a pseudo-first-order rate constant K'_N.

Equation (14) indicates that the total rate of hydrolysis in water (K_h) is pH dependent, unless K_A and $K_B = 0$. This has been reported by Mabey and Mill (1978) who have also reviewed kinetic data for the hydrolysis of a variety of organic chemicals in aquatic systems (see Table 2.5). Some chemicals, including alkyl halides, were found to have hydrolysis rates which are independent of pH over the usual environmental pH range of 4–9, while others such as carboxylic acid esters are acid and base catalyzed with a minimum hydrolysis rate at pH 4–5. At any specific pH, the half-life of an ester can be obtained assuming first-order kinetics from

$$t_{1/2} = \frac{0.693}{K_h} \quad \text{[see Equation (5)]}$$

Calculated half-lives of esters of carboxylic acids in water at 25°C and pH 7 range from 38 min ($Cl_2CH_2COOCH_3$) to up to 140 yr ($CH_3COOC(CH_3)_3$) and some organophosphate insecticides, at pH 7.4 and 20°C, extend from hours up to 130 days (parathion). Also the stability of organophosphorus compounds in aqueous solution rapidly decreases with increasing temperature and at extreme pH values (Korte, 1978).

The mechanisms of hydrolysis, including structure–activity relationships, and predictive test methods have been used to develop estimates of kinetic rates for various compounds (Mabey et al., 1978). For example, molecular structure and substituents have an important influence on hydrolysis rates in certain pollutant chemical classes, for example, carbamates (Wolfe et al., 1978; Tinsley, 1979).

Photolysis

The photolysis of environmental pollutants depends upon the energy of incident sunlight, the absorption spectrum of the molecule, and the presence of photosensitizers in the environment. Direct absorption of energy in the near ultraviolet–visible electromagnetic range (240–700 nm) may result in cleavage of bonds, dimerization, oxidation, hydrolysis, or rearrangement (Mill, 1979; Tinsley, 1979).

Photochemical changes initiated by solar radiation involve three stages: (1) the absorption of radiation of certain wavelengths and the production of an excited state; (2) the primary photochemical process that involves the transformation of the electronically excited state and its de-excitation; and (3) the secondary reactions of the various chemical species that may be produced by the primary photochemical process (Tinsley, 1979). Absorption in the ultraviolet and visible range is generally associated with the presence of unsaturation or aromatic rings. But for photochemical breakdown to occur, sufficient radiation energy should be absorbed to overcome the bond energies and allow dissociation to proceed.

Direct photolysis of chemicals in water, in the atmosphere, or on soil surfaces can be represented by the same general kinetic relationships (Mill, 1980). At low concentrations, the rate of light absorption, I_A, by a chemical at wavelength λ can be given as

$$I_A(\lambda) = \epsilon_\lambda I_\lambda[C] = K_a(\lambda)[C]$$

where ϵ_λ is the molar absorbance; I_λ, the intensity of incident light at wavelength λ; [C], the concentration of the chemical; and K_a the rate constant for light absorption. The rate of direct photolysis can be represented by a simple first-order kinetic expression:

$$R_P = -\frac{dC}{dt} = K_a(\lambda)\phi_\lambda[C] = K_P(\lambda)[C]$$

or

$$K_P = K_a\phi$$

where ϕ_λ is the quantum yield, that is, the efficiency for converting the absorbed light into chemical reaction, measured as the ratio of moles of substrate transformed to einsteins of photons absorbed; and K_P is the first-order rate constant for direct photolysis. Photolytic half-lives of chemicals can be estimated according to the expression

$$t_{1/2} = \frac{0.693}{K_a\phi} \quad \text{[see Equation (5)]}$$

Experimental estimates of environmental photolysis rates in sunlight include laboratory measurements of ϕ at a single wavelength and ϵ_λ. Average intensity (I_λ) data are available in the literature as a function of time of day, season, and latitude. The rate constant in sunlight $K_{P(S)}$ is then given as

$$K_{P(S)} = \phi \sum_\lambda \epsilon_\lambda$$

and the half-life in sunlight is

$$t_{1/2} = \frac{0.693}{K_{P(S)}} \quad \text{[see Equation (5)]}$$

Oxidation

Photochemical oxidation processes may occur in natural waters. The average effective concentrations of two oxidant species, alkylperoxyl radicals and singlet oxygen, were estimated to be $10^{-9}\,M$ and $10^{-12}\,M$, respectively (Mill et al., 1979).

The rate of loss of a chemical by reaction with an oxidant [Ox] can be represented by

$$-\frac{dc}{dt} = K_{ox}[Ox][A]$$

where K_{ox} is a second-order rate constant for reaction of the oxidant with chemical A, and [Ox] and [A] are the concentrations of oxidant and chemical, respectively. Mill (1979) and Mill et al. (1979) have reviewed the available data which are applicable only to the specific cases in which the data have been measured and are not of more general utility.

Redox Systems

The chemical forms of many environmental pollutants are modified by (1) their oxidation–reduction characteristics and (2) the oxidizing and reducing capabilities of their environment. The redox status of an area is defined as the tendency to donate or accept electrons. This can be measured as the activity of electrons present (a_{e^-}) generally expressed as pϵ, where

$$p\epsilon = -\log(a_{e^-})$$

As many redox systems involve both electron and proton transfer it is necessary to take into account both the pϵ and the pH characteristics to describe the various species of a chemical within that system. Usually pϵ–pH diagrams are constructed to indicate the electron activity boundaries for related redox or species environments. For example, Figure 2.3 outlines the redox boundaries for water and different types of natural aquatic environments. Similar diagrams have been developed for a number of inorganic systems, especially metals (Pourbaix, 1974) (e.g., Figure 2.4). However, a similar analysis of organic systems is not available although redox conditions strongly influence many organic degradation processes (Tinsley, 1979). Redox systems in many natural and modified environments have been described (Stumm and Morgan, 1970; Guenther, 1975; and Tinsley, 1979).

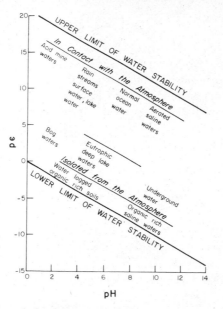

Figure 2.3. Stability boundaries for water and $\rho\epsilon$ vs. ρH characteristics of natural aquatic environments. [From Tinsley (1979). Reprinted with permission from John Wiley & Sons, Inc.]

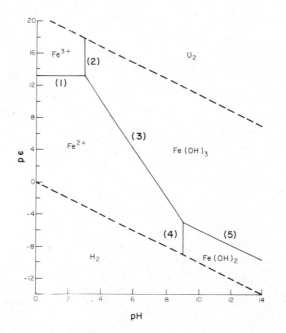

Figure 2.4. A simplified $\rho\epsilon$ vs. ρH diagram for iron at a concentration of $1.0 \times 10^{-5}\,M$. [From Tinsley (1979). Reprinted with permission from John Wiley & Sons, Inc.]

Biodegradation Processes

Biotransformation and biodegradation of pollutants by microorganisms (bacteria, fungi, protozoa, and algae) are important removal and modification processes in waters, sediments, and soils. Reactions include oxidation, reduction, hydrolysis, and sometimes rearrangements, and are influenced by molecular structure and concentration of the pollutant, the nature of the microorganisms, environmental conditions, and temperature.

The Monad kinetic equation is usually used to describe the relation between growth-limiting substrate concentration [A] and the specific growth rate in a well-mixed system:

$$\mu = \frac{d[X]/dt}{[X]} = \frac{\mu_{max}[A]}{K_S + [A]}$$

where μ is the specific growth rate, [X] the biomass per unit volume, μ_{max} the maximum specific growth rate, and K_S the concentration of substrate supporting half-maximum specific growth rate ($0.5\mu_{max}$).

The rate of loss of chemical A (or substrate) is

$$-\frac{d[A]}{dt} = \frac{\mu[X]}{X} = \frac{\mu_{max}[A][X]}{YK_S + [A]} \tag{15}$$

where $\mu_m/Y = K_b$ and Y the biomass produced from a unit amount of chemical or substrate consumed. The constants μ_{max}, K_S, and Y are dependent on the characteristics of the microorganisms, pH, temperature, nutrients, and so on. When the chemical or substrate concentration is high and $[A] \gg K_S$, then the above expression [Equation (15)] becomes simply

$$-\frac{d[A]}{dt} = K_b[X]$$

However, for many pollutants in the environment, concentrations are low, such that $[A] \ll K_S$. Equation (15) then becomes

$$\frac{-d[A]}{dt} = \frac{K_b[A][X]}{K_S} = K_{b2}[A][X]$$

where K_{b2} is a second-order rate constant.

In many environmental conditions, where [X] is relatively large and [A] is low, microbial populations will remain relatively constant when the chemical is consumed. Under these circumstances, the degradation rate can be assumed to be pseudo-first-order as given by

$$-\frac{d[A]}{dt} = K_b'[A]$$

where K_b' is a pseudo-first-order rate constant which is dependent on cell concentration $[X_0]$. Therefore,

$$\frac{K_b'}{X_0} = K_{b2}$$

The half-life of a chemical at a given X_0 will then be

$$t_{1/2} = \frac{0.693}{K_{b2}X_0} \quad \text{[see Equation (5)]}$$

Limitations in the application of Monad kinetics to mixed culture and environmental systems are discussed by Mill (1979) and Callahan et al. (1979). When a chemical is initially introduced into the environment there is often a lag or acclimation period between exposure of the chemical to the unacclimated microorganisms and the initiation of microbial action. The time required to reduce 50% of the original concentration ($T_{1/2}$) is represented by the sum of the time required to reach acclimation (t_0) and the half-life of transformation ($t_{1/2}$). Monad kinetics can apply when acclimated organisms are present.

At present there appear to be few general principles available about useful correlations between structure and reactivity (Dagley, 1975) except the thermochemically limited rate of oxidative biodegradation observed among highly halogenated compounds (Mill, 1979). Microbially mediated reactions with certain metals, for example, Hg, As, and Pb, can result in the formation of metallocompounds such as methyl mercury (Craig, 1980).

POLLUTANT BEHAVIOR IN THE ATMOSPHERE

Transport Processes

The vapor pressure of a pollutant is a useful parameter for predicting the extent to which the substance will be transported into the atmosphere (Thibodeaux, 1979; Tinsley, 1979). The equilibrium vapor pressure of a gas can be viewed as the solubility of the substance in air and the vapor pressure of a liquid or solid is the pressure of the gas in equilibrium with the liquid or solid. For a pure liquid (or solid) in equilibrium with air at constant total pressure ($P_T = 1$ atm), Equation (10) yields the air space mole fraction of chemical A:

$$Y_A = \frac{P_A^\circ}{P_T}$$

where P_A° is the vapor pressure of chemical A and P_T the total pressure in the air space above the interface between the two phases.

The thermodynamic expression describing this equilibrium can be derived from the Clausius–Clapeyron equation (Thibodeaux, 1979):

$$\ln\left[\frac{P_{Aa}^\circ}{P_{Aa0}^\circ}\right] = \frac{\lambda_A}{R}\left(\frac{1}{T_0} - \frac{1}{T}\right)$$

where R is the universal gas constant and λ_A is the molal heat of vaporization. From this equation the vapor pressure of a chemical P_{Aa}° at temperature T can be calculated if the vapor pressure, P_{Aa0}°, at another temperature T_0 is known together with the latent heat of vaporization. Hence, plots of $\ln P_A^\circ$ versus $1/T$ are commonly used for correlating vapor pressure data.

Vapor pressure can provide information on the potential transport of chemicals into the atmosphere via volatilization which has been discussed previously.

The presence of sorption surfaces in water and soil may strongly reduce vapor loss to air. Conversely, sorption surfaces, for example, particulates in air, may also increase the concentration of a chemical in this compartment.

Other important atmospheric transport phenomena (Haque et al., 1980) involve dry and wet deposition, gaseous diffusion, horizontal mixing in the troposphere, and vertical mixing between the troposphere and the stratosphere. The particle size and the physical form of the pollutants (e.g., aerosols, solids adsorbed on solids or liquid aerosols) may affect dispersion and removal processes. Rainout and washout usually occur in only the lowest 5 km of the troposphere. Little is known about dust deposition but in some cases it may lead to deposition of amounts comparable to wet deposition (Korte, 1978).

Transformation Processes

Transformation processes occurring in the atmosphere are complex and generally photochemically induced by sunlight, assisted by catalytic chemical species, for example, trace elements. Reaction steps in these processes usually consist of parallel, sequential, and competitive reactions. Transformations can be divided into two major categories: (1) photolysis and (2) reaction of the chemical with other species. These are considered below (Haque et al., 1980).

1. *Photolytic Action.* This is responsible for the degradation of many organic chemicals that absorb ultraviolet–visible radiation. Many of these molecular species have bond energies that are approximately equivalent to the energy of the sea level solar cutoff. At high altitudes, the total solar flux is greater, resulting in an increased amount of photolytic and excited state reactions.

2. *Reaction with Other Chemical Species.* This includes the reaction of environmental chemicals with hydroxyl radicals ($\cdot$OH) and ozone.

Major sources of hydroxyl radicals in the lower atmosphere are given by the following reactions:

$$H_2O + O(^1D) \longrightarrow 2OH$$

$$NO + HO_2 \longrightarrow OH + NO_2$$

$$H_2O_2 \xrightarrow{+ h\nu\,(\lambda\, <\, 370\,\text{nm})} 2OH$$

Hydroxyl reactivity with chemicals may be represented by a second-order rate mechanism. Reported rate constants average from about 10^9 to 10^{10} mol^{-1} sec^{-1} for hydrocarbons (Lloyd et al., 1976) and reduce to about $10^5 - 10^8$ mol^{-1} sec^{-1} for halocarbons containing labile hydrogen atoms or double bonds (Cox et al., 1976).

In polluted atmospheres, ozone is produced by the following reactions:

$$NO_2 \xrightarrow{h\nu(\lambda < 430\,nm)} NO + O^*$$

$$O^* + O_2 + M \longrightarrow O_3 + M$$

M is a nonreactive third body, for example, N_2 or H_2O.

In the presence of olefins, for example, C_3H_6,

$$O^* + C_3H_6 \longrightarrow products$$

$$O_3 + C_3H_6 \longrightarrow products$$

General reactions of organic compounds, for example, arenes and olefins, with active oxygen $O(^3P)$ and singlet molecular O_2, have been reviewed by Korte (1978). Olefins react much faster with ozone than do most other organics, with rate constants between 10^2 and 10^5 mol^{-1} sec^{-1}.

Overall, the reactions of hydroxyl radicals control rates of oxidation of most chemicals. Ozone is confined to oxidation of some olefins in the atmosphere, as in the case of photochemical smog, and possibly oxidation of some sulfur and phosphorus compounds exposed on surfaces in smoggy areas (Mill, 1979).

SOIL AND SEDIMENT INTERACTIONS WITH POLLUTANTS

Soil and sediment play major roles in the transport and removal of environmental pollutants by (1) providing sorption surfaces, (2) acting as buffer systems, and (3) serving as sinks for pollutants. The dominant transport processes related to soil and sediment are adsorption and leaching.

Adsorption Processes

Factors influencing the adsorption process are (1) structural characteristics of the chemical; (2) the organic content of the soil; (3) the pH of the medium; (4) particle size; (5) ion exchange capacity; and (6) temperature. Most adsorption processes take a comparatively short time to reach equilibrium but the desorption rate is much slower. The adsorption process is usually expressed as an adsorption isotherm. Two widely used relationships for adsorption are set out below.

1. *Langmuir Isotherm.* The moles of solute adsorbed per gram of adsorbent, X, are expressed as a function of the equilibrium concentration of solute in solution C:

$$X = \frac{X_m bC}{1 + bC}$$

where X_m is the number of moles of solute adsorbed per gram of adsorbent in forming a complete monolayer; C the equilibrium concentration of the chemical; b a constant related to energy of adsorption.

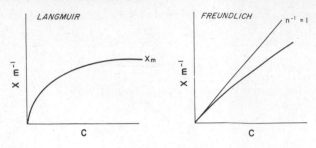

Figure 2.5. Comparison of Langmuir and Freundlich isotherms. Langmuir: linear at low concentration; $x = x_m bC$; limiting adsorption at high concentration. Freundlich: does not approach any limiting value; linear when $1/n = 1$. [From Tinsley (1979). Reprinted with permission from John Wiley & Sons, Inc.]

This isotherm (see Figure 2.5) is applicable to the adsorption of gases on solids, but not for the absorption of chemicals from solution, particularly with heterogeneous adsorbents such as soils (Tinsley, 1979).

2. *Freundlich Isotherm.* This is an empirical relationship (see Figure 2.5) and is expressed as follows:

$$\frac{X}{m} = KC^{1/n}$$

where X is the amount of chemical adsorbed per mg of adsorbent (m); C the equilibrium concentration of the chemical; K the equilibrium constant indicative of the strength of the adsorption ($K = X/m$ when $C = 1$); and $1/n$ is a constant describing the degree of nonlinearity.

A linear form of this relationship is used in the analysis of experimental data:

$$\log \frac{X}{m} = \log K + \frac{1}{n} \log C \tag{16}$$

Haque and Schmedding (1976) report that for several polychlorinated biphenyl-type compounds on several surfaces the adsorption increases as water solubility decreases, and also increases with the organic content of the soil.

For many nonionic organic compounds in solution the slope of the Freundlich isotherm ($1/n$) approaches unity. Equation (16) then becomes

$$\frac{X}{m} = K_S C \tag{17}$$

where K_S is the sorption or distribution constant.

Several empirical relationships have also been observed between distribution phenomena as represented by K_S, n-octanol–water partition coefficient and the organic or carbon content (Haque et al., 1980). The adsorption of a chemical to soil is highly correlated with the proportion of organic carbon (OC) present. It has been found that Equation (17) can be more usefully expressed as

$$\frac{X}{m} = K_{OC} C$$

where the concentration of the adsorbed chemical is expressed per unit of organic carbon in the soil, rather than unit mass of soil. For a given chemical, K_{OC} tends to be more constant among soils than K_S (Tinsley, 1979).

The above relationships, however, do not hold for ionic-type compounds or neutral organic chemicals capable of becoming charged species in solution. The adsorption of these substances on soils and sediments depends on mineral and organic fractions present and properties of the individual chemicals (e.g., molecular size, functional groups, solubility, and partition coefficients). Thus, these interactions will vary significantly for different classes of chemical compounds and adsorbing surfaces. Important adsorptive mechanisms include van der Waal's forces, hydrophobic bonding, hydrogen bonding, ligand exchange, electrostatic attractions—ion exchange, dipole–dipole interactions, and chemisorption (see Tinsley, 1979).

Leaching

The movement of pollutants through soil involves two basic mechanisms: (1) diffusion of chemical species, primarily in gas and liquid phases, and (2) mass transport. The latter mechanism involves water as a carrier, and the movement results from an externally applied force, such as gravity. Significant movement results in leaching which may reduce pollutant concentrations in soil and sediment and may cause groundwater contamination problems, for example, leaching of ions and organic compounds from landfill sites. The adsorption coefficient of a chemical gives an indication of leaching potential (Haque et al., 1980). In general, polar species are more mobile in soils. Thus, among organic compounds adsorbed in soil and sediment increasing mobility occurs in the following order: hydrocarbons < ethers < tertiary amines < nitro compounds < esters < ketones < aldehydes < amides < alcohols < acids (Korte, 1978; see Chapter 7 on pesticides).

Volatilization

Volatilization processes in soils have been reviewed by a number of authors (e.g., Spencer et al., 1973; Tinsley, 1979). Measurement and interpretation of the rates of evaporation of a pollutant from soil surfaces is complicated by factors such as volatilization of the chemical from water at the surface, evaporation of the water itself, prevailing atmospheric conditions, chemical–soil interactions, and concentration gradients in the soil column (Mill, 1979; Tinsley, 1979). At present, experimental estimates are considered inadequate for predicting the fate of chemicals, such as pesticides, under field conditions (Mill, 1980).

Transformation Processes

The major transformation pathway for chemicals in soil and sediment is microbial degradation although photolysis is probably also significant (Haque et al., 1980).

Metabolic transformations by soil microorganisms depend largely upon such factors as concentration of the chemical, temperature, moisture, anaerobic conditions, and organic content of the soil which cannot be simply extrapolated from one region to another (Haque et al., 1980; Tinsley, 1979). Most experimental data on microbial degradation rates, in the form of half-lives, have been calculated using first-order kinetics (Haque et al., 1980).

BIOLOGICAL INTERACTIONS WITH POLLUTANTS

The interactions between environmental processes and physicochemical properties of pollutants determine their distribution, fate, and effect on living organisms. Entry and possible bioaccumulation are controlled by a number of key transport and transformation processes (see Figure 2.6). The net effect of these processes determines the exposure level or dose against which biological response can be measured.

Uptake Processes

Entry of a pollutant into the ecosphere results in uneven dispersion between environmental compartments. For biological systems, this can occur at various levels of organization and between different sets of compartments: fish and water, blood and lipid tissue, microsomes and cytosol. Before a pollutant can be distributed in an organism it must pass across a membrane and enter into cellular spaces (Tinsley, 1979). Membranes play a vital role by controlling the movement of pollutants and other chemicals through concentration gradients on either side of

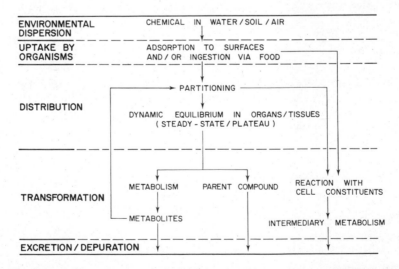

Figure 2.6.   Uptake and retention of chemicals. [From Esser and Moser (1982). Reprinted with permission from Academic Press, Inc.]

membrane boundaries. These processes are necessary for the normal sequence of functions, particularly metabolism. The various transport mechanisms include passive diffusion, filtration, active transport, facilitated diffusion, and pinocytosis.

Passive Diffusion

One of the most important uptake mechanisms for chemicals is passive diffusion. Through the lipid phase of a membrane, this can be defined by Fick's laws of diffusion (Tinsley, 1979). When the concentration gradient is constant, the steady-state flux, F, is

$$F = \frac{DC_0}{h} \tag{18}$$

where C_0 is the concentration on the donor side of the membrane surface, h the membrane thickness, and D the coefficient of diffusion. On the receptor surface of the membrane the substance is instantaneously removed giving an effective concentration of zero. C_0' is the concentration in the contiguous phase to the membrane (see Figure 2.7) so that

$$C_0 = KC_0'$$

where K is the partition coefficient. Since C_0' is usually known, Equation (18) becomes

$$F = \frac{KDC_0'}{h} \tag{19}$$

Equation (19) can be further refined to take into account a factor referred to as the permeability coefficient, P, defined as

$$P = KD \tag{20}$$

Permeability is often expressed in terms of unit membrane thickness giving dimensions of velocity, commonly cm sec^{-1}. This analysis can also be extended to a situation where the concentration in the donor compartment, C_0, decreases as it moves through the membrane into a much larger reservoir or compartment.

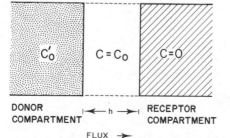

Figure 2.7.   Passive diffusion across a membrane. [From Tinsley (1979). Reprinted with permission from John Wiley & Sons, Inc.]

If the diffusion of the substance is considered as a simple first-order process, the concentration, C', in the donor compartment of volume V can be expressed as

$$C' = C_0' e^{DKt/hV} \tag{21}$$

The coefficient of diffusion, D, can be expressed as a function of parameters that influence the solute mobility including properties of the diffusion substance. One relationship indicates that the coefficient of diffusion is inversely related to the radius of the molecule, r, which in some cases can be expressed as the square root of the molecular weight (MW). Thus, in the absence of viscosity effects within the biological membrane,

$$D = \frac{\text{constant}}{r}$$

or

$$D = \frac{\text{constant}}{(MW)^{1/2}}$$

By combining Equations (20) and (21), the permeability coefficient of a chemical for a given concentration gradient may be expressed as

$$P = \frac{K \times \text{constant}}{(MW)^{1/2}}$$

Thus the following factors have been found: (1) There exists a direct relation between uptake and partition coefficient. (2) The more hydrophobic the compound, the greater is its tendency to move through biological membranes. (3) If the rate of absorption is controlled by the partition coefficient of the compound, the tendency of acids and bases to absorb is influenced by their pK_a and the pH of the environment in which absorption is taking place, for example, water or stomach.

Filtration

Another important transport mechanism across porous membranes involves the passage of solutes through membrane channels as part of the bulk water flow due to hydrostatic or osmotic force. Selective transfer of solutes through membranes is determined by the size of the chemical species and the membrane channels.

Special Transport Systems

In a number of cases membrane transfer of chemicals that are too lipid insoluble to dissolve in cell membranes (passive diffusion) or too large to flow through membrane channels (filtration) occurs. To explain this phenomenon special transport systems have been postulated: These have been summarized by Klaassen (1980).

An active transport system contains the following characteristics: (1) chemicals are moved against electrochemical gradients; (2) at high substrate concentrations the transport system is saturated and a transport maximum exhibited; (3) the transport system is selective, including competitive inhibition; and (4) the system requires energy expenditure so that metabolic inhibitors block the transport process.

Essentially active transport refers to a carrier transport system in which it is presumed a chemical forms a complex with a macromolecular carrier capable of diffusing through cell membranes. This may include cellular adsorption or elimination of xenobiotics.

Similarly, facilitated diffusion involves carrier transport except that the substrate does not move against a concentration gradient, and the process is not energy dependent. Additional specialized forms of membrane transport include pinocytosis and phagocytosis processes in which the cell membrane incorporates particles.

Bioaccumulation Processes

The uptake and retention of a pollutant by an organism results in the development of elevated concentrations which may have deleterious effects. This process can occur by direct absorption from the ambient environment or by absorption of a pollutant in food organisms. Pollutants in a food organism can arise from similar sources. Thus, in a natural food chain, pollutants could be transferred from one trophic level to another.

The retention of a pollutant depends on its biological half-life [see Equation (5)]. Thus, a pollutant must exhibit a comparatively high resistance to breakdown or excretion by the organism to allow a sufficient uptake period for an elevated concentration to occur.

A wide variety of terms are used in an inconsistent and confusing manner to describe organism uptake and retention of pollutants by different paths and mechanisms. Terms often used are: bioamplification, bioaccumulation along food chains, cumulative transfer, trophic contamination, biological magnification, trophic magnification, storage, and uptake. However, three terms can be most usefully applied to these processes and the following definitions are now widely accepted:

1. *Bioaccumulation* is the uptake and retention of pollutants from the environment by organisms via any mechanism or pathway.
2. *Bioconcentration* is uptake and retention of pollutants directly from the water mass by organisms through such tissues as the gills or epithelial tissues.
3. *Biomagnification* is the process whereby pollutants are passed from one trophic level to another and exhibit increasing concentrations in organisms related to their trophic status.

Models for Bioaccumulation

Only a few theoretical models attempt to explain uptake and elimination processes for persistent pollutants in organisms. Of these, the compartmental model (Atkins, 1969), as applied in pharmocokinetics, is a useful approach (Moriarty, 1975). In this model a compartment is defined as a quantity of pollutant that has uniform kinetics of transformation and transport, and whose kinetics are discrete from other compartments.

Single Compartment Model. The bioaccumulation process can be seen as a balance between two kinetic processes, uptake and depuration (see Figure 2.8) as

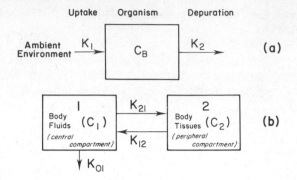

Figure 2.8. Compartment models for bioaccumulation of a pollutant chemical: (a) single-compartment; (b) two-compartment under depuration conditions. [From Connell and Miller (1981). Reprinted with permission from CRC Press.)

quantified by first-order rate constants K_1 and K_2, respectively. The rate of change of pollutant concentration in an organism is given by

$$\frac{dC_B}{dt} = K_1 C_M - K_2 C_B \tag{22}$$

where C_B is the biotic concentration, C_M the concentration in the ambient environment, water, and t the time (see Figure 2.9).

Upon integration of Equation (22), from an initial $C_B = 0$ and $t = 0$, the concentration C_B at time t is

$$C_B = \frac{K_1}{K_2} C_M (1 - e^{-K_2 t})$$

When the concentration in the organism approaches a steady state the uptake

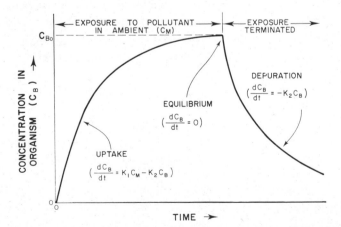

Figure 2.9. Uptake and depuration of a pollutant chemical by an aquatic organism. [From Esser and Moser (1982). Reprinted with permission from Academic Press, Inc.]

and depuration processes will be in equilibrium and

$$\frac{dC_B}{dt} = 0 = K_1 C_M - K_2 C_B$$

and

$$K_1 C_M = K_2 C_B \qquad (23)$$

(see Figure 2.9).

If exposure to the pollutant is terminated then uptake ceases and $K_1 C_M$ is zero. Thus, for depuration

$$\frac{dC_B}{dt} = -K_2 C_B$$

(see Figure 2.9). Upon integration,

$$C_B = C_{B0} e^{-K_2 t}$$

where C_{B0} is the concentration in the organism at the start of the depuration process (see Figure 2.9). Thus,

$$\log C_B = \log C_{B0} - \frac{K_2 t}{2.303}$$

Therefore a semilogarithmic plot of C_B versus time will be linear. The biological half-life of the pollutant in the organism can be derived from the relationship

$$t_{1/2} = \frac{0.693}{K_2} \quad \text{[see Equation (5)]}$$

This model assumes that an organism consists of a single homogeneous compartment, but, in fact, many exist. This provides a reasonable approximation in many cases but has limitations.

Two-Compartment Model.   This model incorporates a peripheral compartment consisting of body fluids (e.g., blood) and a central compartment which includes all other tissues (Moriarty, 1975) (see Figure 2.8). At the steady state, there is a dynamic equilibrium between the external medium and body fluids, and body fluids and tissues. Such a model can be used to explain the initial high rate of loss of a pollutant observed in experimental studies. This phenomenon is attributed to the initial loss of pollutant from the "nonassimilated" pool (e.g., blood) within an organism.

Thus the primary process in uptake and elimination should be seen as equivalent to a simple partitioning between the body fluids and the external medium. The secondary process, when levels in the body fluids exceed a certain threshold limit, results in temporary storage of lipophilic pollutants within cells if metabolic capacities are exceeded.

Moriarty (1975) has described the assumptions of steady-state systems with two or more compartments and derived the equations for the loss of pollutants (see Figure 2.8). Equations for a two-compartment system are of the same general form as for a one-compartment system except that there are two exponential terms instead of one. Therefore, semilogarithmic plots of concentration versus time are not linear. These plots have the general form of an initial rapid pollutant loss phase followed by a slower elimination phase. The equations take the general form

$$C_1 = X_1 e^{-\lambda_1 t} + X_2 e^{-\lambda_2 t}$$
$$C_2 = X_3 e^{-\lambda_1 t} + X_4 e^{-\lambda_2 t}$$

where C_1 and C_2 are the concentrations of pollutants in compartments 1 and 2, respectively, X_1, X_2, X_3, and X_4 are complex coefficients, and λ_1 and λ_2 exponential constants incorporating the three rate constants. For concentration terms, the rates of change of pollutant in compartments also depend on rate constants and the relative volumes of the two compartments (Moriarty, 1975).

This model has been applied to a few experimental studies where two exponential terms account for the loss of pollutant after exposure and allow biological half-lives for rapid and slow elimination periods to be estimated from the exponential constants (e.g., Robinson et al., 1969).

For vertebrates, a three-compartment model can be used in which the liver, as principal site of metabolism, is represented as a discrete metabolic compartment (Moriarty, 1978). The behavior of pollutants in other tissues of the body may be represented by additional compartments.

Differential Behavior Model.   This model has been described by Fagerstrom and Ansell (1974) as a generalized model for bioaccumulation of a trace substance. In any system the trace substance is "carried" to the individual, population, and so on, by a substance that is part of the normal metabolic processes of the system, for example, lipids, proteins, calcium, or any other natural dynamic component of the system. The trace substance chemically resembles the carrier and flows to the system as an analogue to the carrier, where it is stored in a similar manner to the carrier. The trace substance could be a pesticide, such as DDT, associated with lipids, or a metal such as strontium 90 associated with calcium.

In this model, bioaccumulation occurs where the trace substance is more efficiently retained in tissue than the carrier.

A mathematical expression for this model is given by

$$\frac{q_{ts}(t)}{q_c(t)} = f(t) \frac{Q_{ts}(t)}{Q_c(t)}$$

where q_{ts} and q_c are the rates of flow of the trace substance and carrier, respectively, and Q_{ts} and Q_c are the respective pool sizes in the donor compartments of the trace substance and carrier.

The complex function $f(t)$ is considered crucial to each biological system where it may be a constant or a system variable. If $f(t)$ is significantly different from unity, bioaccumulation is implied.

Fagerstrom and Ansell (1974) used the model to explain a relationship between the biological half-life ($Tb_{1/2}$) of a trace substance and body weight (W) based on the common weight dependence of metabolic rates. This is expressed as

$$Tb_{1/2} \propto W^{1-b}$$

where b is a constant usually between 0.7 and 0.9. These authors quote experimentally obtained values of the entity $1 - b$ from studies with radioactive elements.

Bioaccumulation Factor

The bioaccumulation factor, K_B, is the ratio of the pollutant concentration in an organism to that in the ambient environment when the concentration in the organism is in a steady state. At equilibrium $dC_B/dt = 0$ and Equation (23) gives

$$K_B = \frac{C_B^*}{C_M^*} = \frac{K_1}{K_2} \tag{24}$$

where C_M^* and C_B^* are the equilibrium concentrations of a pollutant. Thus, the bioaccumulation factor is also the ratio of the uptake and depuration rate constants. Experimental procedures used to estimate this factor are described by Esser and Moser (1982).

Applications of the Partition Coefficient (n-Octanol–Water)

The bioaccumulation factor, K_B, can be related to physicochemical equilibrium properties such as the n-octanol–water partition coefficient, P_{ow} (Veith et al., 1979; Neely, 1980; Mackay, 1982). This parameter can be measured accurately and comparatively simply and provides a measure for assessing the bioaccumulation potential of a pollutant. This simplified approach assumes that the entry of pollutants into an organism is a partition process between two phases, the organism and water. All organisms, in fact, consist of multiple phases but, for hydrophobic pollutants, the lipid phase is the dominant-concentrating phase, because of its relatively low activity coefficient. With most organisms, n-octanol approximates the solubility properties of the lipid phase.

The relationship between P_{ow} and K_B for nonionic compounds has been tested by regression analysis and found to take the general form (see Figure 2.10)

$$\log K_B = n \log P_{ow} + b$$

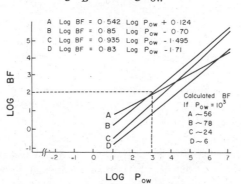

Figure 2.10.   Relationship between octanol–water partition coefficient (P_{ow}) and bioaccumulation factor (BF). [From Esser and Moser (1982). Reprinted with permission from Academic Press, Inc.]

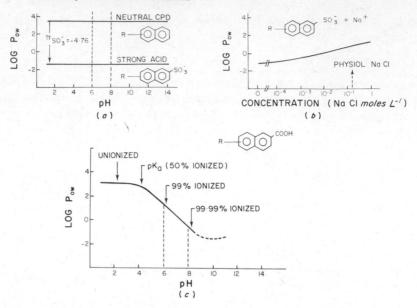

Figure 2.11. Examples of physicochemical factors influencing octanol–water partition coefficients (P_{ow}): (a) effect of charge; (b) effect of neutral salt; (c) effect of pH. [From Esser and Moser (1982). Reprinted with permission from Academic Press, Inc.]

Esser and Moser (1982) report that the slopes of lines B, C, and D are parallel and almost unity, indicating that partitioning is the major process in bioaccumulation. Mackay (1982) has found that a one constant correlation can be shown to exist between K_B and P_{ow} for available bioaccumulation data on fish:

$$\frac{K_B}{P_{ow}} = A \quad \text{or} \quad \log K_B = \log K_{ow} + \log A$$

This correlation would probably be improved by calculating the concentrations in biota in terms of lipid content rather than whole weight.

Nevertheless, there are a number of significant deviations in the data which affect the generalization of this approach for predicting bioaccumulation (e.g., Mackay, 1982). Correlations between K_B and P_{ow} are best observed between P_{ow} values of 10^2 and 10^6, which correspond to K_B values from about 10^1 to 10^4. But within this range equations for predictive purposes should only be applied to lipophilic organic compounds similar to those used to develop the correlation. Chemicals with low P_{ow} values or relatively water soluble may be more susceptible to biodegradation or excretion such that accumulation may be overestimated. Mackay (1982) suggested that compounds with values of K_B below 10 should be treated separately, and that correlations of this type will be invalid if applied to compounds undergoing biodegradation in the biotic phase.

For high P_{ow} values (i.e., $> 10^6$) the bioaccumulation potential is also uncertain. For example, the Neely regression (line A in Figure 2.10) underestimates K_B and

P_{ow} values having logarithms greater than about 6, presumably because the time to reach steady-state conditions is extremely long, which means growth and lipid deposition have to be considered in the analysis (see Mackay, 1982).

Mackay (1982) considers direct equilibrium measurement of partition coefficients in excess of 10^6 difficult since the water phase concentrations are extremely low. Also, equilibrium conditions for measurements of K_B for such compounds may not be reached because of membrane permeability resistance due to molecular size (Mackay, 1982).

Shaw and Connell (1980, 1983) have found that with the PCBs the stereochemistry of the molecule has a marked effect. An empirical steric effect coefficient was developed from stereochemical considerations and when combined with $\log P_{ow}$ an accurate evaluation of bioaccumulation was obtained. It was suggested that this steric factor was related to membrane permeability.

For ionized compounds the value of P_{ow} may be significantly affected by (1) the charge on the molecule, (2) the presence of a neutral salt, such as NaCl, (3) the pK_a value of the compound, and (4) the pH of the aqueous solution (see Figure 2.11) Esser and Moser (1982) have recommended that the P_{ow} of weak acids or bases be measured by using a buffer or physiological pH and sodium chloride solution as the aqueous phase.

Biomagnification

This mechanism has been used to explain the high concentrations of some pollutants, such as DDT, in top carnivores, such as eagles. The mechanism involves uptake of pollutants in the food of organisms but, on metabolism of the food, the pollutant, if it is resistant to degradation, is preferentially retained. This process results in a loss of energy and food matter as a result of respiration and non-assimilated energy, and consequent concentration of the pollutant. The energy content at any trophic level is approximately 90% less than the level below. Thus, an approximately ten-fold concentration of a pollutant would be expected for each trophic level through which a pollutant passes. Alternatively, a step-wise increase in organism concentration related to trophic status could result from the following process. At the lowest trophic level, phytoplankton take up a pollutant in equilibrium with water according to the following relationship derived from Equation (24):

$$C_{PP} = K_B C_W \tag{25}$$

where C_{PP} is the concentration of pollutant in phytoplankton (or detritus), K_B the bioaccumulation factor, and C_W the pollutant concentration in water. When the phytoplankton are consumed by a herbivore the following relationship applies if: (1) the food, phytoplankton, is the only source of pollutant; (2) the pollutant has not been chemically modified; (3) sufficient time has elapsed for equilibrium to be attained; and (4) K_B is constant:

$$C_H = K_B C_{PP} \tag{26}$$

where C_H is the concentration in the herbivore. It would be expected that concentrations would be somewhat lower than C_H due to losses from the organism by establishment of an equilibrium between the organism and the water mass. Substituting Equation (25) into Equation (26) we obtain the following:

$$C_H = (K_B)^2 C_W$$

Similarly, the following general relationship can be established for the conditions outlined above:

$$C_{TR} = (K_B)^{tr} C_W$$

where C_{TR} is the concentration in organisms at a particular trophic level, tr the number of transfers of the pollutant in the food web. In this situation the equilibrium will exist between the stomach contents and the organism rather than the external environment and the organism.

Thus a sequential increase in concentration could be expected in a food chain, with organisms at the highest levels exhibiting the highest concentrations. However, in many natural food chains the conditions necessary for this to occur may not exist. For example, direct uptake of pollutant from the environment may be the most important source, the pollutant may not persist for sufficient time for this comparatively lengthy process to operate, and the bioaccumulation factor may exhibit large variations between organisms. In the natural environment, the degree of biomagnification is usually a complex function of: (1) the number of links in the food chain; (2) kinds of organisms in the food chain; (3) the nature of the compound being bioaccumulated; (4) the dose of substance at each level of the food chain; and (5) the time in contact with the pollutant (Ray and Trieff, 1980). The potential for food chain increases of pollutant concentrations within an ecosystem can be indicated by the relative values of Cs/K ratio in organisms. Increases in the ratio of cesium to potassium over known food chain links can be expected because cesium has been found to have a biological half-life that is generally 2–3 times that of potassium (Young et al., 1980). Using this approach, these workers found that only organic mercury and high molecular weight chlorinated hydrocarbons (total DDT and PCB 1254) appear to increase with trophic level in coastal marine ecosystems off Southern California.

The overall significance of biomagnification in natural ecosystems is somewhat uncertain. In aquatic systems direct uptake from water and sediments appears to be the dominant process for many persistent pollutants. The retention of pollutants by different organisms may depend more on differing rates of metabolism and excretion rather than position in the food web (Moriarty, 1975). In unstructured food webs, consisting primarily of opportunistic, multidirectional feeders, there is little evidence that feeding relationships control the different concentrations of pollutants in member organisms. On the other hand, carnivorous birds often exhibit evidence that pollutants present in muscle tissue are due to biomagnification (e.g., Shaw and Connell, 1983).

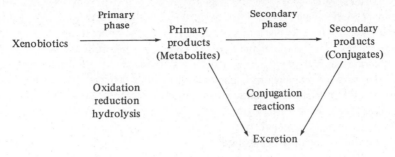

Figure 2.12.   Generalised biotransformation pathways for xenobiotics.

Biotransformations

Many organisms, particularly vertebrates, possess metabolic pathways capable of deactivation and/or activation of pollutants. The general mechanism for the biotransformation of many xenobiotics consists of two phases: formation of metabolites, usually by mixed-function oxidations, and transformations into less toxic and more hydrophilic conjugates (see Figure 2.12). Variations from this scheme may include excretion of some compounds and/or their metabolites without conjugation, by diffusion across cell membranes, and sometimes by direct conjugation.

Oxidation of foreign compounds, including pollutants, is considered the most important primary reaction. The oxidation metabolism of xenobiotic compounds is carried out by a cytochrome P-450 dependent mixed-function oxidase (MFO) system found in the endoplasmic reticulum of the cell. These enzymes exhibit a broad spectrum of reactivity and a predilection for lipophilic compounds which they metabolize through reactions involving functional groups (Wilkinson, 1980). Various researchers, for example, Estabrook et al. (1971), White et al. (1973), Hodgson (1976), and Malins (1977), have proposed schemes or reviewed mechanisms and pathways involved in the bioconversion of exogenous compounds by these enzymes (oxygenases).

The NADPH-dependent oxygenases mediate the introduction of molecular oxygen into organic substrates, for example, aromatics in conjunction with an electron transport system in which cytochrome P-450 complexes both substrate and molecular oxygen. The electrons involved are derived from reduced NADP and the overall reactions takes the form:

$$RH + O_2 + NADPH + H^+ \longrightarrow ROH + H_2O + NADP^+$$

Although the majority of these reactions constitute important detoxification pathways, some can form reactive metabolites, for example, the oxidation of aromatic hydrocarbons and olefins to form intermediate arene or alkene oxides. Certain of these electrophilic metabolites chemically interact with cellular macromolecules, such as DNA and RNA or with genetic substrates, and are implicated as primary carcinogens, mutagens, and cytotoxins. In some cases, microsomal epoxide hydrases can convert the epoxides to diols.

Primary metabolism (phase I) reactions, usually involving an oxidative, reductive, or hydrolytic biotransformation, introduce more polar substituents into the molecule. In some cases, nonpolar metabolites again may be produced by several secondary reactions in higher organisms and plants. These result from the conjugation of the substrate with endogenous substances. Enzymic systems (transferases) located in the cytosol, mitochondria, and endoplasmic reticulum of animal cells are capable of conjugating a wide variety of xenobiotics that contain either electrophilic or nucleophilic centers. Animal systems can form (1) glucuronic acid and related carbohydrate conjugates, (2) sulfate ester conjugates, (3) glutathione and related mercapturic acid conjugates, (4) methylated products, (5) acylated products, (6) amino acids and related conjugates, and (7) a few other miscellaneous conjugates. Plants may also form conjugates with proteins. In animals conjugates are generally more hydrophilic and therefore are more readily secreted or excreted than the parent compound (Paulson, 1980).

REFERENCES

Atkins, G. L. (1969). *Multicompartment Models for Biological Systems*. Methuen, London.

Callahan, M. A., Slimak, M. W., Gabel, N. W., May, I. P., Fowler, C. F., Freed, J. R., Jennings, P., Durfee, R. L., Whitmore, F. C., Maestri, B., Mabey, W. R., Holt, B. R., and Gould, C. (1979). Water-related Environmental Fate of 129 Priority Pollutants. Vol. 1, US-EPA report, EPA-440/4-79-029a.

Cox, R. A., Derwent, R. G., Eggleton, A. E. J., and Lovelock, J. E. (1976). Photochemical oxidation of halocarbons in the troposphere. *Atmos. Environ.* **10**, 305.

Craig, P. J. (1980). "Metal Cycles and Biological Methylation." In O. Hutzinger (Ed.), *The Natural Environment and the Biogeochemical Cycles*. Springer, Berlin, p. 169.

Dagley, S. (1975). "A Biochemical Approach to Some Problems of Environmental Pollution." In P. N. Campbell and W. N. Aldridge (Eds.), *Essays in Biochemistry*, Vol. II. Academic Press, New York, p. 81.

Esser, H. O. and Moser, P. (1982). An appraisal of problems related to the measurement and evaluation of bioaccumulation. *Ecotoxicol. Envion. Safety* **6**, 131.

Estabrook, R. W., Baron, J., Peterson, J., and Ishimura, Y. (1971). Oxygenated cytochrome P-450 as an intermediate in hydroxylation reaction. *Biochem. J.* **125**, 3P.

Fagerstrom, T. and Ansell, B. (1974). "Realism and Generality of Bioaccumulation Models. Towards a General Methodology." In *Proceedings of the International Conference on Transport of Persistent Chemicals in Aquatic Ecosystems IV*, Ottawa, Canada, May 1–3, p. 11.

Guenther, W. B. (1975). *Chemical Equilibrium*. Plenum Press, New York, p. 207.

Haque, R., Falco, J., Cohen, S., and Riordan, C. (1980). "Role of Transport and Fate Studies in the Exposure Assessment and Screening of Toxic Chemicals." In R. Haque (Ed.), *Dynamics, Exposure and Hazard Assessment of Toxic Chemicals*. Ann Arbor Science, Ann Arbor, Michigan, pp. 47–67.

Haque, R. and Schmedding, D. (1976). A method of measuring the water solubility of very hydrophobic chemicals. *Bull. Environ. Contam. Tox.* **14**, 13.

Hodgson, E. (1976). "Comparative Toxicology: Cytochrome P-450 and Mixed-Function Oxidase Activity in Target and Nontarget Organisms." In W. J. Hayes, Jr. (Ed.), *Essays in Toxicology*, Vol. 7, Academic Press, New York, Chap. 3.

Klaassen, C. D. (1980). "Absorption, Distribution, and Excretion of Toxicants." In J. Doull, C. D. Klaassen, and M. O. Amdur (Eds.), *Toxicology – The Basic Science of Poisons*. 2nd ed. MacMillan, New York.

Korte, F. (1978). "Abiotic Processes." in G. C. Butler (Ed.), *Principles of Ecotoxicology*, SCOPE 12. John Wiley & Sons, New York, p. 11.

Liss, P. G. and Slater, P. G. (1974). Flux of gases across the air–sea interface. *Nature (London)* **247**, 181.

Lloyd, A. C., Darnall, K. R., Winer, A. M., and Pitts, Jr., J. N. (1976). Relative rate constants for reactions of hydroxyl radical with a series of alkanes, alkenes and aromatic hydrocarbons. *J. Phys. Chem.* **80**, 789.

Mabey, W. M. and Mill, T. (1978). Critical review of hydrolysis of organic compounds in water under environmental conditions. *J. Phys. Chem. Ref. Data* **7**, 383.

Mabey, W. M., Mill, T., and Hendry, D. G. (1978). Test Protocols for Environmental Processes: Hydrolysis. US-EPA report. Contract 68-03-2227.

Mackay, D. (1982). Correlation of bioconcentration factors. *Environ. Sci. Technol.* **16**, 274.

Mackay, D. and Leinonen, P. J. (1975). Rate of evaporation of low solubility contaminants from water bodies to atmosphere. *Environ. Sci. Technol.* **9**, 1178.

Mackay, D. and Wolkoff, A. (1973). Rate of evaporation of low-solubility contaminants from water bodies to atmosphere. *Environ. Sci. Technol.* **7**, 611.

Malins, D. C. (1977). "Biotransformation of Petroleum Hydrocarbons in Marine Organisms Indigenous to the Arctic and Subarctic." In D. A. Wolfe (Ed.), *Fate and Effects of Petroleum Hydrocarbons in Marine Organisms and Ecosystems*. Pergamon Press, New York, p. 47.

Mill, T. (1979). Structure Reactivity Conditions for Environmental Reactions. US-EPA report, US-EPA 560/11-79-012.

Mill, T. (1980). "Data Needed to Predict the Environmental Fate of Organic Chemicals." In R. Haque (Ed.), *Dynamics, Exposure and Hazard Assessment of Toxic Chemicals*. Ann Arbor Science, Ann Arbor, Michigan, p. 297.

Mill, T., Mabey, W. R., and Hendry, D. G. (1979). Test Protocols for Environmental Processes: Oxidation in Water. EPA Contract 68-03-2227.

Moriarty, F. (Ed.) (1975). *Organochlorine Insecticides: Persistent Organic Pollutants*. Academic Press, London.

Moriarty, F. (1978). "Terrestrial Animals." In G. C. Butler (Ed.), *Principles of Ecotoxicology*, SCOPE 12. John Wiley & Sons, New York, p. 169.

Neely, W. B. (1980). "A Method for Selecting the Most Appropriate Environmental Experiments on a New Chemical." In R. Haque (Ed.), *Dynamics, Exposure and Hazard Assessment of Toxic Chemicals*. Ann Arbor Science, Ann Arbor, Michigan, p. 287.

Paulson, G. D. (1980). Conjugation of foreign chemicals by animals. *Residue Reviews,* **76**, 31.

Pourbaix, M. (1974). *Atlas of Electrochemical Equilibria in Aqueous Solutions.* National Association of Corrosion Engineers, Houston, Texas.

Ray, S. M. and Trieff, N. M. (1980). "Bioaccumulation of Anthropogenic Toxins in the Ecosystem." In N. M. Trieff (Ed.), *Environment and Health.* Ann Arbor Science, Ann Arbor, Michigan, p. 93.

Robinson, J., Roberts, M., Baldwin, M., and Walker, A. I. T. (1969). The pharmacokinetics of HEOD (Dieldrin) in the rat. *Food Cosmetic Toxicol.* **7**, 317.

Shaw, G. R. and Connell, D. W. (1980). Relationships between steric factors and bioconcentration of polychlorinated biphenyls (PCB's) by the sea mullet (*Mugil cephalus* Linnaeus). *Chemosphere* **9**, 731.

Shaw, G. R. and Connell, D. W. (1983). Factors influencing concentrations of polychlorinated biphenyls in organisms from an estuarine ecosystem. *Aust. J. Mar. Freshwater Res.* **33**, 1057.

Smith, J. H., Mabey, W. R., Bohonas, N., Holt, B. R., Lee, S. S., Chou, T. W., Bomberg, D. C., and Mill, T. (1978). Environmental Pathways of Selected Chemicals in Freshwater Systems: Part II. Laboratory Studies. EPA Report 600/7-78-074.

Spencer, W. F., Farmer, W. and Cliath, M. M. (1973). Pesticide volatilization. *Residue Rev.* **49**, 1.

Stumm, W. and Morgan, J. J. (1970). *Aquatic Chemistry.* Wiley-Interscience, New York, pp. 300–382.

Thibodeaux, L. J. (1979). *Chemodynamics.* John Wiley & Sons, New York.

Tinsley, I. J. (1979). *Chemical Concepts in Pollutant Behavior.* John Wiley & Sons, New York.

Veith, G. D., DeFoe, D. C., and Bergstedt, B. V. (1979). Measuring and estimating the bioconcentration factor of chemicals in fish. *J. Fish Res. Board Can.* **36**, 1040.

White, A., Handler, P., and Smith, E. M. (1973). *Principles of Biochemistry,* 5th ed. McGraw-Hill, New York.

Wilkinson, C. F. (1980). "The Metabolism of Xenobiotics: a Study in Biochemical Evolution." In H. R. Witschi, (Ed.), *The Scientific Basis of Toxicity Assessment.* Elsevier, Amsterdam, p. 251.

Wolfe, N. L., Zepp, R. G., and Paris, D. F. (1978). Use of structure reactivity relationships to estimate hydrolytic persistence of carbamate pesticides. *Water Res.* **12**, 561.

Young, D. R., Mearns, A. J., Tsu-Kai Jan, Husen, T. C., Moore, M. D., Eganhouse, R. P., Hershelman, G. P., and Gossett, R. W. (1980). Trophic Structure and Pollutant Concentrations in Marine Ecosystems of Southern California. CalCOF1 Rep. Vol. XXI, 197.

3

ENVIRONMENTAL TOXICOLOGY

The assimilation of xenobiotic chemicals within organisms at sublethal or lethal levels may induce a sequence of biological effects. These range from molecular interference with biochemical mechanisms and interactions with cellular organelles (e.g., DNA and RNA molecules), through to pathological changes at the cellular, tissue, and organ levels. Finally, these result in an integrated functional or behavioral response, experienced at the whole organism level, which may be reversible or irreversible.

There is a need to extend our understanding of the significance of such toxic effects from the individual level to higher levels of biological organization and complexity. Consequently, toxicological assessment of chemicals has shifted toward studies concerned with determining the effects at the population, community, and ecosystem levels (Auerbach and Gehrs, 1980).

On the other hand, new insights into biochemistry, cellular biology, cytogenetics, and pharmacology have focused increased attention on the ability of toxic agents to induce responses and pathological changes at the cellular level through interaction with cellular organelles and enzymes (Jernelöv et al., 1978). Concurrent with this trend, an extensive array of sublethal effects in organisms subjected to environmental stresses or physical and chemical agents has been discovered.

The problem confronting environmental toxicologists is the complex task of identifying (1) the consequences to individuals in the natural environment of effects generally demonstrated in laboratory studies, and (2) the ecological significance of these often subtle effects experienced by an individual. Approaches to this task are based on the structural and functional interrelationships existing between each succeeding level of biological organization. Thus it is important to determine the relationships between effects demonstrated at the macromolecular or cellular level and the ultimate response of the organism. A thorough knowledge of basic mechanisms by which toxic agents impair organisms is fundamental to any ecological assessment. But, in addition to stress–effect relationships, it is important to evaluate the interaction of environmental factors which may modify responses observed in organisms.

TOXICANT BEHAVIOR IN ORGANISMS

Study of the relationships between chemical structures and the biological activities of compounds allows mechanisms to be developed explaining the interaction of

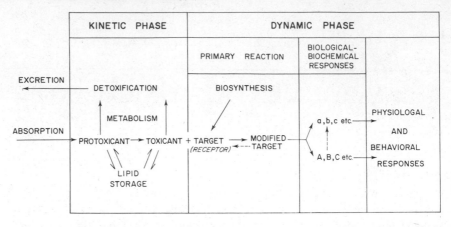

Figure 3.1. Schematic illustration of different phases of toxicity. [Adapted from Aldridge (1980). Reprinted with permission from Elsevier Biomedical Press.]

certain chemical structures with specific cellular processes. This approach also permits the development of predictive relationships between structure and toxic action (Shank, 1980).

Current knowledge of toxic mechanisms offers a basis to formulate generalized concepts of toxicity. Figure 3.1 illustrates a basic toxicological framework which involves a succession of phases from uptake or absorption of the toxicant to the physiological or behavioral response of the exposed organism. The initial phase covers those biological processes that affect the absorption, distribution, and metabolism of the chemical agent, as discussed in Chapter 2. It is this kinetic phase that determines the chemical form and transport of the active chemical agent to its primary site of action. At this point, there is a toxodynamic phase where the proximal toxic chemical interacts with the primary target or receptor (enzyme, lipid, membrane, nucleic acid, etc.). The ligand–receptor complex formed in the primary reaction, or, alternatively, any reactive free radicals (e.g., peroxides or hydroxyl ions) may initiate a complex sequence of biological effects that transpose into a lethal or sublethal response.

Table 3.1 summarizes some of the possible mechanisms through which toxic agents can impair important biochemical processes and physiological functions in living organisms (Ariens et al., 1976; Jernelöv et al., 1978; De Bruin, 1976). The degree of response will depend on the actual concentrations or doses that reach the receptor or target tissues in the dynamic phase.

Metabolic Aspects of Toxicity Mechanisms

Metabolic pathways and the formation of transformation products from xenobiotics were considered in Chapter 2. Within the context of this chapter, it is pertinent to examine the role of toxification and detoxification processes in toxicity mechanisms. Increasing mechanistic evidence on the relationship between xenobiotic metabolism

Table 3.1.   Summary of Some Important Biochemical and Physiological Effects of Toxic Agents

1. *Cellular Membranes*

    Function: Membranes form barriers with specific permeability properties which, together with the mediated transport systems, control the rate and degree of transfer of substances into and out of a cell and cell organelles.

    Toxic Response: (1) Disruption or modification of membrane permeability; (2) disturbance of mediated transfer systems by interference with carriers and ATP production.

2. *Enzymes*

    Function: Highly specialized class of proteins that catalyze the many chemical reactions involved in the intermediate metabolism of cells.

    Toxic Response: Reversible or irreversible inhibition of enzymes (coenzymes, substrates, or metal activators) by chemical agent.

3. *Lipid Metabolism*

    Function: Lipids are important cellular constituents which are involved in the structure of cell membranes and many cellular functions and metabolic processes.

    Toxic Response: Disturbances of lipid metabolism may result in impaired liver function, including pathological lipid accumulation in the liver. The capacity of the liver to synthesize cholesterol may also be impaired.

4. *Protein Biosynthesis*

    Function: The biosynthesis of a specific cellular protein, including enzymes, involves the arrangement of its constituent amino acids in a characteristic sequence in the polypeptide chain. The proper sequence is genetically determined by the structure of nuclear DNA. The actual synthesis is achieved by initial transcription of the base sequence of template DNA to that of messenger RNA, serving as an intermediate carrier of the genetic code, which translates the information stored in DNA for the amino acid sequences of proteins to be synthesized.

    Toxic Response: Protein synthesis can be influenced by a large number of exogenous substances, mainly through depression of protein synthesizing capacity located in the rough endoplasmic reticulum of the cytoplasm within the cell. In some cases, a stimulatory effect develops through enhancement of microsomal protein synthesis.

5. *Microsomal Enzyme Systems*

    Function: Metabolic multienzyme system located in hepatic microsomal fractions. This system is also responsible for biotransformation of exogenous compounds.

    Toxic Response: Alteration in microsomal enzyme function — stimulation or inhibition induced by many environmental chemicals.

6. *Regulatory Processes and Growth*

    Function: Metabolic pathways and rates of biosynthesis and catabolism of cell components are controlled by hormones and other regulatory systems.

Table 3.1.   (continued)

Toxic Response: (1) The structure or activity of regulatory enzymes may be altered and the synthesis, storage, release, or sequestration of hormones may be impaired by toxic agents in many ways. (2) Decrease in growth rate may follow chemical interference with metabolic pathways and rates.

7.   *Carbohydrate Metabolism*

Function: The intermediatory metabolism of carbohydrates involves the following major biochemical processes: (i) glycolysis − the breakdown of glucose to pyruvate or lactate; (ii) citric acid − or tricarboxylic acid cycle − major metabolic route for the aerobic oxidation of acetyl CoA, derived from the decomposition of carbohydrate, lipid and protein, into CO_2 and water; (iii) hexose monophosphate (HMP) or pentose phosphate cycle − alternative degradative route of glucose, leading to its complete oxidation into CO_2 and water; (iv) glycogenolysis − the breakdown of storage glycogen, mainly into glucose.

Toxic Response: Impairment of oxidation and glycolytic processes: biochemical lesions caused by chemical agents may interfere with the normal processes of carbohydrate biosynthesis and breakdown, leading to derangements in respiratory chain reactions.

8.   *Respiration*

Function: Biological oxidation proceeds through a sequence of intermediatory carriers which transfer electrons from substrates to molecular oxygen. This mitochondrial system, known as the respiratory chain, is of vital significance for supplying energy to the living cell. A concomitant process to electron transfer through the mitochondrial chain is oxidative phosphorylation which involves the generation of ATP from ADP through close coupling with oxidative processes in the respiratory chain. This process allows the energy arising from the respiratory process to be chemically captured and stored as high-energy phosphate (e.g., ATP).

Toxic Response: (1) Electron transport along the respiratory chain may be inhibited at specific sites by different toxic agents. (2) Uncoupling and inhibition of oxidative phosphorylation.

and toxicity is available (Khan and Bederka, 1974; Gut et al., 1981; Witschi, 1980; Ciba Foundation Symposium 76, 1980). Any mechanistic hypothesis of toxicity needs to recognize the activities of the parent compound plus metabolites. Boyd (1980) has proposed two general classes of toxicity mechanisms that distinguish between extrahepatic (nonliver) toxicity due to active parent compounds and active metabolites. But both mechanisms can be extended to cover hepatic and extrahepatic toxicity. Figure 3.2 illustrates schematically, the potential toxification and detoxification pathways involving these two classes of toxicity mechanisms.

1.   *Mechanism A.*   The parent compound is the ultimate toxic agent responsible for damage to hepatic or extrahepatic tissues and organs. Metabolism serves the detoxification function only. The site of toxic action may be specific or non-

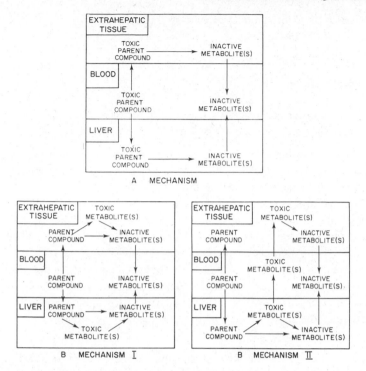

Figure 3.2. Schematic representation of extrahepatic and hepatic metabolic mechanisms of toxic substances. (*a*) Toxicity mechanism A. Parent compound is ultimate toxin; metabolism serves detoxification function only. (*b*) Mechanism B showing potential relationships between toxification and detoxification pathways in extrahepatic toxicity produced by active metabolites generated *in situ* (B mechanism I) or formed in the liver and delivered by the circulation (B mechanism II). [From Boyd (1980). Reprinted with permission from Excerpta Medica.]

specific. It may depend upon factors such as selective exposure, preferential uptake, and accumulation of the toxic agent in the target tissue, or presence of specific receptors or other highly susceptible sites of action.

2.   *Mechanism B.*   Toxicity is induced by active metabolites and not the parent compound. Active metabolites may be generated *in situ* in extrahepatic or hepatic cells and tissues. In relation to extrahepatic toxicity, Boyd (1980) has identified two types of mechanisms: "B" mechanism I where the toxic metabolites are formed *in situ* in extrahepatic tissues or cells and "B" mechanism II where they are formed in the liver and circulated to extrahepatic tissues. Again the degree of damage caused in organs and tissues will depend on factors such as those mentioned with Mechanism A. In addition, the distribution and activities of metabolic systems will also be critical determinants of toxic action.

Furthermore, a number of other mechanistic combinations or hybrid mechanisms can be derived from the two essentially polarized schemes outlined above. These are important in situations with multiple toxicity exposures, for example, uptake of toxic metals and chlorinated hydrocarbon insecticides where hybrid categories, such as A/B, or BI/II would occur.

Toxicological Reactions

Two basic types of toxic reactions can be distinguished (see Ariens et al., 1976) which involve the parent compound or a metabolite as the active agent.

1.   Chemical lesions which are caused by irreversible covalent binding between the chemical agent and the biological substrate or receptor, for instance, the co-valent binding of toxic metals such as lead, cadmium, and mercury, to − SH groups in enzymes and other important proteins, as well as the binding of HCN and H_2S to the iron groups in cytochromes. Other types of lesions include carcinogenic and mutagenic reactions involving DNA molecules and reactive intermediates such as biological aklylating agents and microsomal metabolites (e.g., epoxides).

2.   Reversible interactions which occur between the exogenous substance and biological substrate or receptor. These substances undergo reversible interactions with specific receptors, usually leaving the substance and receptor unaltered. In some instances the receptor may experience temporary alterations in its confor-mations during formation of the ligand–receptor complex.

Further resolution of reaction mechanisms associated with toxic reactions, such as cellular necrosis, has resulted in the postulation of several possible molecular mechanisms of toxicity (e.g., Gillette, 1980; Farber, 1980).

In Figure 3.3, a conceptual model is presented which attempts to incorporate

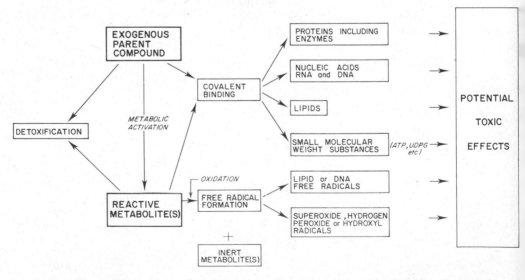

Figure 3.3.   Possible mechanisms of toxicity induced by an exogenous parent compound and/or its chemically reactive metabolites.

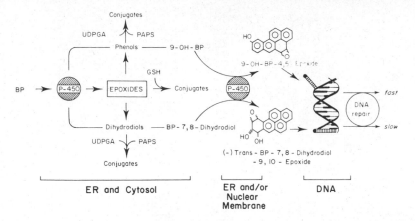

Figure 3.4.   Possible mechanisms involved in the formation and regulation of DNA-binding products of benzo[a]pyrene in hepatocytes isolated from 3-methylcholanthrene-pretreated rats. BP, benzo[a]pyrene; UDPGA, uridine-5'-diphosphoglucuronic acid; PAPS, 3'-phospho-adenosine-5'-phosphosulphate; GSH, glutathione, reduced form; ER, endoplasmic reticulum. [From Orrenius et al. (1980). Reprinted with permission from Excerpta Medica.]

postulated mechanisms of toxicity caused by both chemically reactive parent compounds and/or metabolites. As described by Gillette (1980), the target of a reactive agent could be (1) an intracellular enzyme (or its substrates) required for cellular processes, (2) a phospholipid in cellular membranes which control the compartmentalization of intracellular components, (3) substrates involved in the synthesis of proteins required for the normal replacement of intracellular enzymes, or (4) the DNA required for cellular replication. For example, the metabolic activation of benzo[a]pyrene in hepatocytes involves the formation of DNA-binding products, principally the highly carcinogenic benzo[a]pyrene-7,8-dihydrodiol-9,10-epoxide and a less carcinogenic metabolite, probably 9-hydroxybenzo[a]pyrene-4,5-epoxide (see Figure 3.4) (Orrenius et al., 1980). It is also apparent that toxic effects may depend on the impairment of several of these targets simultaneously (Gillette, 1980).

Toxic responses may be induced by the formation of free radicals and not mechanisms in which the reactive agent or metabolite is covalently bound to the target substance. In certain cases, metabolites can react to form lipid or DNA free radicals or undergo oxidation to initiate superoxide, hydrogen peroxide, or hydroxyl free radicals (Gillette, 1980).

The generation of free radicals in biological systems is usually detrimental to the viability of biological membranes, due to the initiation of destructive auto-oxidation reactions of lipoprotein membrane constituents. Furthermore, potentially toxic polyunsaturated fatty acid peroxides could accumulate in tissues and possibly inactivate sulfhydryl groups in enzymes, or other key bioconstituents, by free radical attack.

Certain types of halogenated hydrocarbons are considered to initiate hepatotoxicity through lipid peroxidative processes linked to a process of free radical

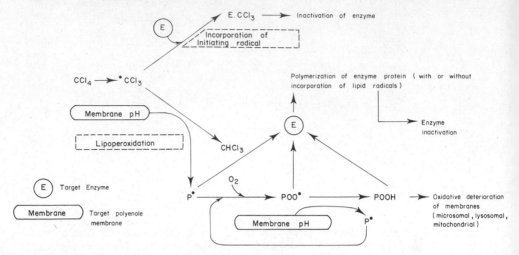

Figure 3.5.   Free radical attack and lipoperoxidation of target biomolecules — enzymes and polyenoic fatty acids. [From De Bruin (1976). Reprinted with permission from Elsevier Biomedical Press.]

generation according to Figure 3.5 for carbon tetrachloride induced lipoperoxidation (see De Bruin, 1976). For example, carbon tetrachloride binds to microsomal cytochrome P-450 in target cells and is rapidly metabolized to metabolites which bind covalently to cell components, lipid peroxidation takes place, and intracellular calcium homeostasis is altered. Subsequently, protein synthesis is inhibited and the endoplasmic reticulum is disrupted, possibly leading to cell impairment or death, for example, liver necrosis (McLean et al., 1980). It is suggested that the initial step in liver necrosis by the protoxin carbon tetrachloride involves the reaction of the free radical metabolite, trichloromethyl, which reacts with lipid to form chloroform and free radical lipids.

The association of reactive substances and covalent binding, or free radical mediated reactions, with the development of toxicity is complex. Experimental studies, particularly on carcinogenesis, indicate that reactive metabolites usually undergo a multiplicity of reactions with cellular components. The rates of these processes may be critical to the emergence of toxicity, and depend on factors such as affinity for certain nucleophilic groups, for example, amino, sulfhydryl, and thiol groups, and stability of ligand–receptor complexes. Covalent binding, for example, indicates the generation of reactive molecules in cells, but the production of cell injury depends on the size and shape of the reactive molecule and on the particular cell macromolecule involved (McLean and Nuttall, 1978).

Basic mechanisms of cell necrosis have not been elucidated (Mclean et al., 1980) although the site of initial interaction is reasonably well established for some toxic substances. For example, in the inhibition of pseudocholinesterase by organophosphorus and carbamate insecticides, stable reactive metabolites combine reversibly with certain sites on proteins and this complex rearranges to form covalently bound material.

Useful biochemical information on toxicity mechanisms has arisen from the investigations of TCDD toxicity. Poland and Glover (1980) have proposed a model for the mechanism of toxicity with dibenzo-p-dioxin congeners, dibenzofurans, azo- and azoxy-benzene, and halogenated biphenyls. A correlation was observed between their potency to induce AHH (aryl hydroxylase) activity and their cytosol binding affinity (see Figure 3.6). It was postulated that the toxicity of these chlorinated aromatic hydrocarbons is mediated by the initial stereospecific recognition of and binding to hepatic cytosol. The ligand–receptor complex so formed controls a battery of genes, and the expression (or repression) of one or more of these genes results in the observed toxic syndrome.

Poland and Glover (1980) have pointed out that their model relates to toxicity mediated through binding to the receptor but it does not explain the ultimate biochemical lesions responsible for the toxicity of these compounds. How this model accounts for the behavior of nonhalogenated compounds (e.g., polycyclic aromatic hydrocarbons and β-naphthoflavone), which also bind to the induction receptor but do not produce the same toxic responses, is not fully clear. However, Poland and Glover (1980) suggest that toxicity differences are related to the sustained occupancy of the receptor by the halogenated aromatic compounds in comparison to the rapidly metabolized polycyclic aromatic hydrocarbons.

It is evident from the preceding discussion that the molecular or biochemical approach to understanding toxicology has identified the significance of metabolic processes in forming reactive metabolites, usually within target cells, which exert biological or toxic responses through specific binding activities with cellular macromolecules. Farber (1980) has stressed, however, the inherent difficulties in attempting to develop chemical reactions into mechanisms of cell injury. Little insight into the sequence of events which presumably exists between these biochemical interactions and cell injury or death is available.

RELATIONSHIPS BETWEEN DOSE OR ENVIRONMENTAL CONCENTRATION AND RESULTANT EFFECTS AND RESPONSES

In considering the relation between the dose or concentration of an environmental toxicant and the resultant reaction in biological systems, it is useful to distinguish between responses and effects. Usually, dose–response refers to the relation between dose and the nature of the effect. Brown (1978) has defined dose–response, in general terms, as the relation between any measurable stimulus, physical, chemical, or biological, and the response of living matter in terms of the reactions produced over the same quantitative range. The reactions to any one stimulus may be multiple in nature with each reaction having its own unique relation with the degree of stimulus. A specific reaction can be quantitatively assessed either by the magnitude of the effect produced, including whether the effect is produced or not, or of the time required for the appearance of the specific effect (Brown, 1978). Time effects can be classified, nominally, as acute or chronic in terms of an organism's lifespan or even multigenerational for certain mutagenic effects. The intensity of abnormal responses to stimuli will determine lethal or sublethal effects.

Relative binding affinity	Relative biological potency	Structure	Relative binding affinity	Relative biological potency
100	100		Inactive (5.4×10^{-7})	Inactive (9.4×10^{-8})
167	100		Inactive (2.7×10^{-8})	Inactive (9.4×10^{-8})
43	43		Inactive (5.4×10^{-8})	Inactive (9.4×10^{-8})
20	22		Inactive (5.4×10^{-8})	Inactive (4.7×10^{-7})
16	8		Inactive (5.4×10^{-8})	Inactive (9.4×10^{-8})
13	3		Inactive (2.7×10^{-8})	Inactive (9.4×10^{-8})

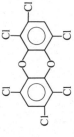

14   0.06   Inactive (1.1×10^{-8}) Inactive (9.4×10^{-8})

Figure 3.6. The cytosol binding affinities and biological potencies of dibenzo-*p*-dioxin congeners relative to TCDD. The binding affinities of the dibenzo-*p*-dioxin congeners for hepatic cytosol were estimated by the capacity of these compounds to compete with (^{3}H)-TCDD for specific binding sites in C57BL/6J liver cytosol. The binding affinities are expressed relative to TCDD which is assigned a value of 100. For inactive analogues, the highest concentration tested that was judged to be soluble is given in parentheses. The biological potency (ED$_{50}$) of each congener was the dose that produced one-half the maximal induction of hepatic AHH activity in the chicken embryo, and the potency was expressed relative to TCDD (TCDD = 100). For inactive compounds, the highest dose tested (in mol kg^{-1}) is given in parentheses. This assumes the weight of an average chicken egg is approximately 50 g. [From Poland and Glover (1980). Reprinted with permission from Elsevier Biomedical Press.]

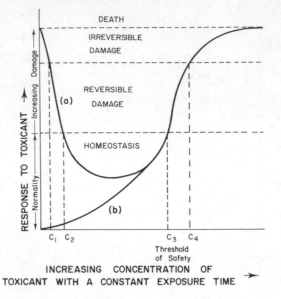

Figure 3.7.   The hypothetical relationship between observed effects and toxicant concentration, that is, dose–response curve: (*a*) the observed effects of a substance essential to life at low concentrations; (*b*) the effects of a nonessential substance. [From Duffus (1980). Reprinted with permission from Edward Arnold (Publishers), Inc.]

Unless exact mechanisms or modes of toxic action within organisms are known, it is common to measure the relation between different dose levels of toxicants and induced effects on biological systems that can be either readily observed in nature or experimentally tested under laboratory or field conditions. Relationships are generally expressed in the form of dose–response or effect curves.

Hypothetical relationships between observed effects and toxicant concentration, that is, a dose–response curve, are represented in Figure 3.7 where two basic types of effects are represented: (a) the effects on an organism exposed to a substance essential for life and (b) the effects induced by a nonessential substance. In case (a), homeostasis is maintained between the concentration ranges C_2 and C_3. Below the concentration C_2, a deficiency in the essential substance exists, a nutritional deficiency, while at excessive concentrations above C_3, the substance acts as a toxicant. Irreversible damage and ultimately death may occur at concentrations below C_1 and above C_4. In case (b), the nonessential substance acts as a toxicant at concentrations above C_3, which represents the "no effect" level or threshold of safety. For low dose–response of nonessential substances the existence of a threshold level is equivocal and will be discussed in the section entitled Dose–Response and Dose–Thresholds.

Dose–response relationships describe what percentage of a population, for example, organisms or receptor sites, exhibit a defined effect at a certain dose or environmental concentration (see Figure 3.7). In comparison, dose–effect relations are usually explained on the basis of interactions between the reactive substance A and molecular sites of action, the receptors R (Ariens et al., 1976). It is assumed that these interactions, according to the Law of Mass Action, can be described by a simple equilibrium reaction:

$$A + R \rightleftharpoons RA$$

The induced effect E is a simple function of the quantity of ligand–receptor complex formed:

$$E = f[RA]$$

or a fraction of the total number $[r]$ of receptors occupied, $[RA]/[r]$.

When this quotient equals 1, that is, all receptors are occupied, substance A induces its maximal effect E_{max} (100% effect).

As described by Ariens et al. (1976), the principle behind the dose–effect concept rests with the affinity, expressed as the equilibrium constant, of a substance for specific receptor sites and its intrinsic ability to cause changes in receptor molecules, for example, conformation of the receptor which initiates a sequence of events that leads to the effect. The characteristics of affinity and intrinsic activity can be compared to the affinity constant for the interaction between enzymes and substrates and the V_{max} value for the conversion of a substrate by an enzyme (conversion rate reached when the enzyme system is saturated by the substrate), respectively. The magnitude of the effect will depend on the dose of the active substance, for example, reactive metabolite, and its affinity and intrinsic activity in competition with other substances, for example, essential endogenous substances or antagonists for receptor sites. From dose–effect curves the dose can be determined at which a certain effect (e.g., 50% of the maximal effect) is produced and the magnitude of the maximum effect that can be reached with the particular substance.

Dose–Response Models

Dose (or environmental concentration)–response expressions refer to the relationships between the characteristics of exposure to a toxic agent and the spectrum of observed effects. The relationships imply a reasonable presumption that the observed effects are induced by a known toxic agent (Klaassen and Doull, 1980). Numerical and graphical expressions of the dose–response relationship are based on assumptions that: (1) the response is a function of the concentration of a toxic agent at a site of action; (2) the concentration at a site is a function of the dose; and (3) response and dose are causally related (Klaassen and Doull, 1980). That is, the probability of a test organism having some type of response is a function of the dose (exposure level) or, more appropriately, the effective dose D of the toxicant. Mathematical models for estimation of dose–response relationships have been reviewed recently by Brown (1978) and Van Ryzin and Rai (1980).

In toxicological studies, or bioassays, a sensitive and unequivocal measure or index of toxicity is necessary. This may be related, directly or indirectly, to a biochemical, physiological, or behavioral function such that a causal relationship can be established or reasonably presumed. One type of response commonly measured is the quantal ("all or nothing") response, for example, death, particularly where quantification of a response is difficult or impossible. In practice, lethality represents a precise and unequivocal toxicity measure that is widely used to estimate dose–response relationships within an exposed test population of organisms.

When the response is quantal, there will be some level of stimulus, dose, or concentration, at which, under constant environmental conditions, the response will not occur and above which it will. This level is referred to as the organism's tolerance. However, it is more useful to consider the distribution of tolerances over a population of test organisms. This situation can be represented in the form of a tolerance distribution, $f(D)dD$, along with its corresponding cumulative distribution $P(D)$, as

$$P(D_1) = \int_0^D f(D)dD$$

when a population is exposed to a dose of D_1. The function $P(D)$ represents the dose–response relationship for the population when the response is quantal in nature and the parameter values are assumed to be $P(0) = 0$ (no responders for zero dose) and $P(\infty) = 1$ (all respond to some high dose).

Quantal responses such as lethality exhibit a range of differences in susceptibility among individuals within a population. Usually, these responses follow a normal gaussian distribution in relationship to log (dose) as represented by the frequency histogram, given in Figure 3.8A. More conveniently, the data in Figure 3.8A can be replotted in the form of a sigmoidal cumulative dose–response curve, as in Figure 3.8B. A probit transformation is frequently used to plot mortality, or other quantal data, against dose since in a normal population this results in a straight line (Finney, 1971).

In a normally distributed population, the mean ± standard deviation (SD) represents 68.3% of the population, the mean ± 2 × SD represents 95.5% of the population, and the mean ± 3 × SD equals 99.7% of the population. Since quantal dose–response phenomena are usually normally distributed, the percent response can be converted to units of deviation from the mean or normal equivalent deviations (NED). Thus the NED for a 50% response is zero; a NED of + 1 is equated with 84.1% response, and − 1 represents a 15.9% response. Units of NED are usually converted by the addition of 5 to the value to avoid negative numbers and these converted units are called "probit" units. The probit, then, is a NED + 5. Thus a set of relationships between percent response, normal equivalent deviation, and probit can be established, for example, 0.1, − 3, 2; 2.3, − 2, 3; 97.7, + 2, 7; and 99.9, + 3, 8. Using probit units normal distributions can be plotted as straight lines as shown for mortality data in Figure 3.8C (Finney, 1952).

The dose–response relationship can be simply described by the LD_{50} (or LC_{50}) value, which is the dose determined from the line of best fit at the 50% mortality value (see Figure 3.8B) or equivalent probit unit for 50% mortality (i.e., 5) (see Figure 3.8C). Although probit analysis is a commonly used parametric technique

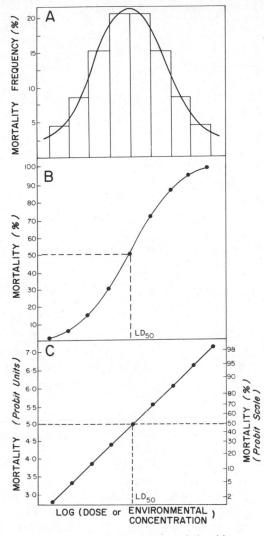

Figure 3.8.   Dose–response relationships.

for handling toxicity data, significant deviations from the log probit model can occur, for example, when data are not normally distributed. Buikema et al. (1982) have recently reviewed statistical and experimental approaches used in acute toxicity testing for aquatic organisms, including a discussion on departures of experimental data from log probit models.

Other mathematical models of tolerance distributions which lead to the sigmoid appearance of the corresponding dose–response curves are the logistic and sine curves (see Brown, 1978). Several major mathematical models proposed in the literature, including the probit model, are given in Table 3.2. It should be noted that these models refer only to dose–response and do not consider time.

Table 3.2.   Mathematical Dose–Response Models

Models	$P(D)$	Parameter Values
Probit	$\Phi(\alpha + \beta \log D)$	$-\infty < \alpha < \infty$
	$\Phi(t) = \displaystyle\int_{-\infty}^{t} (2\Pi)^{-1/2}\, e^{-u^2/2}\, du$	$\beta > 0$
One-hit	$1 - e^{-\beta D}$	$\beta > 0$
Multihit	$\displaystyle\int_0^{\beta D} \frac{u^{k-1} e^{-u}}{\Gamma(k)}\, du$	$k > 0$
	$\Gamma(k) = \displaystyle\int_0^{\infty} u^{k-1} e^{-u}\, du$	$\beta > 0$
Multistage	$1 - \exp\left(-\displaystyle\sum_{i=1}^{k} \alpha_i D^i\right)$	k a positive integer $\alpha_i \geqslant 0$

Source: Van Ryzin and Rai (1980). Reprinted with permission from Elsevier Biomedical Press.

Apart from the probit model the other three models in Table 3.2 are stochastic biological models derived from what has been called "hit theory." These models do not assume a dose–tolerance distribution to produce a dose–response curve, but rather general mechanistic dose–response assumptions are made (see Turner, 1975). In the one-hit and more general multihit models, a toxic response is conceived as the result of one or more hits to the target organ which results in a toxic response, when the number of hits is a homogeneous Poisson process in dose with intensity parameter β_k (Van Ryzin and Rai, 1980). For the multihit model, the parameter k represents the minimum number of "hits" on a receptor which results in a toxic response. In the low-dose range, as $P(D)$ becomes approximately equal to D, the one-hit model takes a linear form passing through the origin.

The multistage model was originally proposed as a mechanism in carcinogenesis (see Van Ryzin and Rai, 1980). In this model, the positive-integer k refers to the number of stages in the process affected by the toxic agent. When $k = 1$ in the multihit or multistage models, these models take the form of the one-hit model.

A useful environmental application of these models includes the incorporation of a nonzero background response, $p > 0$, which may simulate natural or spontaneous responses. For models in Table 3.2, where $P(0) = 0$, the extended model with background response is given by

$$P^*(D) = p + (1-p)P(D)$$

where $P(D)$ is as in Table 3.2. The additional risk over background due to an added dose D can be defined as

$$P(D) = \frac{P^*(D) - p}{1 - p}$$

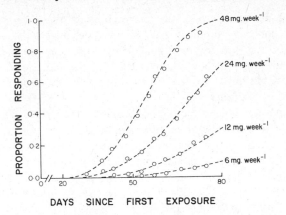

Figure 3.9. Comparison of estimated (−−−) and observed (ooo) proportions of animals with skin tumors at four dose levels. [From Lee and O'Neill (1971). Reprinted with permission from H. K. Lewis and Company, Ltd.]

Time–Effect Relationships

Apart from dose (environmental concentration–effect relationships), time–effect relationships are also important in understanding toxic effects. Figure 3.9 gives an example of the relationships between time, dose, and response. The time taken for the induction of the effect or exposure time without evidence of any effects are important aspects. An example of a "toxicity curve" is shown in Figure 3.10

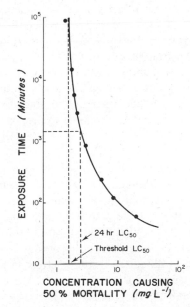

Figure 3.10. Toxicity curve of log exposure time vs. log LC_{50} with the "incipient" or "threshold" LC_{50} indicated when the curve becomes asymptotic to the time axis.

where LC_{50} is plotted against the corresponding exposure time for aquatic organisms. The concentration at which the curve becomes asymptotic to the time axis is referred to as the "asymptotic," "threshold," or "incipient" LC_{50}. Concentrations below this will not cause lethality in the short term. This measure is often used in establishing risk criteria for natural populations of organisms exposed to chronic levels of toxicants.

Thus the exposure time, or duration of a bioassay experiment, is a critical factor in deriving "safe" levels of exposure. For example, in lethal and sublethal tests with marine invertebrate species exposed to the water soluble fraction of No. 2 fuel oil, Anderson (1977) was able to demonstrate the significance of time–concentration factors by discriminating between short-term physiological responses such as respiration (10 hr or less) and long-term growth and reproduction effects (72–672 hr). The animals were able to adapt to short-term physiological stress and recover upon transfer to clean seawater. Juveniles and larvae tolerated constant exposure to water-soluble hydrocarbons over the long term but the appearance of significant reduction in growth became more obvious when concentration and time were taken into consideration together.

Various mathematical models have been developed which attempt to relate dose levels or environmental concentrations, and time–effect relationships. Brown (1978) has reviewed the use of these models, all of which have direct correspondence with quantal dose–response models (Chand and Hoel, 1974).

Dose–Response and Dose Thresholds

In quantal dose–response models it is assumed that for any dose $D > 0$, $P(D) > 0$, that is, there is no absolute threshold level below which the probability of responses is zero (Brown, 1978). Many experimental studies exist, however, where an apparent threshold level can be: (1) extrapolated from the dose–response curve to correspond to a dose level at which there appears to be no added response over controls or (2) simply indicated by dose levels at which there are no abnormal responses over controls. It is a common toxicological practice to predict risks associated with low doses by extrapolation from high dose–response curves, particularly with stochastic effects such as in tumorigenesis where it is experimentally difficult or costly to observe low-dose responses.

With this work it is important to determine whether the dose–response curve displays a threshold or quasithreshold, or is linear (see Figure 3.11). Bridges (1980) suggests threshold or quasithreshold curves originate via a progressively saturating system of detoxification, an inducible activation pathway, or by differential repair processes. Brown (1978) points out that downward extrapolation of the observed dose–response relation, often linear, can be erroneous when the curve is convex. Conversely, it is possible to obtain a saturating effect in the dose–response curve at high dose levels (see Figure 3.11). In this case, the curve has a steep slope at low doses but tends to flatten out at higher levels. The saturating curve could reflect several mechanisms: an inducible pathway of detoxification, a progressively saturating activation pathway, or differential repair processes.

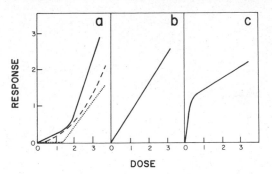

Figure 3.11. Hypothetical dose–response curves: (*a*) threshold (···), quasithreshold (–––), or cumulative (——); (*b*) linear; (*c*) saturating. [From Bridges (1980). Reprinted with permission from Excerpta Medica.]

Statistical prediction of the existence of threshold levels appears to be inconclusive. Brown (1976) has shown that statistical analysis of bioassay results cannot discriminate between mathematical models which assume the existence or nonexistence of an actual threshold. Furthermore, experimental observations of "no effect" levels do not validate that the probability of response is equal to zero. If zero organisms respond out of an exposed population N, then the risk is consistent at the 5% significance level, with an actual response rate between zero and approximately $3/N$ (Brown, 1978).

Consequently, at low doses, the shape of the dose–response relationship cannot be established with statistical confidence; hence it is usual to assume a linear dose–response relationship without threshold. This assumption is a useful approximation in the absence of a knowledge of exact toxicity mechanisms. In terms of risk assessment, the linearity assumptions can be extended to cover small increments in doses above already existing or "background" doses such as, for example, natural sources of radiation. However, for large increments, particularly at high doses, nonlinear dose–response relations are more likely.

INTERACTIONS BETWEEN POLLUTANTS AND MULTIPLE TOXICITY: CONCEPTUAL ASPECTS

In most polluted systems individuals are exposed to a diverse range of toxicants rather than a single specific agent. Interactions between these pollutants may mutually enhance, or alternatively, inhibit toxic responses in exposed organisms. Mechanisms of interaction between constituents of pollutant mixtures and models for assessing multiple toxicity responses have been described by Anderson and D'Apollonia (1978). Interactions can be either chemical or physiological within (1) the kinetic phase, by altering mechanisms of toxicant uptake, distribution, deposition, metabolism, and excretion; and (2) the dynamic phase, by altering toxicant–receptor binding affinity and activity.

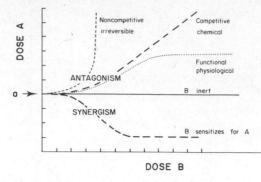

Figure 3.12.   Isoboles (curves of equal biological response) for combinations of a substance A that is active on its own and a substance B that is inactive when given alone, but influences the action of A. Synergism (increased sensitivity for A by the action of B) and antagonism (decreased sensitivity for A by the action of B) are indicated. [From Ariens et al. (1976). Reprinted with permission from Academic Press, Inc.]

Mechanisms of interaction which may induce multiple toxicity, as proposed by Anderson and D'Apollonia (1978), are as follows:

Environmental Phase.   Chemical interactions between pollutants which produce new compounds, complexes, chelates, or changes of chemical state; interactions between pollutants and substrates which alter physicochemical forms of pollutants and their toxicities;

Dynamic Phase.   Multipollutants that act at the same site(s) in target tissue(s); multipollutants that act at different sites possibly in different target tissues but which contribute to a common adverse response; mixture of pollutants where one is normally inactive as a toxic agent but in combination changes the response of an organism to one or more of the other toxic constituents; multipollutants that mutually produce a toxic response different from the response induced by each toxicant alone;

Kinetic Phase.   Multipollutants that alter toxicant availability to the target tissues; multipollutants that enhance or induce (e.g., by the mixed oxidase system in liver) the production of metabolites more than the original pollutants.

Some important types of pollutant interactions and related responses in the kinetic and dynamic phases are illustrated in Figures 3.12 and 3.13. These involve synergistic and antagonistic interactions for combinations of two substances. In Figure 3.12, the types of antagonisms indicated are "competitive antagonism," in which the antagonist displaces the agonist from its site of action, for example, displacement action of additional oxygen uptake in carbon monixide intoxication; "chemical antagonism," in which the antagonist inactiviates the agonist through chemical interaction, for example, chelation of some metals; and "noncompetitive antagonism," in which the antagonist interferes with the induction of the effect by the agonist without reacting with the agonist itself or with the specific receptors of

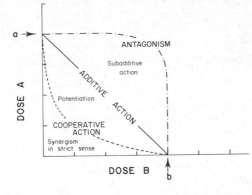

Figure 3.13.   Isoboles for combinations of two substances A and B, each of which is active by itself. Addition forms the boundary between the synergism (potentiation) and antagonism. [From Ariens et al. (1976). Reprinted with permission from Academic Press, Inc.]

the agonist, for example, atropine blocks the specific receptors of acetylcholine which accumulates after inactivation of acetylcholinesterase by organophosphorus compounds (Ariens et al., 1976). In functional antagonism, there is antagonism between two agonists that act on the same cell system, but contribute in opposite ways to a certain response in the cell. Where agonists act on different cell systems and produce opposing effects in these systems this is known as physiological antagonism (Ariens et al., 1976).

LETHAL AND SUBLETHAL EFFECTS

Pollution stresses extend from effects on individual species through to successively higher levels of biological organization incorporating populations, communities, and ecosystems. Analyses or predictions of any systematic effects or changes depend on an adequate data base on the individual organisms.

Under increasing degrees of environmental pollution these effects may transcend the normal range of adaptation and tolerance exhibited by living organisms. There usually exists a range of sublethal responses, where long-term survival potential of the organism is reduced and ultimately a level of pollution is reached which results in comparatively rapid death (Rosenthal and Alderdice, 1976). Effects on organisms are generally categorized into those causing (1) direct lethal toxicity and (2) sublethal disruption of physiological or behavioral activities.

In identifying these effects, it is pertinent to recognize not only the direct effects of a pollutant on an organism but the effects due to interactions of pollutants with the organism's environment (e.g., physical alteration of habitat by oil spills; loss of food organisms; increased competition) and modification to the pollutant caused by environmental factors (e.g., photochemical decomposition) or the organism (e.g., metabolism) (see Auerbach and Gehrs, 1980).

Of particular interest is the distinction between lethal and sublethal effects of

Table 3.3.   Selected Parameters for Sublethal Studies

Uptake, Accumulation, and Excretion

Complexation and storage, distribution within tissues and organs, kinetics of uptake and release, bioconcentration, bioaccumulation.

Physiological Studies

Metabolism, photosynthesis and respiration, osmoregulation, feeding and nutrition, heart beat rate, blood circulation, body temperature, water balance.

Biochemical Studies

Carbohydrate, lipid and protein metabolism, pigmentation, enzyme activities, blood characteristics, hormonal functions.

Behavioral Studies (Individual Responses)

Sensory capacity, rhythmic activities, motor activity, motivation and learning phenomena.

Behavioral Studies (Interindividual Responses)

Migration, intraspecific attraction, aggregation, aggression, predation, vulnerability, mating.

Reproduction

Viability of eggs and sperm, breeding/mating behavior, fertilization and fertility, survival, life stages, and development.

Genetic

Chromosome damage, mutagenetic and teratogenic effects.

Growth Alterations and Delays

Cell production, body and organ weights, developmental stages, e.g., larval and juvenile stages.

Histopathological Studies

Abnormal growths, respiratory and sensory membranes, tissues and organs, e.g., reproductive organs.

Interactions

Environmental or ecospecific factors with pollutants, multiple pollutant combinations.

---

pollutants on organisms. Qualitatively, lethal effects can be defined as those responses that occur when physical or chemical agents interfere with cellular and subcellular processes in the organism to such an extent that death follows directly. In severe cases this may take the form of smothering and suffocation, interference with movements to obtain food or escape predators, or destruction of habitat (e.g., for sedentary organisms). In comparison, sublethal effects are those that disrupt physiological or behavioral activities but do not cause immediate mortality, although death may follow because of interference with feeding, abnormal growth or behavior, greater susceptibility to predation, lesser ability to colonize, or other indirect causes. These effects may not only lead to changes in populations of individual species but may also result in shifts in species composition and diversity (see GESAMP, 1977).

Hence, Rosenthal and Alderdice (1976) have described sublethal effects as those (1) elicited through application of pollution stress at one level of organization or development, (2) recognized through the appearance of altered structure or function at a later stage of development, and (3) whose significance is fully manifested as lower survival potential at a further stage of development. Sublethal responses can be broadly subdivided into a number of categories according to effects on the organisms: (1) physiology, (2) biochemistry/cell structure, (3) behavior, and (4) reproduction (Waldichuk, 1979).

Observations can reveal empirical evidence of effects on physiological functions, such as growth and reproduction, or changes in behavioral patterns which influence chances of survival. Alternatively, effects on specific systems such as enzyme activity, tissues, and organs require laboratory examination. For example, Table 3.3 outlines a number of important laboratory tests used to measure sublethal responses in marine organisms exposed to pollutants.

A critical aspect of these investigations is the distinction between an adverse effect and an adaptive response. Thus interpretation may be confused by adaptive changes, such as acclimation, in the short term (less than one generation) or by genetic changes in populations in the long term. Furthermore, sublethal changes in organisms must be evaluated in terms of several levels of organization of the organism, which integrate stepwise from within a cell to, eventually, the success or otherwise of groups of organisms in an ecosystem.

Assessment of sublethal effects depends on two important factors: (1) the selection of physiological and behavioral parameters which predict ecologically significant responses; and (2) the experimental measurement of sublethal responses. These factors must include the recognition of critical effects which occur at different life stages (see Table 3.4). The task of extrapolating significant responses to the ecosystem level is complex and currently uncertain. For most organisms there is an inadequate understanding of natural variations in physiological responses and interactions with environmental parameters. Also, natural perturbations may quite easily mask any physiological changes induced by long-term exposure to low levels of exogenous agents.

Sublethal and Lethal Relationships

Measurements of lethality are frequently used to derive "safe" levels of exposure to toxicants. This includes, for example, the use of "application factors" (e.g., 1% or 0.01 of 96 hr LC_{50}) to calculate "safe" levels which may also function as water quality criteria for specific toxicants. The assumptions adopted in the application factor approach are not well supported empirically, and as an alternative, the use of

Table 3.4.   Some of the Sublethal Effects of a Pollutant on Various Life Stages of a Marine Organism

Life Stage	Vital Life Processes	Critical Effects of Pollutants
Egg	Meiotic division of cells; fertilization; hatching enzyme activity; cleavage mitoses of fertilized egg; respiration	Reduced fertility; gene damage; chromosome abnormalities; damage to egg's membrane; reduced hatching enzyme activity; direct toxicity to embryo from pollutant; impaired respiration
Larva	Metamorphosis; morphological development; feeding; growth; avoidance of predators; parasites and disease	Toxicity from bioaccumulated poisons in yolk sac during early feeding; biochemical changes; physiological damage; deformities; behavioral alterations
Juvenile	Feeding; growth; development of immune systems; endocrine glands; avoidance of predators; parasites and disease	Direct toxicity; reduced feeding and growth; altered predator–prey relations; impaired chemoreception; reduced resistance to parasites and disease
Adult	Feeding; growth; sexual maturation (development of gonads with male and female gametes)	Direct toxicity; adverse alteration of environmental conditions, e.g., dissolved oxygen; physiological and biochemical changes; behavioral and alterations
Reproduction	Spawning migration; spawning act (fertilization); successful completion of life cycle	Avoidance reaction on spawning migration, destruction of spawning grounds; direct toxicity; reduced fertility

Source: Waldichuk (1979). Reprinted with permission from The Royal Society.

chronic, sublethal tests may be more appropriate. Sublethal measurements are considered suitable for predicting "safe" or "ecologically insignificant" levels of toxicants if responses are quantified and statistic dose–response relationships are derived. Conceptually, this approach appears feasible but selection of sublethal criteria, test organisms, and experimental conditions are major experimental obstacles.

Relationships between sublethal and lethal toxicities continue to be of major importance. The general dose–response relationship for biological assays on the effect of pollutants on experimental organisms is schematically represented in Figure 3.14. A dichotomy of dose–responses can be postulated, linear and non-

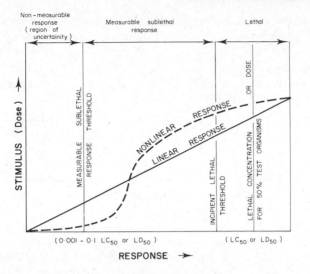

Figure 3.14. Hypothetical relationship of concentration of pollutant to response of organisms, showing some significant points and regions on the curve. [From Waldichuk (1979). Reprinted with permission from The Royal Society.]

linear, to illustrate the transition from sublethal to lethal responses. Hypothetical threshold levels are used to classify response phases in relation to the median lethal dose or concentration.

For most pollutants, and experimental circumstances, a region of uncertainty, equivalent to a nonmeasurable response, can be postulated for very low doses. As doses tend toward lethality, a nonlinear response is generally observed (e.g., sigmoidal). In toxicity testing, measurement can be either by *in situ* or field experiments and survey, model ecosystem testing or by laboratory bioassays. The latter technique is generally favored because experimental conditions can be controlled and the response of test organisms observed or monitored to a greater degree. Responses of test organisms can be placed in the following categories:

1. An acute effect, which is the organism's response to a condition severe enough to bring about a rapid response usually within 96 hr, for example, 96 hr LC_{50}.
2. A subacute effect, which is the organism's response to a less severe condition than present in (1) and usually after a longer term.
3. A chronic effect, which is the organism's response to a continuous condition maintained for at least 10% of the organism's life span.

In practice, most bioassay results are expressed in terms of acute lethal toxicities (e.g., 96 hr LC_{50}) or measure certain sublethal responses in test organisms over a specified period of time. For sublethal measurements, the results may be expressed as the "median effect concentration," or EC_{50}, which is the concentration at which 50% of the test organisms display the response being measured. The EC_{50} usually exhibits similar patterns of relationships to those for the LC_{50} as indicated in Figures 3.7, 3.8, and 3.9. Toxicity tests such as acute lethality tests, embryo/larval toxicity tests, chronic toxicity tests for reproduction effects, and tests on bioconcentration/bioaccumulation are useful for assessing chemical hazards to aquatic life (Cairns et al., 1978).

REFERENCES

Aldridge, W. N. (1980). "The Need to Understand Mechanisms." In H. R. Witschi (Ed.), *The Scientific Basis of Toxicity Assessment.* Elsevier/North Holland Biomedical Press, Amsterdam, pp. 305–319.

Anderson, J. W. (1977). "Responses to Sub-lethal Levels of Petroleum Hydrocarbons: Are They Sensitive Indicators and Do They Correlate with Tissue Contamination?" In D. A. Wolfe (Ed.), *Fate and Effects of Petroleum Hydrocarbons in Marine Organisms and Ecosystems.* Pergamon Press, New York, p. 95.

Anderson, P. D. and D'Apollonia, S. (1978). "Aquatic Animals." In G. C. Butler (Ed.), *Principles of Ecotoxicology*, SCOPE 12. John Wiley & Sons, New York, p. 187.

Ariens, E. J., Simonis, A. M., and Offermeier, J. (1976). *Introduction to General Toxicology.* Academic Press, New York.

Auerbach, S. I. and Gehrs, C. W. (1980). "Environmental Toxicology: Issues, Problems, and Challenges." In H. R. Witschi (Ed.), *The Scientific Basis of Toxicity Assessment.* Elsevier/North Holland Biomedical Press, Amsterdam.

Boyd, M. R. (1980). "Effects of Inducers and Inhibitors on Drug-Metabolizing Enzymes and on Drug Toxicity in Extrahepatic Tissues." In Ciba Foundation Symposium 76, *Environmental Chemicals, Enzyme Function and Human Disease.* Excerpta Medica, Amsterdam, pp. 43–66.

Bridges, B. A. (1980). "Induction of Enzymes Involved in DNA Repair and Mutagenesis." In Ciba Foundation Symposium 76, *Environmental Chemicals, Enzyme Function and Human Disease.* Excerpta Medica, Amsterdam, pp. 67–81.

Brown, C. C. (1976). Mathematical aspects of dose–response studies in carcinogenesis – the concept of thresholds. *Oncology* **33**, 62.

Brown, C. C. (1978). "The Statistical Analysis of Dose–Effect Relationships." In G. C. Butler (Ed.), *Principles of Ecotoxicology*, SCOPE 12. John Wiley & Sons, New York.

Buikema, Jr., A. L., Niederlehner, B. R., and Cairns, Jr., J. (1982). Biological monitoring. Part IV – Toxicity testing. *Water Res.* **16**, 239.

Cairns, Jr., J., Dickson, K. L., and Maki, A. W. (Eds.) (1978). *Estimating the Hazard of Chemical Substances to Aquatic Life.* ASTM, Philadelphia.

Chand, N. and Hoel, D. G. (1974). A Comparison of Models for Determining Safe Levels of Environmental Agents. In *Reliability and Biometry.* SIAM, Philadelphia, pp. 681–700.

Ciba Foundation Symposium 76 (1980). *Environmental Chemicals, Enzyme Function and Human Disease*. Excerpta Medica, Amsterdam.

De Bruin, A. (1976). *Biochemical Toxicology of Environmental Agents*. Elsevier/North Holland Biomedical Press, Amsterdam.

Duffus, J. H. (1980). *Environmental Toxicology*. Edward Arnold Ltd., London.

Farber, J. L. (1980). "Molecular Mechanisms of Toxic Cell Death." In H. R. Witschi (Ed.), *The Scientific Basis of Toxicity Assessment*. Elsevier/North Holland Biomedical Press, Amsterdam, pp. 201–210.

Finney, D. J. (1952). *Statistical Methods in Biological Assay*. Hafner Publishing, New York.

Finney, D. J. (1971). *Probit Analysis*, 3rd ed. Cambridge University Press, Cambridge.

GESAMP, IMCO/FAO/UNESCO/WMO/WHO/IAEA/UN (1977). Joint Group of Experts on the Scientific Aspects of Marine Pollution (GESAMP), Impact of Oil on the Marine Environment, Rep. Study No. 6. Food and Agriculture Organization, Rome.

Gillette, J. R. (1980). "Pharmacokinetic Factors Governing the Steady-State Concentrations of Foreign Chemicals and Their Metabolites." In Ciba Foundation Symposium 76, *Environmental Chemicals, Enzyme Function and Human Disease*. Excerpta Medica, Amsterdam, pp. 191–217.

Gut, I., Cikrt, M., and Plaa, G. L. (Eds.) (1981). *Industrial and Environmental Xenobiotics Metabolism and Pharmokinetics of Organic Chemicals and Metals*. Springer-Verlag, Berlin.

Jernelöv, A., Beijer, K., and Soderlund, L. (1978). "General Aspects of Toxicology." In G. C. Butler (Ed.), *Principles of Ecotoxicology*, SCOPE 12. John Wiley & Sons, New York, pp. 151–168.

Khan, M. A. Q. and Bederka, Jr., J. P. (Eds.) (1974). *Survival in Toxic Environments*. Academic Press, New York.

Klaassen, C. D. and Doull, J. (1980). "Evaluation of Safety: Toxicologic Evaluation." In J. Doull, C. D. Klaassen, and M. O. Amdur (Eds.), *Toxicology – The Basic Science of Poisons*, 2nd ed. MacMillan Publishing, New York, pp. 11–27.

Lee, P. N. and O'Neill, J. A. (1971). The effect of both time and dose applied on tumor incidence rate in benzopyrene skin painting experiments. *Br. J. Cancer* **25**, 759.

McLean, A. E. M. and Nuttall, L. (1978). Paracetamol injury to rat liver slices and its subsequent prevention by some anti-oxidants. *Biochem. Pharmacol.* **27**, 425.

McLean, A. E. M., Witts, D. J., and Tame, D. (1980). "The Influence of Nutrition and Inducers on Mechanisms of Toxicity in Humans and Animals." In Ciba Foundation Symposium 76, *Environmental Chemicals, Enzyme Function and Human Disease*. Excerpta Medica, Amsterdam, p. 275.

Orrenius, S., Thor, H., and Jernstrom, B. (1980). "The Influence of Inducers on Drug-Metabolizing Enzyme Activity and on Formation of Reactive Drug Metabolites in the Liver." In Ciba Foundation Symposium 76, *Environmental Chemicals, Enzyme Function and Human Disease*. Excerpta Medica, Amsterdam, pp. 25–42.

Poland, A. and Glover, E. (1980). "2,3,7,8-Tetrachlorodibenzo-*p*-Dioxin: Studies on the Mechanism of Action." In H. R. Witschi (Ed.), *The Scientific Basis of Toxicity Assessment.* Elsevier/North Holland Biomedical Press, Amsterdam, pp. 223–239.

Rosenthal, H. and Alderdice, D. F. (1976). Sub-lethal effects of environmental stressors, natural and pollutional, on marine fish eggs and larvae. *J. Fish. Res. Board Can.* **33**, 2047.

Shank, R. C. (1980). "Chemical Structure, Metabolic Pathways and Toxicity: Introduction," In H. R. Witschi (Ed.), *The Scientific Basis of Toxicity Assessment.* Elsevier/North Holland Biomedical Press, Amsterdam, p. 221.

Turner, M. E. (1975). Some classes of hit-theory models. *Math. Biosci.* **23**, 219.

Van Ryzin, J. and Rai, K. (1980). "The Use of Quantal Response Data to Make Predictions." In H. R. Witschi (Ed.), *The Scientific Basis of Toxicity Assessment.* Elsevier/North Holland Biomedical Press, Amsterdam, pp. 273–290.

Waldichuk, M. (1979). Review of the problems. *Phil. Trans. R. Soc. Lond. Ser. B* **286**, 399.

Witschi, H. R. (Ed.) (1980). *The Scientific Basis of Toxicity Assessment.* Elsevier/North Holland Biomedical Press, Amsterdam.

4

SOME PRINCIPLES OF
POLLUTION ECOLOGY
AND ECOTOXICOLOGY

The environment of an organism contains a wide variety of physical, chemical, and biological factors controlling its development. The set of organisms comprising an ecosystem are genetically adapted to these factors as part of their natural environment. In addition, the organisms are adapted to the natural changes which occur throughout the seasons, diurnally, and in other ways. Pollution introduces a change into the set of factors operating and controlling a natural ecosystem.

For example, an ecosystem is subjected to the natural ranges of temperature which occur in its environment. However, thermal pollution can lead to excessively low or excessively high environmental temperatures. Also, the pH of water may be altered by acid rain or discharges to give acid or alkaline conditions beyond the normal range. Some substances or environmental factors may cause pollution only when they extend beyond the normal range in one direction and not the other. For example, plant nutrients in excess lead to pollution but if plant nutrient concentrations are too low, this is usually not due to a pollution situation.

Each organism in the system has a natural ability to tolerate these changes and exhibits a characteristic optimal activity and range of changing activities in relation to the concentration or factor. Thus, many environmental changes which result in a shift toward the optimum will be beneficial to some particular organisms in the system.

Figure 4.1A shows how hypothetical populations of two organisms could respond in individual and characteristic ways to changes in concentrations or environmental factors of the general type discussed above. Thus, at level x, organism a will be favored by the conditions which exist, and at level y, organism b will be favored. Thus, there will be a change in the composition in the community of a and b related to the concentration or factor operating in the system.

Another general type of change is that due to concentrations or factors to which the organisms are not specifically adapted, generally the introduction of exogenous substances to the system for example, synthetic pesticides, PCBs, some petrochemicals, and some heavy metals. In this situation, the organisms show a characteristic response related to metabolic, physiological, and other characteristics. Usually,

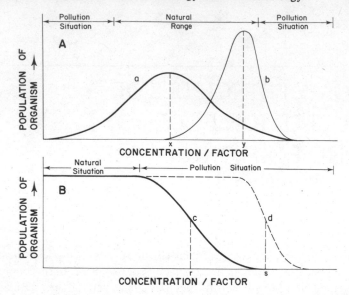

Figure 4.1. Generalized patterns of change in a population of organisms induced by pollution. (*A*) Population response produced by naturally occurring concentrations of substances and factors essential to the survival of organisms but which are detrimental outside the range encountered in natural ecosystems. (*B*) Response induced by substances or factors which are not part of the natural ecosystem and are detrimental at some level.

there are no observable effects on organisms or the ecosystem if the concentrations or factors are sufficiently low, but at higher levels, characteristic effects on the population are observed. Figure 4.1B shows the hypothetical response of the populations of two organisms *c* and *d* to different concentrations or environmental factors in this situation. Thus, conditions at *r* will favor *c* and conditions at *s* will favor *d*. So there will be a set of changes in the community structure of *c* and *d* related to the level of the concentration or factor involved.

Furthermore, factors such as competition, food, and a variety of other ecological factors have an impact on organisms in the system and will also be affected by environmental conditions. However, in a pollution situation the magnitude of the concentration or environmental factor change is often sufficient to give a set of related, comparatively easily identified, effects. Nevertheless, a clear understanding of more subtle and long-term effects in pollution situations is not usually available.

In polluted systems, organisms are selected for tolerance, which can lead to genetic adaptation over time with successive generations becoming more tolerant to the pollution conditions. The occurrence of this depends on a variety of factors, including the proportion of interacting and reproducing population affected by the pollution, reproduction rate, and genetic variability. If only a small proportion of the reproducing population is affected, as often occurs in pollution situations, little adaptation can be expected.

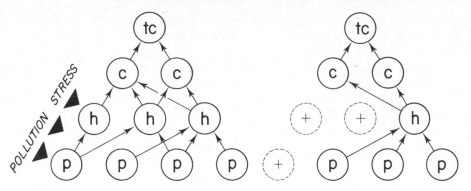

Figure 4.2.   A hypothetical food web illustrating modification as a result of pollution where tc = top carnivore, c = carnivore, h = herbivore, and p = producer.

In natural ecosystems there are sets of animals and plants adapted to the particular environmental conditions which exist, for example, temperature, availability of oxygen, and availability of nutrients. Usually there are a wide range of interacting organisms present including microorganisms, plants, invertebrates, vertebrates, and so on. But the interrelationships within the groups are not particularly well understood, which can lead to difficulties in the understanding of pollution effects. However, in many situations, the pollution effect is large compared with variations experienced in the natural system and this leads to a larger and more obvious biological effect.

Some pollutants such as toxicants may act directly on organisms while others may alter the environment to give an effect on the ecosystem. But, since all organisms are affected in different ways by pollution, it can be expected that there will be an alteration to the structure of a food web resulting from pollution (see (Figure 4.2).

With some exceptions the general effects of pollution on an area can be summarized as follows:

1. There is a decrease in the suitability of the area as a habitat for the living components of the ecosystem which have been naturally established and adapted to the area.

2. There is a detrimental impact on certain species and groups related to the intensity and type of pollution.

3. An alteration to the community structure occurs and, as a general rule, the number of species present declines.

4. The flows of energy and matter in the ecosystem are changed.

5. Removal of larger organisms with longer life spans occurs (FAO, 1972).

6. There is the appearance of opportunistic species with short life spans exhibiting large population fluctuations in time and space (FAO, 1972).

Patterns of change may be apparent in pollution situations since pollutants discharged to the environment are acted on by a variety of physical, chemical, and biological forces which generally lead to systematic loss and dilution. Thus, in relation to a point discharge there is a sequence of physicochemical and related biological changes. These patterns can occur in space or in time.

CLASSES AND GENERAL EFFECTS OF POLLUTANTS

The deleterious effects of pollutants can be generally related to three environmental factors:

1. Excess plant production.
2. Deoxygenation.
3. Toxic or similar deleterious physiological effects on organisms.

Factor 2 mentioned above applies exclusively to aquatic systems and factor 1 almost so, whereas factor 3 applies to all systems. The overall effects of all different types of pollutants can be conceptually divided into the three factors listed above and weighted accordingly. The three weighted properties of each pollutant can then be plotted to give a point on a triangular graph. Figure 4.3 shows the points plotted for a wide variety of different pollutants from waste waters to pathogenic microorganisms. A limited number of pollutants do not fit this scheme particularly well,

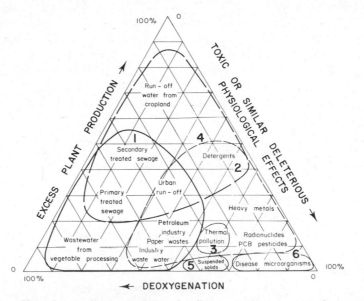

Figure 4.3. Plot of generalized properties of pollutants and waste waters in terms of three common properties (i) excess plant production; (ii) deoxygenation and (iii) toxic or similar deleterious physiological effects. Six classes of pollutants can be delineated: (1) organic matter, (2) plant nutrients, (3) thermal wastes, (4) toxicants, (5) suspended solids, and (6) disease microorganisms (pathogens).

for example, thermal pollution. Nevertheless, it provides a reasonable basis for comparison and contrast of pollution effects. For example, pathogenic micro-organisms lead to direct toxic or similar deleterious physiological effects but this group does not have an impact on dissolved oxygen or stimulate excess plant production. Suspended solids can be classified as having toxic or similar deleterious physiological effects and causing reduced plant production with a resultant impact on dissolved oxygen. Another example are detergents which contain phosphates, causing stimulation of plant growth, and surfactants with a toxic impact. Petroleum can lead to dissolved oxygen reduction due to microbial degradation and also toxic effects due to the presence of aromatic hydrocarbons.

Using the plots shown in Figure 4.3, it can be seen that there are related groups of pollutants. These are set out below:

1. *Organic Matter.* This consists principally of carbohydrates, proteins, and fat and leads to dissolved oxygen reduction by stimulating the growth of microorganisms.

2. *Plant Nutrients.* These substances are usually rich in nitrogen and phosphorus and stimulate excess plant growth.

3. *Toxic Substances.* These are substances which interfere with metabolism and physiological activity in a detrimental manner in low concentrations.

4. *Suspended Solids.* These substances have similar effects to toxic substances but act by physical interaction at comparatively high concentrations.

5. *Energy.* Energy pollution is mainly due to thermal discharges. The effects are similar to toxic effects but this activity is due to thermal energy inputs.

6. *Pathogenic Microorganisms.* These exhibit a toxic effect on organisms but the effects are due to organisms rather than chemical substances.

These classes provide the basis for discussion of the overall interaction of pollutants with ecosystems.

ECOLOGY OF DEOXYGENATION AND NUTRIENT ENRICHMENT

On a worldwide scale, these two processes are among the most important pollution effects in aquatic areas. Deoxygenation and nutrient enrichment are due to the two classes, organic matter and plant nutrients, respectively, and can be discussed together since many interrelated processes occur in both situations. Deoxygenation is caused by the addition of organic matter, rich in chemical energy (carbohydrate, protein, and fat), to a water body. If these substances are in a form whereby they can be readily attacked by microorganisms, they are often referred to as putrescible matter. A wide variety of microorganisms are present in all water bodies and some of these can readily utilize organic matter and are capable of rapid population expansion. Thus, respiration occurs and dissolved oxygen in the water mass is consumed in this process. Dissolved oxygen is essential for the survival of most

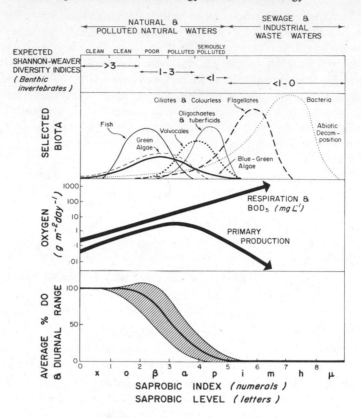

Figure 4.4.   Some generalized characteristics of waters at different saprobic levels. [Compiled from Sladecek (1979), Tittizer and Kothe (1979), Odum (1956), and Welch (1980).]

aerobic aquatic organisms. Reduction in dissolved oxygen leads to a change in the ecological structure of the affected community.

Figure 4.4 shows how increasing respiration and bacterial numbers are accompanied by a decrease in primary production and dissolved oxygen in an aquatic system, leading to a set of changes in the associated biota. Fish, which require reasonably high dissolved oxygen levels, survive best in the range indicated. Benthic invertebrates, such as oligochaetes and chironomids, need less dissolved oxygen and so occur in a lower range.

The addition of organic matter therefore causes a corresponding set of changes in dissolved oxygen, primary production, respiration, and biota which are reasonably consistent and related to a particular level of pollution. These changes can be classified into a sequence of states, zones, or classes containing a particular range of chemical and related biological characteristics. This can be done in a variety of ways (see Persoone and De Pauw, 1979) but one of the simplest is clean, poor, polluted, seriously polluted (Figure 4.4). Other classifications use terms such as clean, active decomposition, septic, degradation, and so on. The classification with the longest history is the saprobic system which uses a saprobic level or index (see

Figure 4.4) and has been recently reviewed by Sladecek (1979). All the schemes are similar in principle but differ in nomenclature and methods of zone classification. Considerable controversy surrounds the validity of these classifications and their application to pollution situations (Persoone and De Pauw, 1974). Nevertheless, the basic principle of a related sequence of chemical and related biological changes is sound and has been used as a framework for the following discussion.

The initial increase in primary production which occurs with increasing organic pollution is due to the release of plant nutrients (e.g., nitrogen and phosphorus) by the decomposition of organic matter containing these substances. The absorption of solar radiation initiates photosynthesis during daylight hours, producing an input in photosynthetic oxygen. On the other hand, respiration occurs throughout the daily cycle and this leads to a diurnal cycle of dissolved oxygen with a maximum during the day and a minimum at night (see Tittizer and Kothe, 1979). But overall, the rate of respiration exceeds primary production which leads to a deficit in dissolved oxygen and a reduction in the dissolved oxygen present in the water mass with increasing organic pollution (see Figure 4.4). When the dissolved oxygen falls to low levels, hydrogen sulfide and other toxic substances are produced which inhibit plant growth, and thus there is a decrease in primary production at higher saprobic levels. In addition, there is a steady fall in the species diversity of benthic invertebrates due to increasing divergence in environmental conditions from the natural situation to which the system has adjusted (see Figure 4.4).

The higher saprobic levels do not occur in natural systems which have been polluted. These are confined to waste waters but can be placed in this saprobic sequence as has been done by Sladecek (1979) (see Figure 4.4).

When organic matter is discharged to a stream a sequence of characteristic changes occurs. The dissolved oxygen content of the water decreases due to increased microbial respiration and replenishment occurs by solution of atmospheric oxygen at the air–water interface. This leads to a characteristic dissolved oxygen sag shown in Figure 4.5. The pollution level rises and falls in relation to the dissolved oxygen decrease and subsequent increase. Other related changes in biota, BOD, dissolved salts, saprobic index and diversity follow. This leads to a characteristic set of changes related to the discharge point (see Figure 4.5).

Some of the features of nutrient enrichment are related to the factors outlined above. A discharge can consist of water containing high concentrations of plant nutrients but low in organic matter, and thus little direct demand on dissolved oxygen occurs. The receiving waters are enriched with plant nutrients, usually nitrogen and phosphorus compounds, and there is an increase in photosynthesis and primary production. Changes in phytoplankton biomass can be detected by measuring the concentration of chlorophyll a, the green pigment in plants. Also, an increase in respiration occurs due to the increase in organic matter present from plant matter production. A resultant set of changes in environmental conditions and biota results, including such factors as water turbidity, type and density of plants, and type and density of animals (see Figure 4.6). The available data suggest that, as a general rule, the number of species present decreases with increasing enrichment in natural systems. However, Weiderholm (1980) reports that it would

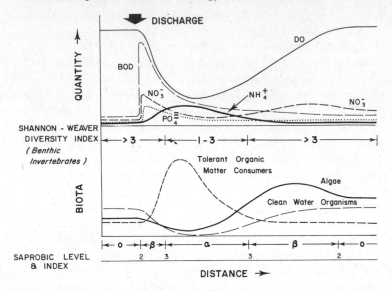

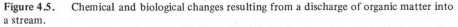

Figure 4.5.   Chemical and biological changes resulting from a discharge of organic matter into a stream.

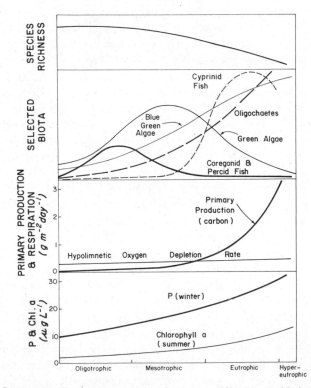

Figure 4.6.   Some generalized characteristics of water at different trophic levels. [Data from Welch (1980), Weiderholm (1980) and Mason (1981).]

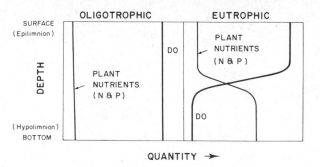

Figure 4.7. Generalized vertical distribution patterns of some water constituents in oligo-trophic and eutrophic conditions with stratification.

be expected that a maximum in species would occur at a particular enrichment level. Lower species numbers than this would occur with water extremely low in nutrients, for example, ultraoligotrophic, while a decrease in species would occur at higher levels.

Many aquatic areas, from small pools to the continental shelves, exhibit summer stratification due to heating of the surface water. This can increase plant pro-duction in the epilimnion and, upon death, the plant matter falls to the bottom increasing microbial respiration, dissolved oxygen consumption, and nutrient concentrations. In oligotrophic waters, there is little change from surface to bottom in dissolved oxygen and plant nutrients, but in eutrophic conditions, distinct vertical profiles are developed (see Figure 4.7). Thus, the depletion of hypolimnetic dissolved oxygen increases with increasing nutrient enrichment.

In addition, there are a related set of seasonal changes in the epilimnion of many larger water bodies (see Figure 4.8). In winter, the surface and bottom waters are mixed but the conditions of temperature and solar radiation are un-favorable for the growth of phytoplankton. With the onset of summer, there is the development of good growth conditions and also the stratification of the

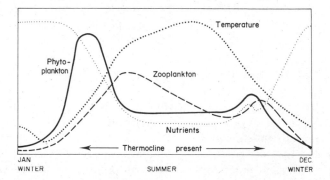

Figure 4.8. Seasonal variation in the cycle of nutrients, phytoplankton, and zooplankton in the temperate North Atlantic. [From Meadows and Campbell (1978). Reprinted with per-mission from Blackie and Son, Ltd.]

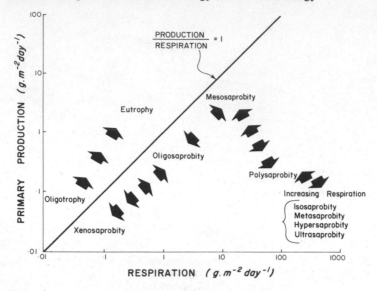

Figure 4.9.   Plot of primary production vs. respiration for a variety of saprobic and trophic conditions in aquatic areas and the sequence of interrelationships.

water column. Rapid growth of phytoplankton proceeds with the consequent consumption of waterborne plant nutrients, and, due to the isolation of the nutrient-rich bottom waters, there is depletion of plant nutrients in the epilimnion ultimately leading to reduced plant growth. Mixing of the water column can occur in autumn leading to a small growth of phytoplankton while the temperature conditions and solar radiation are still favorable. With the onset of unfavorable winter conditions low phytoplankton numbers once again are established. Generally, zooplankton growth follows the phytoplankton pattern but is displaced in time.

There have been many attempts to relate saprobity to trophy. These are usually based on the two major processes which occur in both situations, primary production and respiration. There is disagreement on the primary production and respiration characteristics of different trophic and saprobic classes (Odum, 1956; Rodhe, 1970; Sladecek, 1979; Tittizer and Kothe, 1979). However, Figure 4.9 represents an evaluation of the position of the various classes of saprobity and trophy on a primary production versus respiration graph derived from the available data.

The two basic processes of nutrient enrichment (trophy or eutrophication) and organic enrichment (deoxygenation and saprobity) are distinctively different but related. In the nutrient-enrichment process additions of plant nutrient into the water mass lead to acceleration of the primary production process. Accordingly, the nutrient enrichment process is controlled by primary production which would be expected to exceed respiration at all stages (see Figure 4.9). On the other hand, saprobity is due to the addition of decomposable organic matter (putrescible matter) which leads to deoxygenation. Consequently, this process and its impact are controlled by respiration which would be expected to exceed primary production at all stages (see Figure 4.9).

The development of the various classes of saprobity and trophy has proceeded independently and according to different criteria. Eutrophication classes have been developed according to plant growth and related characteristics, whereas the saprobic classes have been developed according to deoxygenation effects. Thus, it would be expected that the various classes of saprobity and trophy having the same prefix would not necessarily be equivalent. The relationships between the various classes are shown approximately in Figure 4.9. But, as indicated previously, there are differences between the characteristics used by various authors to define the various classes. It can be seen from this figure that the physicochemical and biological criteria are different for all saprobic and trophic classes.

An important difference between nutrient enrichment and organic enrichment is the reversibility of the process. Organic enrichment is a reversible process which can be seen in a forward and reverse sequence in a stream receiving an organic-rich discharge. In this way, the sequence xenosaprobic through to polysaprobic and the reverse can be seen in stream situations. On the other hand, nutrient enrichment is a process which proceeds only in one direction — toward increasing eutrophication.

The closest relationship between the classes of saprobity and trophy occurs when the levels of plant nutrients and organic matter, that is, primary production and respiration, are very low. Hence oligotrophic and xenosaprobic conditions are somewhat similar.

The interrelationships between the various classes are demonstrated by following a number of hypothetical sequences in water bodies receiving various discharges. Figure 4.10A shows the pattern of eutrophication of a lake or similar water body. Figure 4.10B illustrates the sequence in a stream receiving a discharge rich in organic matter containing chemically bound plant nutrients. There is a clean upstream section of the stream (xenosaprobic), then after discharge a progression through oligosaprobic, mesosaprobic, and polysaprobic conditions. Following the decomposition of organic matter and consequent reduction in respiration, there is a rise in the dissolved oxygen content in the water mass and the reverse sequence occurs. At this stage, the water is enriched with nutrients produced by the degradation and decomposition of the nutrient-rich organic matter. These newly available nutrients give an increase in primary production at a higher level than that on the forward path. Thus, the reverse sequence exhibits characteristics somewhat different from those in the forward sequence. When the original organic content is completely removed from the water mass, oligosaprobic conditions develop and the complete release of all the chemically bound nutrients has occurred. At this stage, depending on the environmental conditions, phytoplankton or other plants may increase and there is a development of conditions along the trophic path (see Figure 4.10B). This can only proceed to a limited extent since excessive growth of plants and accumulation of nutrients does not usually occur in a flowing stream.

Another similar sequence can be visualized with an organic matter discharge which is poor in chemically bound plant nutrients, that is, low in nitrogen and

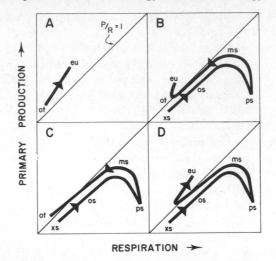

Figure 4.10.   Generalized sequence of changes in saprobic and trophic state in streams and lakes resulting from various discharges: (*A*) A lake receiving plant nutrients without any organic matter over time; (*B*) heavy pollution of a clean stream with waste waters containing nutrient-rich organic matter, e.g., sewage discharge, sequence from upstream of the discharge; (*C*) heavy pollution of a clean stream with organic matter poor in nutrients, sequence upstream to down-stream of the discharge; (*D*) heavy pollution of a clean stream with organic matter rich in nutrients and discharging into a lake, sequence upstream of the discharge to the stream and, finally, to the lake. (eu = eutrophy; ot = oligotrophy; xs = xenosaprobity; os = oligosaprobity; ms = mesosaprobity; ps = polysaprobity.)

phosphorus. This sequence would be expected to follow a similar path to that shown in Figure 4.10B, but in this case the lack of released nutrients at the oligo-saprobic stage inhibits plant growth so the path cannot continue toward eutro-phication but proceeds toward xenosaprobic conditions (see Figure 4.10C).

Another situation can occur with a stream receiving nutrient-rich organic matter and discharging into a lake or similar body of water with little flushing. In this case, the sequence would be expected to follow that indicated in Figure 4.10B, but when the stream discharges into the lake, the hydrological conditions allow an accumulation of plant nutrients and primary production to occur. The sequential development of eutrophic conditions results (see Figure 4.10D). Thus, both pro-cesses can occur in sequence and the sequence which occurs in any situation is dependent on the nature of the discharge and the hydrological features of the receiving system.

ECOTOXICOLOGY OF TOXIC SUBSTANCES

Toxic substances can affect ecosystems in a number of different ways, but in its simplest form, two basic types of effect are possible:

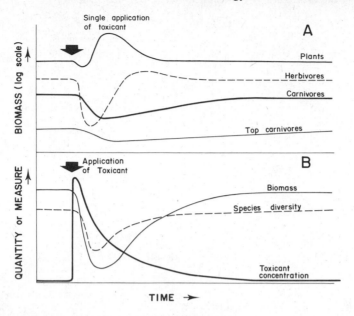

Figure 4.11.   Some generalized characteristics of the effects of a toxicant on an ecosystem. [Data from Eschenroeder et al. (1980) and Hynes (1974).]

1. Acute lethal toxicity over a short time period due to a discharge of a toxic substance, or treatment of an area with a toxic material on a single occasion.

2. Chronic sublethal effects can occur in an area due to exposure to sublethal concentrations over a longer time period on a continuous or intermittent basis.

Acute lethal toxicity with limited time exposure is quite common. However, the concentration of toxicant would be expected to decrease with time due to processes such as dispersal, dilution, and degradation (see Figure 4.11). Generally, animals are more sensitive to environmental toxicants than plants. But many carnivores are highly mobile and able to avoid the initial toxic discharge or treatment. Good examples of organisms exhibiting this behavior are eagles, hawks, carnivorous fish, and mammals. Consequently, toxic effects are often most acute with herbivores and omnivores.

The general pattern of ecosystem response to a toxicant causing herbivore lethality is shown in Figure 4.11. Parts of this general pattern have been developed as a result of simulation modeling (Eschenroeder et al., 1980). Although the initial impact on carnivores and top carnivores is delayed, there is, nevertheless, a longer-term impact and recovery lags in comparison with the herbivores. Initially both species diversity and biomass fall and then recover as the toxicant concentration decreases.

This pattern of ecosystem response is often observed in streams receiving a toxic discharge where the generalized pattern shown in Figure 4.11 occurs with distance

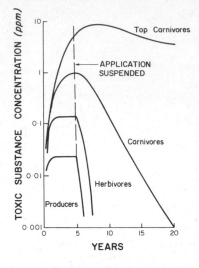

Figure 4.12.   Simulation of uptake and loss of a persistent toxicant in an ecosystem with 5 years of initial constant application. [From Eschenroeder et al. (1980). Reprinted with permission from Ann Arbor Science Publishers, Inc.]

downstream from the discharge point. Similarly, following oil spills in marine ecosystems, this general pattern occurs although changes in the carnivores and top carnivores can be difficult to quantify. However, phytoplankton blooms following oil spills are quite common. Also, this general type of sequence has been observed in the application of insecticides. With this situation, predator organisms are often preferentially removed from the population leading to an outbreak of prey species rather than plants (Dempster, 1975).

The chronic sublethal effects of persistent toxicants are difficult to quantify to enable establishment of a general pattern. Figure 4.12 shows a simulation of the concentration of a persistent toxicant in organisms within an ecosystem over time. The organisms at high trophic levels show a slow buildup in concentrations but with much higher ultimate concentrations and slower rate of loss than other organisms in the ecosystem. In this case, the greatest impact of the toxicant would be expected at the highest trophic levels and there are many examples of population reductions with top carnivorous birds as a result of insecticide usage (see Chapter 7).

SUSPENDED SOLIDS

Suspended solids can be expected to have broadly similar effects to toxicants in aquatic ecosystems. Many organisms show tolerance to comparatively high concentrations of suspended solids. Nevertheless, plants would be expected to show a population decrease due to the reduction in light penetration. A major impact on the ecosystem is the elimination of food organisms. Also, a reduction in the numbers of fish and other organisms using visual means to seek prey occurs as well.

THERMAL ECOLOGY

There is little information of the type required to allow the formulation of general principles for the effects of thermal discharges on aquatic ecosystems. With thermal pollution, aquatic ecosystems are mainly affected because the heat input to a specific system in this situation is much more substantial than with terrestrial organisms.

Heat represents an input of energy into an ecosystem which is not readily assimilated by organisms. However, it can be utilized by organisms in providing a suitable environmental temperature for metabolic processes.

In general terms thermal pollution causes similar changes in organisms and ecosystems to those noted previously with other forms of pollution. Generally, there can be expected to be a systematic distribution of numbers of organisms, number of species, or some other measure of ecosystem success against temperature. Thus, there will be an optimum temperature, T_0, where maximum success would be expected and, on either side of this optimum, a range of temperatures where the population was less successful (see Figure 4.13). In natural ecosystems, there will be a range of temperatures due principally to diurnal and seasonal changes with consequent changes in ecosystem success. Thus, T_u represents the upper limit of the normal range of temperature and T_L represents the lower temperature limit in a normal ecosystem (see Figure 4.13). Any ecosystem contains a diversity of organisms which are naturally adapted to these thermal conditions. The seasonal patterns of change may result in a corresponding change of species related to temperature or other seasonal characteristics. Also, there is an absolute upper and lower limit for the survival of living organisms. So, irrespective of normal environmental temperature, we would expect that all ecosystems will approach extinction at these limits.

For tropical systems with a natural temperature regime, close to the upper limit for all organisms, a skewed ecosystem success curve would be expected with a limited range between the maximum and the upper extinction limit. Constraints on temperatures lower than the optimum are not as acute as the upper temperatures so the ecosystem success curve would extend further in this direction. In

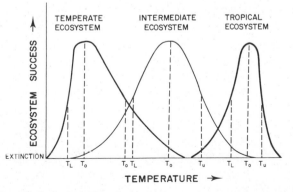

Figure 4.13. Hypothetical plot of ecosystem success against temperature with T_L and T_u representing the natural variation in the ecosystem and T_0 as the optimum.

addition, it would be expected that there would be a limited seasonal range of temperatures and thus the tropical ecosystem would be adapted to a narrow range of temperature tolerance. Similar arguments apply with temperate ecosystems but the temperature range of adaptation would be larger and a lower level of susceptibility to increased temperatures would occur.

Changes in environmental temperatures due to thermal pollution will give a change in ecosystem success. This change will depend on the position in the seasonal and diurnal cycle when the discharge occurs. If it results in temperature movement toward the maximum or optimum it will be beneficial to the ecosystem. On the other hand, if it results in a movement away from the maximum a detrimental impact will occur. Thus, as a general rule, temperatures outside of the natural range (i.e., greater than T_u and less than T_L) would be expected to be detrimental to the particular ecosystem involved. In general terms, on this basis, tropical ecosystems would be expected to be the most susceptible to detrimental effects resulting from thermal additions and temperature elevation. This general conclusion is substantiated by results described by Johannes and Betzer (1975) as well as Zieman and Wood (1975).

Generally, thermal stress leads to a somewhat similar pattern of ecosystem change as that caused by the discharge of toxicants. But in this case it is not clear which groups of organisms will be affected and which will benefit from changes in thermal conditions. Nevertheless, the overall changes indicated in Figure 4.11B would be expected.

Organisms can adapt to thermal stress. An ecosystem subject to thermal stress results in a process of selection and the development of a temperature-tolerant breeding stock. Thus, long-term changes in ecosystems may differ from the short-term effects.

Radionuclides also result in an energy input to organisms and ecosystems, but this energy cannot be utilized in any significant way. In fact, radionuclides act as toxicants by interfering with metabolism and physiological activity in a detrimental manner at low concentrations. Thus, the effects of exposure to radionuclides are of a similar nature to that described for toxicants previously.

PATHOGENIC MICROORGANISMS

Pathogenic microorganisms are a very important water pollution factor, particularly regarding human health. Their impact on natural populations is much less important. Nevertheless, microorganism-borne disease from pollution can possibly be expected to affect natural populations. The effects are probably similar to the general pattern of toxicant impact illustrated in Figure 4.11B.

BIOTIC RESPONSES TO POLLUTION AT DIFFERENT LATITUDES

This topic has been reviewed by Johannes and Betzer (1975). There are basic differences between temperate and tropical regions which lead to difficulties in

Table 4.1.   How the Behavior and Impact of Water Pollutants
May be Expected to Differ Quantitatively in the Tropics[a]

Solubility of gases	Lower
Solubility of liquids and solids	Higher
Biological uptake rates	Higher
Biological release rates	Higher
Rates of physicochemical degradation	Higher
Rates of biological degradation	Higher
Toxicity	Higher
Toxicity thresholds	Lower
Rates of chemical oxygen depletion	Higher
Rates of biological oxygen depletion	Higher
Biological impact of nutrients	Higher
Biological impact of suspended solids	Higher[b]
Rates of oxidation in sewage treatment plants	Higher
Rates of denitrification in sewage treatment plants	Higher
Rates of disinfection in sewage treatment plants	Higher

Source: Johannes and Betzer (1975). Reprinted with permission
from Elsevier Scientific Publishing Company.

[a] There are some exceptions to these generalizations.
[b] In coral reef communities, but not in mangroves.

applying knowledge developed in temperate regions to the tropics. Most of the
background knowledge we have on these differences relates to aquatic systems.
The differences are caused fundamentally by different levels of incoming solar
radiation leading to a variety of different physical, chemical, and biotic characteristics which stimulate different responses of these systems in a pollution
situation.

A fundamental characteristic which results in many differences between tropical
and temperate regions is water temperature. The temperature of water governs the
rate of environmental transformations, solubility of natural substances and
pollutants, stability of pollutants, and the metabolic rate of organisms. Thus, the
behavior and impact of pollutants would be expected to vary with temperature
and, consequently, with latitude. Table 4.1 summarizes the various differences
between tropical and temperate regions.

In the preceding discussion on thermal ecology it was suggested that tropical
ecosystems operate closer to the upper thermal limit and thus would be expected
to be more susceptible to thermal pollution. There is a reasonable body of evidence
and observations to support this (Johannes and Betzer, 1975; Zieman and Wood,
1975).

In laboratory experiments many toxicants have been found to increase in
toxicity with increasing temperature. On the other hand, respiration rates increase
with temperature leading to a higher rate of metabolism and excretion. In this
situation it would be expected that the Q_{10} rule would apply and thus the rates of
metabolism and excretion would double for every 10°C increase in temperature.

Therefore, toxicants would tend to be more toxic but their toxic effects would decrease more rapidly with time.

An important aspect of pollution in the tropics concerns the interaction of dissolved oxygen concentrations with the respiration rates of organisms. With increasing temperature the respiration rate of organisms increases while the solubility of oxygen in water decreases. In addition, as the respiration of microorganisms increases with temperature, an increasing rate of dissolved oxygen consumption occurs. The effects have been summarized by Johannes and Betzer (1975) as follows:

1.   The dissolved oxygen levels in tropical waters are generally lower than temperate waters due to the higher ambient temperatures.

2.   The rate of oxygen depletion of water due to discharges of organic matter will be more rapid due to the higher respiration rates of microorganisms.

3.   Oxygen demand will decrease more rapidly with distance from a discharge point and with time.

4.   Due to higher respiration rates and lower dissolved oxygen levels, tropical aquatic ecosystems will generally be closer to the limiting levels of dissolved oxygen.

Many tropical marine systems have developed over geological time in very stable conditions of temperature, salinity, solar radiation, and other physicochemical factors allowing the development of a rich and diverse ecosystem. Therefore, it is suggested that the tolerance of many of these organisms to environmental fluctuations would be relatively limited. Thus, susceptibility to pollution could be elevated. The concept that increasing trophic complexity confers stability on ecosystems due to the multiplicity of checks and balances does not seem to hold with these communities (Johannes and Betzer, 1975).

BIOTIC INDICES

The previous sections have considered the fundamental interrelationships between pollution factors and the resultant change in biological characteristics. In most pollution investigations of natural environments, concern is mainly centred on the "health" of the living system. Thus, direct measurements of biotic characteristics are of particular importance. The response of biota to pollution in a natural ecosystem is an integral part of a pollution evaluation.

Biotic indices are used in this way and take a variety of forms. Matthews et al. (1982) have divided biotic measures into three classes: (1) those based on changes in the function of an ecosystem; (2) those based on changes in the community structure of biota in the ecosystem; and (3) combinations of functional and community changes.

There are a number of different measures of the functions of the ecosystems which are modified by pollution. One of the most fundamental is the change in

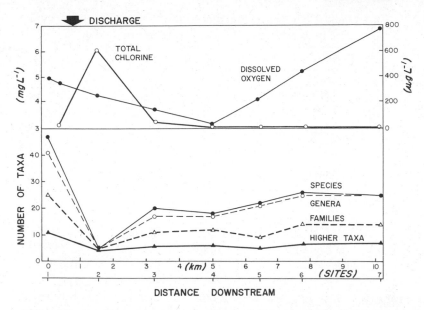

Figure 4.14. Number of species, genera, families, and higher taxa of benthic invertebrates in a stream receiving a discharge from a secondary sewage treatment plant. [From Arthington et al. (1982). Reprinted with permission from the Australian Water Resources Council.]

the respiration to photosynthesis ratio. This has been discussed previously and measures of the respiration to photosynthesis ratio provide an estimation of eco-system function related to enrichment by organic matter and also eutrophication (see Figures 4.9 and 4.10). Cairns (1977) has also suggested such factors as changes in oxygen consumption, ADP and lipid concentrations, filtering rates, reproduction rates, and nutrient cycling rates. The BOD test (see Chapter 5) is a good example of a measure of community function which is widely used. This is a measure of respiration, as oxygen consumption, by the microorganism community in the water mass.

Changes in community structure in an area affected by pollution involves identification, classification, and quantification of biota in the area. This can be applied to a wide range of biota but the most commonly used group are the benthic macroinvertebrates. This group consists of relatively sedentary organisms and therefore would be expected to reflect conditions in the affected area. They are relatively easily sampled and the species show a wide range of tolerance to pollution conditions. However, algae, fish, and other organisms have also been used to determine community structure in areas affected by pollution.

A simple structural biotic index is the number of higher taxa, families, genera, or species in an area. Figure 4.14 shows the variation in numbers of taxa with distance downstream from an organic discharge. The dissolved oxygen profile shows a sag with a minimum at the third site below the discharge. This is due to suppression of microbiological activity by chlorine in the discharge which evapo-rates rapidly from the water, allowing major microbiological activity by the third

Table 4.2.   Some Equations for Various Indices Used in Assessment of Biota in Polluted Areas

Index	Equation[a]
Shannon–Weaver	$H' = -\Sigma \dfrac{n_i}{N} \log_2\left(\dfrac{n_i}{N}\right)$
Margalef	$I = \dfrac{S-1}{\ln N}$
Brillouin	$H = \dfrac{1}{N} \ln \dfrac{N!}{n_1! n_2! \cdots n_s!}$
Pielou	$E = \dfrac{H}{H_{max}} = \dfrac{H}{\log_2 S}$

[a] N = total number of individuals; S = number of species; n_i = number of individuals of the ith species.

site downstream. The data show improved sensitivity of this index by identification to the species level.

The number of species and species diversity drop when the dissolved oxygen plus chlorine have a major impact in the area (see Figure 4.14). A minimum number of species can be seen at site 2, the first sample point below the discharge, where pollution conditions would be expected to be at their worst. The same set of species would be expected in similar conditions in other situations. Thus a limited number of species could be selected as indicators of the presence of this type of pollution condition. Similarly, sets of "indicator species" could be selected to represent different levels of pollution as indicated in Figure 4.14 (Hawkes, 1975). There are a number of limitations with this approach to the biotic evaluation of pollution situations. These are principally due to the presence of indicator species in unpolluted areas, the need to identify animals to the species level due to the different pollution tolerance of closely related taxonomic groups, and the non-quantitative nature of results obtained.

The use of the number of species as a numerical biotic index for pollution is an expression of species diversity in the area. However, this technique also has several limitations. One is that there is no weighting for the different numbers of individuals represented by the different species present. All species are given an equal weighting, irrespective of their numerical representation in an area. This can be overcome by using equations which provide a weighting for the number of individuals of a species and the different relationships between number of individuals, number of species, and number of individuals representing each species. The indices in Table 4.2 present a variety of different ways in which the data can be combined to a single biotic index. Figure 4.15 illustrates the results obtained by applying these indices to a pollution situation resulting from organic enrichment. Arthington et al. (1982) have suggested that Margalef's species richness index and the number of species provided the most satisfactory measure of

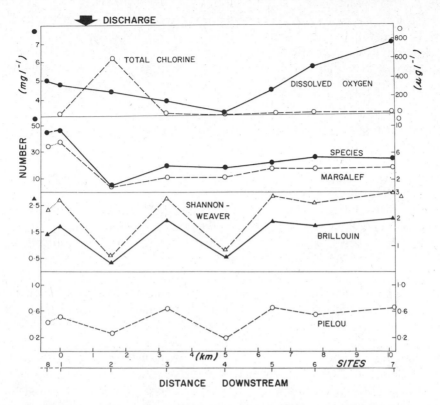

Figure 4.15. Diversity indices of benthic invertebrates, dissolved oxygen, and chlorine in a stream receiving a discharge from a secondary sewage treatment plant. [From Arthington et al. (1982). Reprinted with permission from the Australian Water Resources Council.]

pollution conditions in this situation. These indices have a close relationship to the chemical parameters of dissolved oxygen and total chlorine and also distance downstream from the discharge. Nuttal and Purves (1974), in comparing the different methods for developing a biotic index, found diversity indices to be the most sensitive to changes in biota as a result of pollution.

Species diversity may be affected by factors other than pollution (Hawkes, 1979). For example, stream headwaters with high water velocity may have low diversity but high water quality. In addition, the species composition factors used in the species diversity index take no account of the physiological stress caused by different pollution conditions on the differently susceptible species present. Hawkes (1979) has raised some questions regarding the use of diversity indices in pollution investigations for organic pollution and eutrophication, although he has suggested that this method may be useful for pollution by toxic substances.

The final class of indices which contains combined structural and functional information is not well developed. Matthews et al. (1982) have outlined several possible areas for development of indices of this type.

Indices alone or in combination with other information are a valuable technique in the investigation of pollution situations. But it should be kept in mind that an index represents a simplification of a complex set of data and there is a loss of information in the process. This may introduce errors and inaccuracies into the final result.

INDICATOR SPECIES, ECOLOGICAL INDICATOR SPECIES, AND CHEMICAL MONITOR SPECIES

The expression "indicator species" has been used in two distinctively different ways:

1. To describe particular species that are selectively adapted to certain pollution conditions; for example, heavily polluted or clean. Thus the presence of these particular species can be used to indicate pollution conditions as described in the previous section.

2. Organisms that bioaccumulate toxic substances present in trace amounts in the environment. Chemical analysis of these species then indicates the presence of toxicants in the environment more effectively than direct analysis of an environmental sample, such as water.

Confusion can arise using this terminology. "Ecological indicator species" is probably more appropriate to the first type and "chemical monitor species" for the second.

Phillips (1980) has thoroughly reviewed the use of chemical monitor species. He reports that the basic characteristics of a monitor species are the following:

1. The organism should accumulate the pollutant without being killed by the levels encountered in the environment.

2. The organism should be sedentary in order to be representative of the study area.

3. The organism should be abundant throughout the study area.

4. The organism should be sufficiently long lived to allow the sampling of more than 1 year class if desired.

5. The organism should be of reasonable size, giving adequate tissue for analysis.

6. The organism should be easy to sample and hardy enough to survive in the laboratory, allowing defecation before analysis, if desired, and laboratory studies of pollution uptake.

7. The organism should tolerate brackish water.

8. A simple correlation should exist between the pollutant content of the organism and the average pollutant concentration in the surrounding water.

9. All organisms of a given species used in a survey should exhibit the same correlation between their pollutant content and the average pollutant concentration in the surrounding water at all locations studied under all conditions.

In basic terms the major advantage of this strategy is that low concentrations of toxicants in the environment can be monitored. Also, this technique provides an integrated measure of the concentration encountered in the environment over time whereas analyses of environmental samples provide the concentration at a particular time. The results may give a more relevant assessment of biological exposure to a toxicant present in water and sediment in a variety of different forms with different biological activities. With food organisms as monitors, they provide a direct measure of human exposure. Biological examples are easier to store than water and so analyses can be delayed to a convenient time.

Phillips (1980) has reviewed the influence of such factors as age, lipid content, seasonal variations, sex, shore level, depth, and so on on the use of organisms as monitors. These factors can lead to complications in the use of organisms as chemical monitors and make interpretation difficult.

A wide variety of marine, estuarine, and freshwater organisms have been used in this role. Principal attention has been focused on coastal and estuarine areas. Probably the most widely used group are the bivalves since these satisfy most of the criteria outlined above.

REFERENCES

Arthington, A. H., Conrick, D. L., Connell, D. W. and Outridge, P. M. (1982). The ecology of a polluted urban creek. Australian Water Resources Council Technical Paper No. 68, Australian Government Publishing Service, Canberra.

Cairns, J. (1977). "Quantification of Biological Integrity." In R. K. Ballentyne and L. J. Guaria (Eds.), *The Integrity of Water.* U.S. Environmental Protection Agency, Washington, D.C., pp. 171–187.

Dempster, J. P. (1975). "Effects of Organochlorine Insecticides on Animal Populations." In F. Moriarty (Ed.), *Organochlorine Insecticides: Persistent Organic Pollutants.* Academic Press, London, pp. 231–248.

Eschenroeder, A., Irvine, E., Lloyd. A., Tashima, C., and Khanh Tran (1980). "Computer Simulation Models for Assessment of Toxic Substances." In R. Haque (Ed.), *Dynamics, Exposure and Hazard Assessment of Toxic Chemicals.* Ann Arbor Sciences, Ann Arbor, Michigan, pp. 323–368.

FAO (1972). Summary of discussion, Section 4, "Ecosystem Modifications and Effects on Marine Communities." In M. Ruivo (Ed.), *Marine Pollution and Sealife.* Fishing News (Books) Limited, London, England, pp. 348–349.

Hawkes, H. A. (1975). "River Zonation and Classification." In B. A. Witton (Ed.), *River Ecology.* Blackwell, Oxford, pp. 312–374.

Hawkes, H. A. (1979). "Invertebrates as Indicators of River Water Quality." In A. James and L. Evison (Eds.), *Biological Indicators of Water Quality.* John Wiley & Sons, Chicester, pp. 2-1 to 2-45.

Hynes, H. B. N. (1974). *The Biology of Polluted Waters.* Liverpool University Press, Liverpool.

Johannes, R. E. and Betzer, S. B. (1975). "Introduction: Marine Communities Respond Differently to Pollution in the Tropics than at Higher Latitudes."

In E. J. Ferguson-Wood and R. E. Johannes (Eds.), *Tropical Marine Pollution.* Elsevier Scientific Publishing, Amsterdam, pp. 1–12.

Matthews, R. A., Bukiema, A. L., Cairns, J., and Rodgers, J. H. (1982). Part 2A – Receiving systems functional methods, relationships and indices. *Water Res.* **16,** 129.

Meadows, P. S. and Campbell, J. I. (1978). *An Introduction to Marine Science.* Blackie, Glasgow, p. 176.

Nuttall, T. M. and Purves, J. B. (1974). Numerical indices applied to the results of a survey of the macro invertebrate fauna of the Tamar Catchment. *Freshwater Biol.* **4,** 213.

Odum, H. T. (1956). Primary production in flowing waters. *Limnol. Oceanogr.* **1,** 102–117.

Persoone, G. and De Pauw, N. (1979). Systems of biological indicators for water quality assessment. In O. Ravera (Ed.), *Biological Aspects of Freshwater Pollution.* Pergamon Press, Oxford, p. 39.

Phillips, D. J. H. (1980). *Quantitative Aquatic Biological Indicators.* Applied Science Publishers, London, p. 488.

Rodhe, W. (1970). "Crystallisation of Eutrophication Concepts in Northern Europe. In *Eutrophication – Causes, Consequences and Correctives.* National Academy of Sciences, Washington, D.C., pp. 50–64.

Sladecek, V. (1979). "Continental Systems for the Assessment of River Water Quality." In A. James and L. Evison (Eds.), *Biological Indicators of Water Quality.* John Wiley & Sons, Chichester, pp. 3-1 to 3-32.

Tittizer, T. T. and Kothe, P. (1979). "Possibilities and Limitations of Biological Methods of Water Analysis." In A. James and L. Evison (Eds.), *Biological Indicators of Water Quality.* John Wiley & Sons, Chichester, pp. 4-1 to 4-21.

Welch, E. B. (1980). *Ecological Effects of Waste Water.* Cambridge University Press, Cambridge, p. 337.

Wiederholm, T. (1980). Use of benthos in lake monitoring. *J. Water Pollut. Control Fed.* **52,** 537.

Zieman, J. C. and Wood, E. J. F. (1975). "Effects of Thermal Pollution on Tropical-Type Estuaries with Emphasis on Biscayne Bay, Florida. In J. Ferguson-Wood and R. E. Johannes (Eds.), *Tropical Marine Pollution.* Elsevier Scientific Publishing, Amsterdam, pp. 75–98.

PART TWO

CHEMICAL BEHAVIOR AND ECOTOXICOLOGY OF POLLUTANTS

5

DEOXYGENATING SUBSTANCES

Photosynthesis and respiration are fundamental to life on earth since these processes involve the harvesting and consumption of solar energy by living systems. The chemical processes involved can be simply expressed with glucose as set out below, but a highly complex series of reactions are involved in proceeding from one side of this reaction to the other. Also, the formation and consumption of proteins and fats involve different reaction sequences.

$$6CO_2 + 6H_2O \underset{\text{respiration}}{\overset{\text{photosynthesis}}{\rightleftharpoons}} 6CH_2O + 6O_2$$

These processes involve carbon dioxide and oxygen in the earth's atmosphere, which, in aquatic ecosystems, are dissolved in the water mass. The involvement of the water mass causes a marked difference between aquatic and terrestrial systems which is an important aspect of pollution ecology.

The basic transformations of carbon in natural aquatic ecosystems are shown in Figure 5.1. Carbon from dissolved carbon dioxide is incorporated by photosynthesis into the autotrophic biomass of the ecosystem. Respiration of the heterotrophic biota involves the consumption of autotrophic biomass as food and finally the decomposition of body tissues after death. Both of these processes result in the formation of carbon dioxide and methane which are returned to the water and, ultimately, the atmosphere. Some carbon is permanently incorporated into the sediments and thus is removed from the system.

If substances rich in organic carbon are added to the system some of the pathways shown in Figure 5.1 are increased in magnitude and also some of the pools of organic carbon are increased in size. This results in an increase in respiration, mainly through the respiration of microorganisms, giving rise to increased amounts of carbon dioxide and methane (see Figure 5.1).

The transformations of carbon involve oxygen and the related transformations of oxygen are shown in Figure 5.2. The oxygen needed for aerobic respiration is obtained from the dissolved oxygen in the water mass. This oxygen is usually substantially derived from the atmosphere and converted into carbon dioxide which is discharged into the water mass and ultimately to the atmosphere. Increased respiration, due to an increased amount of organic carbon in the water, results in changes in the magnitude of the pathways and changes in the pools of oxygen involved (see Figure 5.2). Most importantly, there is a demand on the reservoir of dissolved oxygen in the water mass which is comparatively small since oxygen

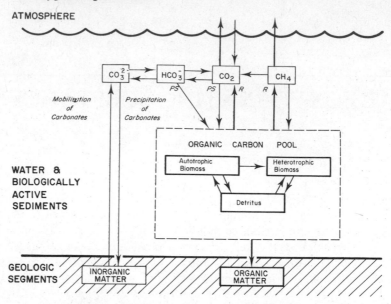

Figure 5.1. Transformations of carbon in aquatic systems with processes accelerated or increased by added organic matter shown in heavy lines. [From Arthington et al. (1982). Reprinted with permission from the Australian Water Resources Council.]

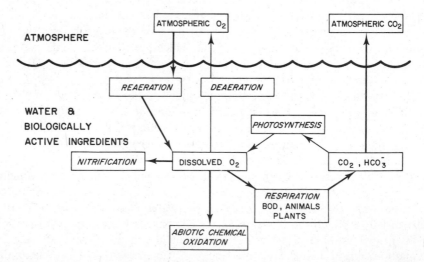

Figure 5.2. Oxygen dynamics in aquatic systems with processes increased or accelerated by organic discharges indicated in heavy lines. [From Arthington et al. (1982). Reprinted with permission from the Australian Water Resources Council.]

has limited solubility in water, usually ranging from about 6 to $14\,mg\,L^{-1}$. Therefore, substantial reductions in dissolved oxygen can occur which have significant implications for aquatic organisms.

Wastes rich in organic carbon are commonly discharged into waterways in the form of sewage, food processing wastes from the fruit, meat, dairy and sugar industries, as well as wastes from paper manufacturing, and a wide variety of other industries. The organic carbon in these wastes usually exists in the form of carbohydrates, proteins, fats, humic substances, surfactants, and a wide variety of related and derived substances. Organic wastes are produced in large quantities in all urban population centers and their discharge to waterways can substantially reduce the dissolved oxygen content of those bodies. Primary, secondary, and tertiary wastewater treatment can reduce these problems but the most common forms of treatment are primary and secondary. The discharge resulting from these treatment techniques can still result in a significant demand on oxygen in a waterway, particularly wastes derived from primary treatment. Secondary treatment usually results in substantial reduction in the oxygen demand.

Organic pollution can also result from urban runoff giving nonpoint discharges to waterways. Also, off-shore dumping sites for sewage wastes are commonly on the continental shelf and elsewhere. In some situations nonurban catchments can give a high organic input where certain types of activities, such as logging, are undertaken.

FACTORS AFFECTING DISSOLVED OXYGEN IN WATER BODIES

Respiration and Nitrification

In water bodies usually the most important heterotrophic organisms, in terms of respiration, are the microorganisms. In well-oxygenated waters, aerobic respiration occurs mediated by aerobic microorganisms. The chemical reaction involved can be simply expressed, for glucose, as

$$6CH_2O + 6O_2 \longrightarrow 6H_2O + 6CO_2$$

Environmental temperature is a strong influence on this process since different species have a narrow range of temperatures for their optimal activity. However, in most aquatic bodies communities of microorganisms exist which cover a wide range of optimal temperatures.

The equation above outlines the reaction of carbon during aerobic respiration, but organic matter in discharges also contains various other elements, particularly nitrogen and sulfur. Nitrogen is present principally in the amino group of the peptide link in proteins. Complete oxidation of the organic nitrogen present leads to the formation of the nitrate ion. Sulfur is present in organic matter in a variety of different forms but principally as the sulfhydryl group, the disulfide group, and the sulfide group. Complete oxidation yields the sulfate ion.

In the absence of oxygen, anoxic or hypoxic conditions develop and anaerobic microorganisms take over the degradation and decomposition of organic matter.

Some are facultative organisms, which can function as either aerobic or anaerobic organisms, while others, particularly the methane-producing group, are obligate anaerobes. In fact, molecular oxygen is toxic to this group. With glucose the anaerobic respiration reaction can be simply expressed as

$$6CH_2O \longrightarrow 3CH_4 + 3CO_2$$

This process makes no demand on the oxygen present in the water mass but it is very important for the removal of oxygen-demanding substances in waterways. The methane and carbon dioxide produced are released into the water mass then to the atmosphere, resulting in the removal of organic carbon and oxygen demand from the system. This process occurs in swamps, organic-rich bottom muds, and water bodies suffering from pollution by organic wastes (Manahan, 1975).

Anaerobic respiration with sulfur- and nitrogen-containing organic substances gives rise to hydrogen sulfide and ammonia, respectively. Ammonia in the presence of oxygen is readily oxidized to nitrate. Nitrite is formed under suitable oxidation and reduction conditions but is less commonly the end product of nitrogen metabolism. The formation of nitrate from organic matter is described as the nitrification reaction and is mediated by a variety of microorganisms. A series of reactions occur which can be simply expressed as follows:

$$2O_2 + NH_4^+ \longrightarrow NO_3^- + 2H^+ + H_2O$$

This process can make an important demand on dissolved oxygen in natural water bodies, especially where sewage contamination containing suitable microorganisms occurs in the temperature ranges between 25 and 30°C (Manahan, 1975).

These oxygen-demanding processes can be measured by the biochemical oxygen demand (BOD test) (APHA, 1975; USEPA, 1974). This test measures the uptake of oxygen by water samples incubated over a period of 5 days under standard conditions, often with seed microorganisms added. In general, clean water has a BOD of 1 or less mg L^{-1} and seriously polluted water contains greater than 10 mg L^{-1}. The nitrification reaction occurs in what is described as the second stage of the BOD test at time periods greater than 5 days (Nemerow, 1974). The precision of the BOD test is about ± 17% and may not satisfactorily reflect the actual conditions existing in a natural water body. The test conditions may differ from those existing in the environment in temperature, microorganisms present, nutrient status, and so on.

The kinetics of BOD changes in waterways has been subject to intense study. Velz (1970) has described one of the principles governing BOD in natural waterways first formulated by Phelps as "The rate of biochemical oxidation of organic matter is proportional to the remaining concentrations of unoxidized substance measured in terms of oxidizability." Thus the oxidation of organic matter follows first-order decay reaction kinetics. From this the following series of mathematical expressions can be derived:

$$-\frac{dL}{dt} = K_1 L$$

which can be integrated to

$$\ln \frac{L}{L_0} = -K_1 t$$

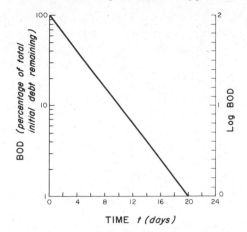

Figure 5.3.   Normal amortization of organic BOD debt at $20°C$ ($k = 0.1$) in an aquatic area. [From Velz (1970). Reprinted with permission from John Wiley & Sons, Inc.]

or

$$\log \frac{L}{L_0} = -0.434 K_1 t = -k_1 t$$

$$\frac{L}{L_0} = 10^{-k_1 t}$$

where L_0 is the total BOD debt at time zero; L the BOD debt at time t; t the time period since time zero; K_1 and k_1 the BOD decay rate coefficients, that is, empirical constants ($0.434 K_1 = k_1$).

By measurements of BOD in waterways over long periods Velz (1970) reported that at $20°C$, $k = 0.1$, if the time period is measured in days. Since first-order kinetics apply it can be shown that $t_{1/2} = 0.301/k$ [see Chapter 2, Equation (5)] and thus the half-life normally exhibited by BOD demanding substances in natural waterways is equal to 3 days. Also, from the mathematical treatment outlined above a plot of log percentage remaining BOD versus time will give a straight line (see Figure 5.3). Thus after 5 days, $\log (L/L_0) = -0.5$ and from this it can be shown that, the total BOD debt or ultimate BOD, $L_0 = BOD_5/0.68$.

The BOD decay rate coefficient (k) varies with temperature. It can be shown that the deviation of this constant from the $20°C$ value can be derived from the equation:

$$k_T = k_{20°C} \times 1.047^{(T-20)}$$

Since the BOD measure has some disadvantages associated with it, such as lack of precision and length of measurement time, other measures have been developed for measuring oxygen demand. One of these is the chemical oxygen demand (COD) which is performed by subjecting the test sample to strong oxidizing agents so that extensive and vigorous oxidation proceeds. While this is more rapid than the BOD test the oxygen taken up by strong oxidizing agents may not reflect the amount of oxygen consumed under natural conditions. Another method measures the total organic carbon (TOC) content of the water, from which the oxygen demand can be estimated. This procedure is rapid and comparatively precise compared with the BOD test. However, once again its relevance to natural systems is not as clear as that of the BOD test.

Temperature and Salinity

The solubility of molecular oxygen in water is affected by (1) the partial pressure of oxygen gas in contact with the water, (2) the temperature, and (3) the salinity. The influence of the partial pressure of the gas on solubility is given by Henry's Law:

$$\text{Solubility (molar concentration)} = H_C P_x$$

where H_C is Henry's Law constant and P_x is the partial pressure.

Ambient temperature has a strong influence on the solubility of oxygen in water as shown by the Clausius–Clapeyron equation:

$$\log \frac{C_2}{C_1} = \frac{\Delta H}{2.303R} \left(\frac{1}{T_1} - \frac{1}{T_2} \right)$$

where C_1 and C_2 are gas concentrations in water; T the absolute temperature in kelvins; H the heat of solution (cal mol^{-1}); and R the gas constant (1.987 cal K^{-1} mol^{-1}).

Thus the solubility of oxygen in water decreases with temperature. Actual figures are shown in Table 5.1. In fact, water at 0°C contains almost twice the concentration of water at 30°C. Dissolved oxygen also decreases with increasing salinity (see Table 5.1).

The Dissolved Oxygen "Sag"

One of the most important sources of dissolved oxygen in waterways is oxygen in the atmosphere which dissolves in the water mass at the water surface. Velz (1970)

Table 5.1.   Solubility of Oxygen in Water Exposed to Water-Saturated Air[a]

| Temperature (°C) | Chloride Concentrations (mg L^{-1}) | | | | |
| | 0 | 5,000 | 10,000 | 15,000 | 20,000 |
	Dissolved Oxygen (mg L^{-1})				
0	14.6	13.7	12.9	12.1	11.4
10	11.3	10.7	10.1	9.5	9.0
20	9.1	8.6	8.2	7.7	7.3
25	8.2	8.0	7.4	7.1	6.7
30	7.5	7.2	6.8	6.5	6.2

[a] At 760 mm Hg.

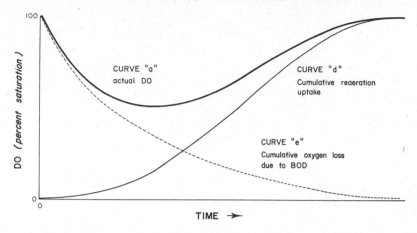

Figure 5.4. Deoxygenation, reaeration, and oxygen sag curves as a result of an increase in BOD at time zero.

and Nemerow (1974) have reported that the rate of reaeration is proportional to the oxygen deficit in the water mass. Thus

$$\frac{dC}{dt} = k_2 D \tag{1}$$

where C is the concentration of oxygen; k_2 the reaeration coefficient; and D the oxygen deficit (i.e., $C_{saturation} - C_{actual}$).

Losses of dissolved oxygen are associated with BOD due to organic discharges. The rate of loss of BOD follows first-order kinetics. Thus the rate of loss of oxygen is proportional to BOD present at the time:

$$\frac{dD}{dt} = k_1 L \tag{2}$$

where D is the oxygen deficit; L the BOD debt or concentration of BOD; and k_1 the BOD decay rate coefficient.

By combining Equations (1) and (2), the actual oxygen deficit can be obtained by subtracting the uptake from the deficit due to BOD. Thus,

$$\frac{dD}{dt} = k_1 L - k_2 D \tag{3}$$

These two processes are illustrated diagrammatically in Figure 5.4. Curve "e" is the cumulative deoxygenation curve which would represent dissolved oxygen values if no reaeration was to occur. However, as soon as an oxygen deficit occurs, soon after zero, reaeration commences (curve "d") and increases as the deficit increases. Curve "d" represents the cumulative oxygen input to the water as a result of deoxygenation. The summation of these curves gives the actual dissolved oxygen profile, the dissolved oxygen "sag" (curve "a"). An expression for this curve is mathematically obtained by integrating Equation (3):

$$D = \frac{k_2 L_0}{k_2 - k_1} (e^{-k_1 t} - e^{-k_2 t}) + D_0 e^{-k_2 t}$$

where L_0 and D_0 are values at time zero.

Critical factors in this expression are k_1 and k_2. The BOD decay coefficient (k_1) can be calculated from experimental results and is usually 0.1 at 20°C with time in days. The reaeration coefficient (K_2 and k_2) is more difficult to measure and several methods have been used. For example, Streeter (see Nemerow, 1974) proposed

$$K_2 = \frac{CV^n}{h^2}$$

where K_2 is the reaeration coefficient (per day); V the mean velocity (ft sec^{-1}); h the mean depth above extreme low water (ft) and C and n constants for the particular stretch of river.

This method has given inconsistent results. O'Connor and Dobbins (see Nemerow, 1974) have found the following expressions satisfactory:

$$k_2 = \frac{480 D_L^{1/2} S^{1/4}}{h^{5/4}}$$

This formula is used for streams showing nonisotropic turbulence, that is, the velocity in different vertical layers is significantly different. For streams with iso-tropic flow the following expression applies:

$$k_2 = \frac{127 (D_L U)^{1/2}}{h^{3/2}}$$

where D_L is the coefficient of molecular diffusion (liquid film ft^2 per day); S the slope of river channel (ft per ft); h the average depth (ft); U the mean velocity of flow (ft sec^{-1}); and k_2 reaeration (per day).

Several other methods of calculation are described by Nemerow (1974). On the other hand Churchill (see Nemerow, 1974) has taken actual measurements of dissolved oxygen in order to calculate reaeration in the field according to the formula

$$k_2 = \frac{\log_{10} D_2 - \log_{10} D_1}{t_2 - t_1}$$

where water flows between two sampling points in the period $t_2 - t_1$ and D_2 and D_1 are dissolved oxygen values at those points. These k_2 values were compared with calculated values using a wide variety of different formulas and the following were found most appropriate:

$$k_2(20°C) = 5.026 \frac{V^{0.969}}{R^{1.673}}$$

where V is velocity (ft sec^{-1}) and R mean depth (ft).

Temperature has a marked impact on the movement of molecules between water and the atmosphere. Churchill (see Nemerow, 1974) has found the following relationship:

$$k_2(t°C) = k_2(20°C) \times 1.0238^{(t-20)}.$$

In the discussion above only oxygen demand due to BOD is considered. This is due to organic matter present as suspended or dissolved matter in the water mass. However, organic matter can be sedimented out of the water mass to form bottom sediments which are rich in organic matter. These organic-rich layers may produce anaerobic degradation in the bottom waters but above this aerobic respiration can occur. This leads to an oxygen demand not measured by BOD which can exceed the BOD in some cases (e.g., Connell et al., 1982).

Photosynthesis and Diurnal Variations in Dissolved Oxygen

Many polluted waterways have very little plant biomass but, in some cases, conditions can be suitable for the growth of aquatic plants. Large rivers and estuaries have significant populations of phytoplankton but many small and shallow streams are dominated by rooted aquatic plants and attached algae (Westlake, 1966). Plants can have a very important influence on dissolved oxygen content through photosynthesis and respiration.

Photosynthesis occurs during the daylight hours but respiration by plants occurs throughout the diurnal cycle. Thus, if significant plant growths are present, this will lead to an input of photosynthetic oxygen during the daylight hours but a continuous consumption of dissolved oxygen by respiration. Figure 5.5 shows these processes cause a rise in dissolved oxygen during the daylight hours with a maximum in the afternoon. At sunset, production of photosynthetic oxygen ceases and so the dissolved oxygen content of the water starts to drop due to respiration by plants and other aquatic organisms. This continues overnight reaching a minimum before dawn when sunlight once again initiates photosynthesis.

These diurnal curves can be used to calculate the total oxygen produced by photosynthesis and the total oxygen consumed by respiration throughout the

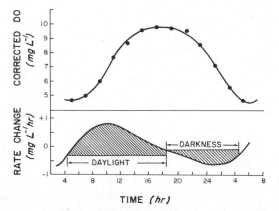

Figure 5.5.   Plot of average variation of corrected dissolved oxygen concentrations and rate of change of dissolved oxygen concentration in a 24 hr cycle in an estuary. [From Connell et al. (1982). Reprinted with permission from the CSIRO Editorial and Publications Service.]

Table 5.2.   Variation in Dissolved Oxygen and Chloride Ion in a Typical Estuary over 24-hour Period

Sample No.	Time	Description of Tide Situation	Chloride Ion (ppm)	Dissolved Oxygen (ppm)
1	6:00 A.M.	Low	1,000	3.0
2	9:00 A.M.	Mean tide	7,500	5.0
3	12:15 P.M.	High	25,000	8.0
4	3:30 P.M.	Mean tide	9,000	5.2
5	6:45 P.M.	Low	1,220	2.8
6	10:00 P.M.	Mean tide	8,000	4.5
7	12:45 A.M.	High	20,000	7.8
8	3:55 A.M.	Mean tide	8,500	5.6

Source: Nemerow (1974). Reprinted with permission from the Hemisphere Publishing Corporation.

diurnal cycle (Odum, 1956). This is done by plotting the rate of change of dissolved oxygen with time and measuring the area under or over the curves (see Figure 5.5).

Dissolved oxygen production and consumption is influenced by light intensity, plant biomass, as well as ambient water temperatures. Generally, Edwards (1968) found that over a diurnal cycle the consumption of oxygen by plant respiration was about 0.75 of the oxygen produced by photosynthesis. During periods of overcast weather, photosynthetic oxygen production can be very low and oxygen consumption due to plant respiration can exceed the photosynthetic oxygen production leading to a decrease in dissolved oxygen in the water mass. In addition, large growths of plants upon death deposit on the bottom and decay, enriching the bottom sediments with organic matter. However, this can lead to anaerobic respiration in bottom waters and sediments but on the surface aerobic respiration occurs. Edwards (1968) has found that a similar situation can occur after the use of aquatic herbicides which cause extensive death and decay of aquatic plants leading to substantial reductions in dissolved oxygen.

Diurnal variations can also occur in aquatic areas, particularly estuaries, as a result of tidal flow. The tides may bring oceanic water rich in dissolved oxygen into an area that has lowered dissolved oxygen levels due to high respiratory activities. This results in a cycle of dissolved oxygen concentrations related to tidal flows (see Table 5.2). Also, changes in the concentrations of BOD can occur in estuarine waters due to the influx of oceanic water often low in BOD. Tidal flows follow a time cycle out of phase with the periods of daylight and darkness so they will introduce patterns out of phase with those produced by photosynthetic activity. These factors can complicate the analysis of dissolved oxygen variation in estuaries (Nemerow, 1974; Velz, 1970).

Seasonal Variations and Vertical Profiles of Dissolved Oxygen

Many environmental factors vary seasonally. For example, the incidence of sunlight has an important effect on photosynthesis and resultant dissolved oxygen concentrations in many aquatic areas. Odum (1956) has found that primary production and thus the production of photosynthetic oxygen is related to the incidence of sunlight, with a maximum occurring in summer and a minimum in winter. Dawes et al. (1977) also found that ambient temperatures affect algal photosynthesis, with extreme temperatures (12, 18, and 42°C) giving the lowest photosynthetic activity.

Turbulent water conditions give well-mixed waters in which vertical profiles of dissolved oxygen are constant from the surface to the bottom waters. However, thermal stratification, due to solar heating of the surface waters, leads to isolation of the bottom waters. If the bottom sediments are enriched with organic matter the bottom waters may become depleted in dissolved oxygen while the surface waters remain unaffected. In this manner, a vertical profile showing considerable variation in dissolved oxygen concentration can occur.

In temperate areas vertical stratification usually occurs in a seasonal pattern with stratification in the summer, and mixing or overturn in the winter. Corresponding to this there is usually a seasonal pattern of dissolved oxygen variation in bottom waters which is common in lakes and relatively small water bodies, but can also occur on the continental shelves of the oceans.

Figure 5.6 shows the seasonal effect of stratification on dissolved oxygen concentration in subsurface waters in the New York Bight located on the continental shelf of the northeastern United States. Armstrong (1977) has suggested that the bottom waters and sediments are enriched by phytoplankton decay, river discharge containing oxygen-demanding substances and nutrients, and dumping of sewage sludge in the area. In 1976 there was an early arrival of spring leading to elevated surface temperatures which coincided with high discharge of water from the adjacent Hudson and Delaware Rivers (see Figure 5.6). This led to a prolonged period of stratification in conjunction with the presence of relatively large quantities of organic matter causing severe depletion of dissolved oxygen in the bottom waters (see Figure 5.6).

Oxygen Balance

In investigating the dissolved oxygen levels and final oxygen balance in a water body it is necessary to identify all the factors contributing to losses and additions of oxygen. These factors have been identified by Connell et al. (1982) as:

Additions

1. *Reaeration.* Solubilization of atmospheric oxygen at the water surface from the atmosphere.

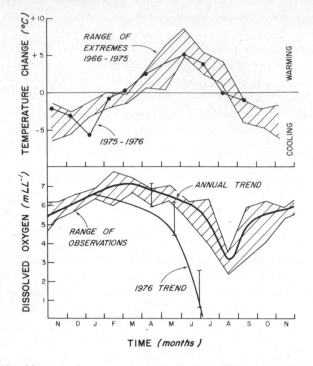

Figure 5.6.   Monthly sea surface temperature change, July 1975–August 1976, and its historic 1966–1975 range at 39–40N, 73–74W, New York Bight (upper). Subsurface (> 20 m) dissolved oxygen as predicted and observed in 1976 and historical range and mean at the same location (lower).
(Values from gulf stream, National Weather Service, NOAA, January 1975–August 1976, and The Gulf Stream Monthly Summary, U.S. Naval Oceanographic Office, January 1966–December 1974). [From Armstrong (1977).]

2. *Photosynthesis.*   Production of photosynthetic oxygen by aquatic plants.

3. *Accrual.*   The addition of dissolved oxygen to the body by discharge of of tributaries.

Losses

1. *Aerobic Respiration.*   The aerobic respiration activities of all organisms present in the water body, including microorganisms in the water mass (measured as BOD), benthic organisms (particularly microorganisms), plants, fish, and other large organisms.

2. *Export.*   The discharge of water containing dissolved oxygen from the water body.

3. *Deaeration.*   If saturation exceeds 100% there is a loss of dissolved oxygen from the water mass to the atmosphere.

The amounts in these various categories would be expected to vary from area to

Table 5.3.   Oxygen Budget for an Urban Estuary

	kg day^{-1}	Standard Deviation	Percent
Additions			
Accrual	76	48	13
Aeration	225	24	45
Photosynthesis	214	–	42
Total	515	–	100
Losses			
Export	30	6	5
Deaeration	89	24	14
BOD	91	35	14
Plant respiration	97	–	15
Benthic respiration	335	~ 45	52
Total	642	–	100

Source: Connell et al. (1982). Reprinted with permission from the CSIRO Editorial and Publications Service.

area. However, Connell et al. (1982) have quantified the amounts for a tidal subtropical creek as shown in Table 5.3.

RESPONSE OF INDIVIDUALS TO REDUCED DISSOLVED OXYGEN AND ASSOCIATED CONTAMINANTS

In natural systems, low dissolved oxygen concentrations in water are usually associated with anaerobic respiration and the production of substances that are toxic to aquatic organisms. Hydrogen sulfide is probably the most important substance in this category. In this case, it can be difficult to distinguish between the effects on organisms caused by low dissolved oxygen concentrations and hydrogen sulfide (Poole et al., 1978). Thus, in considering the response of individuals to reduced oxygen, it is also important to consider the possible effects of other substances, particularly hydrogen sulfide. The effects of both these factors on organisms in aquatic systems are outlined below.

Dissolved Oxygen Reduction

Most aquatic organisms utilize the aerobic respiration process to obtain energy but some invertebrates have a limited capacity for anaerobic respiration (Poole et al., 1978). For example, some polychaetes, bivalves, and even mesopelagic fish are reported as being facultative anaerobes (Poole et al., 1978).

In considering the concentration of dissolved oxygen in water, the most appropriate units to use are concentration, usually as milligram per liter. This is more appropriate than percent saturation since the metabolic demand for oxygen by fish is in terms of actual quantity of oxygen and the percent saturation changes with water temperature (Welch, 1980).

Oxygen uptake by aquatic mammals and reptiles is achieved by direct intake of air at the water surface, for example, porpoises, whales, water snakes, and turtles. Thus for these organisms the concentration of dissolved oxygen in the water mass is of no direct significance. However, many aquatic organisms utilize gills whereby dissolved oxygen is passed from the water mass into the circulatory fluid of the organism. In the circulatory fluid, oxygen is attached to hemoglobin which is then circulated by the heart to the muscles where oxygen is consumed together with carbohydrates to produce carbon dioxide in the aerobic respiration process. Carbon dioxide is respirated to the external environment by the reverse path. Smaller organisms, such as microorganisms and some invertebrates, utilize diffusion through the external body surfaces as a source of oxygen. Circulatory fluids are not necessary since the distances involved are short and uptake can occur relatively rapidly. This process is referred to as cutaneous respiration.

Lowered dissolved oxygen concentrations in the water mass lead to low oxygen uptake by organisms and, consequently, muscles are not fed with sufficient oxygen for aerobic respiration to continue at an optimum rate. This can be compensated for in fish and other organisms by more rapid pumping of water over the gills. But if oxygen uptake is inadequate, insufficient muscle activity will occur and eventually death of the organism will result (Erichsen Jones, 1964). Moderately reduced dissolved oxygen levels reduce physiological activity of aquatic organisms. For example, with fish, food consumption, growth, and swimming velocity all decrease at dissolved oxygen concentrations less than $8-10 \text{ mg L}^{-1}$ (Welch, 1980; Stewart et al., 1967) (see Figure 5.7). Thus adverse effects, although not lethal, can be detected in many fish species when concentrations fall below 100% saturation.

Mobile species of aquatic organisms can exhibit avoidance behavior to low dissolved oxygen concentrations in water. Poole et al. (1978) have reported that several species of fish avoid concentrations of dissolved oxygen ranging from 3 to 5 mg L^{-1} in both field and laboratory experiments. Similarly, invertebrate species have been shown to avoid water with dissolved oxygen concentrations less than full saturation. It has been suggested that trout in freshwater can detect an oxygen gradient and thus move into areas with more favorable oxygen content.

Some aquatic organisms have adapted to continuous low dissolved oxygen concentrations. For example, the respiratory pigments of annelid worms are adapted to take up comparatively high quantities of oxygen by changes in the molecular structures. The proportion of oxygen taken up is directly correlated with the molecular weight of the pigment (Poole et al., 1978). Alternatively, some benthic organisms have developed high concentrations of respiratory pigments in body fluids to compensate for low dissolved oxygen in the water mass, for example, the chironomid larvae which have a highly visible red coloration. Furthermore, there are physical adaptations to low dissolved oxygen concentrations. The larvae of *Eristalis* (the rat-

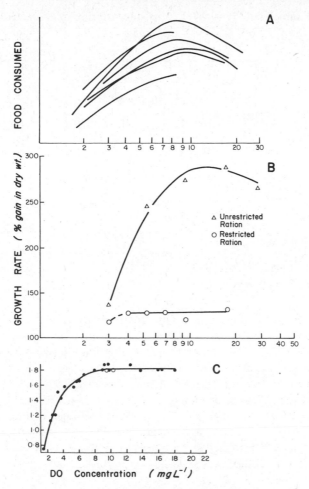

Figure 5.7.   (A) Food consumption in largemouth bass related to D0 concentration. Each curve represents a separate experiment. [From Welch (1980). Reprinted with permission from Cambridge University Press.] (B) The growth rate of juvenile coho salmon fed restricted and unrestricted rations at constant (and diurnally fluctuated) oxygen concentrations. [From Doudoroff and Shumway (1967). Reprinted with permission from the American Fisheries Society.] (C) Relationship between D0 concentration and swimming velocity in coho salmon at 20, 15, and 10°C. Each point represents velocity at which first underling failed to maintain orientation. [From Davis et al. (1963). Reprinted with permission from the American Fisheries Society.]

tailed maggot) has a long telescopic tail with an air tube through which it can breathe (Hynes, 1960).

Different organisms exhibit different rates of oxygen consumption dependent upon size, metabolic characteristics, activity, and many other factors. For example, oxygen consumption rates in terms of cubic centimeters of O_2 per gram of body per hour are 0.0034–0.005 for the jelly fish, 0.055 for the mussel, 0.04 for the eel, and 0.22 for the rainbow trout (Jones, 1964).

Table 5.4.   Examples of Limiting Oxygen Concentrations for Aquatic Organisms[a,b]

Organism	Temperature (°C)	Test	Oxygen mg L^{-1}
Brown trout (*Salmo trutta*)	6.4–24	Limiting concentration	1.28–2.9
Coho salmon (*Oncorhynchus kisutch*)	16–24	Limiting concentration	1.3–2.0
Rainbow trout (*Salmo gairdnerii*)	11.1–20	Limiting concentration	1.05–3.7
Worm (*Nereis grubei*)	21.7–26.3	28 days LC_{50}	2.95
Worm (*Capitella capitata*)	21.7–26.3	28 days LC_{50}	1.50
Amphipod (*Hyalella azteca*)	–	Threshold	0.7
Amphipod (*Gammarus fasciatus*)	–	Threshold	4.3

[a] Compiled from Jones (1964) and Poole et al. (1978).
[b] Limiting values for existence.

Similarly, different organisms show different minimum levels of dissolved oxygen for survival (see Table 5.4). It is interesting to note that often closely relates species (e.g., amphipods in Table 5.4) can show widely different minimum dissolved oxygen levels. Also, many organisms show a different range of responses to lowered dissolved oxygen concentrations in the water mass (see Figure 5.8).

Another important aspect of organism response to reduced dissolved oxygen is the different oxygen requirements of the different life stages of aquatic organisms. For example, the Coho salmon embryo is dependent on an increasing minimum dissolved oxygen as it increases in size (Welch, 1980). Water velocity also has an interactive effect with dissolved oxygen concentration and size as illustrated in Figure 5.9.

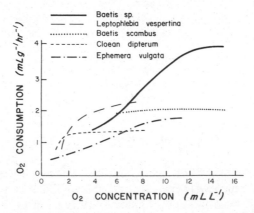

Figure 5.8.   Respiration rate (O_2 consumption) vs. O_2 concentration in five ephemeropterans. [From Fox et al. (1935). Reprinted with permission from Cambridge University Press.]

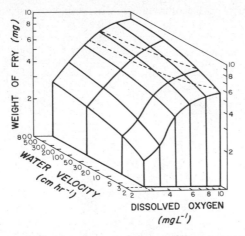

Figure 5.9.   Development of coho salmon embryos at different concentrations of DO and at different water velocity levels. Dashed lines indicate DO and velocity combinations producing 20 and 33% reduction in growth. [From Shumway et al. (1964). Reprinted with permission from the American Fisheries Society.]

Dissolved Oxygen Criteria

Until recently, 5 ppm of dissolved oxygen was considered an acceptable minimum for normal growth and reproduction of fish (Jones, 1964). However, it has now become apparent that there are a number of complicating environmental factors which can have an impact on the minimal dissolved oxygen level required for fish

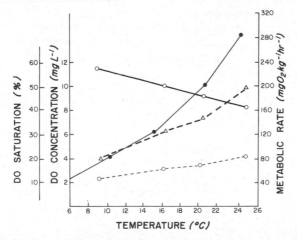

Figure 5.10.   Standard metabolism of *Salvelinus fontinalis* in relation to minimum DO needed in concentration and percent saturation. (●, standard metabolic rate; ○, DO concentration at 100% saturation; △, minimum % saturation allowing standard metabolic rate; ○ ---, minimum DO concentration allowing standard metabolic rate.) [From Welch (1980). Reprinted with permission from Cambridge University Press.]

Table 5.5.   Examples of Recommended Minimum Concentrations of Dissolved Oxygen (DO)

Estimated Natural Seasonal Minimum DO	Recommended Minimum DO for Selected Levels of Fish Protection (mg L^{-1})			
	Yearly Maximum	High	Moderate	Low
5	5	4.7	4.2	4.0
6	6	5.6	4.8	4.0
7	7	6.4	5.3	4.0
8	8	7.1	5.8	4.3
9	9	7.7	6.2	4.5
10	10	8.2	6.5	4.6
12	12	8.9	6.8	4.8
14	14	9.3	6.8	4.9

Source: USEPA (1972).

to complete the normal life cycle. Seasonal variations in the dissolved oxygen requirement of aquatic organisms are known to occur. Also, there is a different susceptibility of aquatic organisms to lowered dissolved oxygen levels at differing latitudes (Johannes and Betzler, 1975). Much of this variation in susceptibility to lowered dissolved oxygen levels is related to water temperature. As water temperature increases, the metabolic rate, and thus oxygen demands, of organisms increases but the solubility of oxygen in the water mass decreases (see Figure 5.10).

The setting of criteria or standards can be approached in a number of different ways. With one method the criterion can be based on the seasonal minimum dissolved oxygen concentrations (see Table 5.5). Thus, this method assumes that organisms in the area are adjusted to these seasonal variations and criteria can be accordingly related to the seasonal minima. Another method is to use the solubility at different temperatures together with the metabolic needs of organisms at those temperatures (see Table 5.6).

Table 5.6.   Example of Minimum Acceptable Concentrations of Dissolved Oxygen in Freshwaters

Temperature (°C)	DO Complete Saturation (mg L^{-1})	Minimal Levels for Protection of Aquatic Life	
		mg L^{-1}	Saturation (%)
36.0	7	5.8	82.9
27.5	8	5.8	72.5
21.0	9	6.2	68.9
16.0	10	6.5	65.0
7.7	12	6.8	56.7
1.5	14	6.8	48.6

Source: USEPA (1972).

Hydrogen Sulfide

Hydrogen sulfide can occur as a significant pollutant in some waste-water discharges. In this present case, hydrogen sulfide is considered as a product of anaerobic respiration produced at reduced dissolved oxygen levels in aquatic areas by organic discharges. Thus natural ecosystems affected by reduced dissolved oxygen will also be influenced by hydrogen sulfide (Poole et al., 1978).

Hydrogen sulfide in aqueous solution undergoes the following dissociation:

$$H_2S \longrightarrow H^+ + HS^- \longrightarrow H^+ + S^{2-}$$

At temperatures of 25°C and pH less than 6, most of the hydrogen sulfide is dissolved as the undissociated form. But at pH greater than 7.8, the bisulfite ion predominates (Poole et al., 1978).

The toxicity of hydrogen sulfide at intermediate pH values is shown in Table 5.7. It can be seen that hydrogen sulfide is very toxic to a wide range of aquatic organisms. Thus in most situations where reduced dissolved oxygen occurs, there will be the two effects acting on organisms in the area, that is, reduced dissolved oxygen and toxic effects due to hydrogen sulfide.

Long-term tests have shown that many adverse effects occur at very low concentrations (Poole et al., 1978). For example, *Gammarus pseudolimnaeus* shows adverse effects on reproduction and growth of young at concentrations as low as $2 \mu g L^{-1}$. It is noteworthy that this figure is 10 times lower than the 96 hr LC_{50}. Similarly, other adverse effects have been found with fish and eggs at low concentrations. Poole et al. (1978) have reported that generally the toxicity resulting from hypoxia and hydrogen sulfide together is less than additive. The toxic effect of hydrogen sulfide is believed to result from inhibition of metalloenzymes by reaction of the substance with the metals present.

Table 5.7.   Lethal Effects of H_2S to Freshwater Aquatic Fauna

Species	96 hr LC_{50} ($\mu g L^{-1}$)	pH	Temperature (°C)
Gammarus pseudolimnaeus	22	7.7–7.9	17.8–18.1
Asellus militaria	1070		
Crangonyx richmondensis laurentianus	840		
G. pseudolimnaeus	59	7.5	15.0
Baetis vagans	20		
Ephemera simulans	316		
Hexagenia limbata	11		
Notropis cornutus	278[a]	6.7–7.9	13.4–14.1
Carassius auratus	110	7.8	15.0

Source: Poole et al. (1978). Reprinted with permission from CRC Press, Inc.

[a] Recalculated from total sulfide concentration.

EFFECTS OF REDUCED DISSOLVED OXYGEN AND ASSOCIATED FACTORS ON COMMUNITIES AND ECOSYSTEMS

There are several factors associated with dissolved oxygen reduction which have an ecological impact. Dissolved oxygen reduction usually results from the addition of organic matter to a water body. However, organic matter itself, apart from its secondary effect of reducing dissolved oxygen, has an ecological impact. It results in an additional supply of food and energy for aquatic organisms and hydrogen sulfide is produced. So the ecological effects usually associated with organic discharges are due to the combined effects of reduced dissolved oxygen, organic enrichment, and the toxic effect of hydrogen sulfide. Although laboratory experiments can be conducted to demonstrate the different effects of these factors, similar distinctions cannot be made in affected natural ecosystems.

It should be noted that in the longer term, the degradation of organic matter releases nitrogen and phosphorus salts. These substances play an important role in the eutrophication process and thus organic enrichment is connected with eutrophication. This aspect is considered in Chapters 4 and 6.

Overall Impact on Aquatic Systems

In broad terms, freshwater, estuarine and marine systems exhibit similar responses to organic enrichment and dissolved oxygen reduction. The basic metabolism of an aquatic system is altered by the addition of organic matter. As discussed previously, photosynthesis and respiration are the fundamental chemical processes occurring in aquatic ecosystems and these processes regulate and control the living components. The discharge of organic matter to an aquatic area results in an increase in respiration and, in addition, may reduce photosynthesis. Thus there will be an alteration in the basic metabolism in the aquatic area as measured by the ratio of photosynthesis to respiration.

Odum (1956) has classified a wide variety of aquatic systems according to their photosynthesis to respiration ratio as shown in Figure 5.11. Thus, water receiving a high level of organic pollution, often described as polysaprobic, is located in the bottom right-hand area of the diagram where respiration greatly exceeds photosynthesis. Respiration in this zone is due to large numbers of bacteria and protozoa and counts in excess of 1 million per milliliter have been obtained (Hawkes, 1962). As the quantity of organic matter decreases, the availability of plant nutrients (nitrogen and phosphorus salts) increases, stimulating photosynthesis and plant growth and leading to the development of a photosynthesis to respiration ratio approaching 1 (see Figure 5.11). When the organic matter has been removed by respiration, the conditions are described as oligosaprobic.

The increase in respiration resulting from additions of organic matter mentioned above is principally due to bacteria and protozoa but there is a large population increase in many other heterotrophic organisms. The major larger heterotrophes include the polychaetes (e.g., *Capitella* sp.) and tuberficids. Overall, there is an

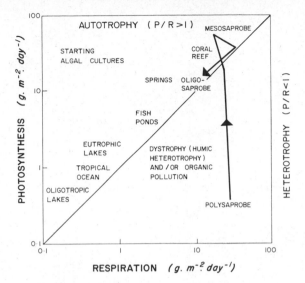

Figure 5.11. A diagram for the functional classification of communities according to the total metabolism and relative dominance of photosynthesis and respiration. [Modified from Odum (1956). Reprinted with permission from the American Society of Limnology and Oceanography, Inc.]

increase in the biomass even when the dissolved oxygen levels drop to zero (Poole et al., 1978).

When the dissolved oxygen reduction extends to lethal levels "kills" of aquatic organisms occur. These are more common in rivers and streams due to the limited volume of water that is present compared to the discharge volume. In oceanic areas the volume of water is usually very large compared with the discharge volume and thus dissolved oxygen reductions are less frequent. Nevertheless, stratification can occur isolating the bottom layers which may then become anoxic (see Figure 5.6).

Changes in the structure of aquatic communities are related, to a certain extent, to the mobility of the organisms in the area. Fish can undertake avoidance movement to remain clear of areas where there are low dissolved oxygen and concentrations of hydrogen sulfide. But large kills of fish often occur in rivers where there may be dissolved oxygen reduction and fish have restricted movement. Fish kills in oceanic areas are less common due to comparatively unrestricted movement, the lengthy period it usually takes to develop anoxic conditions, and the restriction of anoxic conditions to bottom waters. Thus large bodies of water usually exhibit kills of immobile benthic animals (Garlo et al., 1979) and, under extreme conditions, whole groups of organisms (e.g., fish) may be removed from an ecosystem. Also, anoxic conditions on the continental shelf in the northeastern United States have caused avoidance movements which have altered the migration patterns of finfish and lobsters (Steimle and Sindermann, 1978; Garlo et al., 1979).

Table 5.8. Tolerance of Organisms to Organic Pollution[a]

Systematic Group	Saprobic Classification of Members			
	Polysaprobic (grossly polluted)	α-Mesosaprobic (polluted)	β-Mesosaprobic (mildly polluted)	Oligosaprobic (nonpolluted)
Algae				
Cyanophyceae (blue-green algae)		Spirulina	Oscillatoria spp	Chamaesiphon
Bacillariaceae (diatoms)		Nitzschia palea	Gomphonema parvulum, Rhoicosphenia curvata	Cocconeis placentula
Chlorophyceae (green algae)		Stigeoclonium	Chlamydomonas spp., Oedogonium, Ulothrix, Cosmarium, Scenedesmus	Rhizoclonium, Cladophora spp., Chaetophora, Vaucheria, Spirogyra, Closterium, Draparnaldia
Rhodophyceae (red algae)				Batrachospermum, and all other fresh water spp.
Angiosperms			Elodea canadensis (Canadian water weed or water thyme), Glyceria aquatica	Polygonum amphibium

	Potamogeton pectinatus	Potamogeton spp. (pondweeds)	
	P. interruptus		
	Ranunculus fluitans (water crowfoot)		
	Lemna minor (lesser duckweed)		
	Ceratophyllum		
		Nuphar luteum	
		Nymphaea alba	
Platyhelminthes			
Turbellaria (Planarians)	Dendrocoelum lacteum	Planaria alpina	
		P. gonocephala	
	Polycelis nigra	Polycelis cornuta	
Annelida (true worms)	Tubifex tubifex		
	Lumbricus rubellus		
	Limnodrilus		
	Lumbriculus spp.		
	Lumbricillus lineatus		
	Stylaria	Chaetogaster	
	Nais	Gordius	
Hirudinea (leeches)	Erpobdella octoculata		
	Glossiphonia complanta		
	Helobdella stagnalis		
	Haemopis sanguisuga		
	Erpobdella testacea		
	Glossiphonia heteroclita		

(continued on next page)

Table 5.8. (continued)

Systematic Group	Saprobic Classification of Members			
	Polysaprobic (grossly polluted)	α-Mesosaprobic (polluted)	β-Mesosaprobic (mildly polluted)	Oligosaprobic (nonpolluted)
Arthropoda				
Crustacea			Asellus aquaticus (water louse)	Daphnia spp. (water flea) Cypris spp. Cyclops spp. Gammarus pulex (water shrimp) Astacus fluviatilis (freshwater crayfish)
Insecta				
Plecoptera (Stone flies)				All species
Ephemeroptera (May-flies)				All species
Neuroptera (Alder flies)		Sialis lutaria		
Trichoptera (Caddis flies)			Hydropsyche spp. Anabolia sp. Molanna sp.	All other species
Diptera (True flies)	Eristalis tenax (rat-tailed maggot) Chironomus plumosus	Chironomus plumosus	Tanypus spp. Culex spp. Simulium ornatum Simulium reptans Simulium aureum	Chironomus spp. Other Simulium species

120

Mollusca

Sphaerium corneum
Limnaea auricularia
Ancylus fluviatilis
Pisidium spp.
Unio
Planorbis

Pisces
(Fish)

Alburnus alburnus
(Bleak)
Anguilla anguilla
(eel)
Gasterosteus aculeatus
(3-spined stickleback)
Carassius carassius
(Crucian carp)
Carassius auratus
(goldfish)
Nemacheilus barbatula
(stoneloach)
Other species, e.g.,
Salmo fario
(river trout)
Esox lucius
(pike)
Cottus gobio
(Miller's Thumb)
Gobio gobio
(gudgeon)
Perca fluviatilis
(perch)

Source: Hawkes (1962). Reprinted with permission from Butterworth and Company (Publishers) Ltd.

a Based on the Kolkwitz–Marsson classification.

Different organisms exhibit different ranges of oxygen consumption related to the availability of dissolved oxygen in the water mass (see Figure 5.8). Similarly, different species exhibit different lower lethal limits with reduced dissolved oxygen concentrations (see Table 5.4). Therefore, in an area where oxygen reduction occurs, we would expect to see a differentiation of species and a reduction in species diversity related to the different response to reduced oyxgen levels.

The presence of organic matter from a discharge alters the type and availability of food and the community will also respond to this type of change. Welch (1980) has described broad food preferences and indicated how changes in food supply could cause shifts in community structure. The community members can be classified as follows: (1) Detrital feeders which can include the net spinning caddis flies, aquatic sow bug, chironomids, clams, snails, some mayflies, and blackflies. These are largely collectors of fine particulate matter. (2) Grazers include most stone flies, some mayflies, case-building caddis flies, and snails. (3) Predators as represented by the dragonflies, leeches, a few stone flies, beetles, some midges, and a few caddis flies.

An increase in the amount of detritus present through the discharge of organic matter would result in an increase in the detrital feeders, especially at the expense of the grazers. In addition, predators are often reduced or eliminated by low dissolved oxygen concentrations in the water mass due to their generally greater sensitivity.

Hawkes (1979) has summarized the tolerance of a wide variety of freshwater riffle organisms to organic enrichment (shown in Table 5.8). Three major groupings have been devised by Cairns and Dickson (1971) for tolerance of benthic organisms. The tolerant group includes sludge worms, certain midges, leeches, and certain snails; the mildly tolerant group includes most snails, sow bugs, skuds, blackflies, crane flies, fingernail clams, dragonflies, and some midges, whereas the intolerant group includes mayflies, stone flies, caddis flies, riffle beetles, and hellgrammites.

The reproduction rate of organisms is an important factor affecting their abundance in polluted conditions. Welch (1980) has reported that although oligochaete worms are adversely affected by dissolved oxygen reduction, large numbers occur because of the rapid reproduction rate in a wide range of differing environmental conditions.

Biotic Indices

There are a number of measures of the biotic changes that occur in an aquatic ecosystem from the effects described above. Predictably, a relationship should exist between water composition and measures of changes in the aquatic ecosystem. As a result of this, biotic indices (see Chapter 4) are often used together with physico-chemical measures of the quality of water in an area (Sladecek et al., 1982).

The measure with the longest history is the "saprobic" or "saprobien" system, initially developed by Kolkwitz and Marsson (1902) for continental Europe. Sladecek (1979) has defined saprobity as the state of water quality with respect to

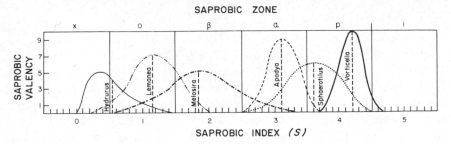

Figure 5.12. Part of the scale of saprobity showing the position of six benthic species (*Hydrurus foetidus, Lemanea fluviatilis, Melosira varians, Apodya lactea, Sphaerotilus natans,* and *Vorticella microstoma*) where o = oligosaprobic; α and β = mesosaprobic; and p = polysaprobic. χ and i are extensions at either end of this scale where χ = xenosaprobic and i = isosaprobic. [From Sladecek (1979). Reprinted with permission from John Wiley & Sons, Inc.]

the content of putrescible organic material as reflected by species composition of the community. This system has been subject to many modifications over time and some of these are described by Sladecek (1979). The method basically consists of dividing the various stages of recovery of a stream polluted by an organic discharge into zones related to the organisms present (see Table 5.8). There are basically three zones: polysaprobic, mesosaprobic, and oligosaprobic, but further subdivision of these zones is commonly used.

The polysaprobic zone is characterized by high concentrations of decomposable organic matter, the absence of oxygen, and the presence of hydrogen sulfide. The biotic community is restricted to a few groups — primarily large numbers of bacteria and protozoa. The mesosaprobic zone has a well-established oxidation process occurring and is often subdivided into two zones. The α-mesosaprobic zone may contain substantial quantities of oxygen, that is, often in excess of 50% saturation, and hydrogen sulfide is not present. Biologically it is rich in bacteria and protozoa. The β-mesosaprobic zone has continued oxidation or mineralization with the oxygen content never being less than 50% saturation. However, in this zone there is a decrease in the number of bacteria and protozoa and an increase in the diversity of plants and animals. In the oligosaprobic zone the oxidation or mineralization processes are complete and the organic content is low. It is biologically characterized by low numbers of bacteria and a high species diversity including fish (Hawkes, 1962).

Sladecek (1979) has described more advanced applications of the saprobien system. The system can be placed on a more quantitative basis by the use of the "saprobic valency." This takes account of the fact that single species rarely are represented in only one zone or one saprobic level. Each species is considered to be distributed in an approximately normal fashion in the saprobic scale and is assigned numerical values according to this (see Figure 5.12). Thus each species has a maximum valency in a particular area of the saprobic scale and a distribution on either side of this. To apply this system requires a detailed knowledge of the characteristics of individual species.

Subjective estimations can be replaced by a "saprobic index" (Pantle and Buck, 1955) which is defined as follows:

$$S = \frac{\sum (hS_i)}{\sum h}$$

where S_i is the individual saprobic index and h is the abundance.

Sladecek (1979) reports that this is the most widely used index in central and eastern Europe. But it should be noted that there are a number of criticisms of the saprobic system. For example, Hynes (1960) concludes that it is only applicable to heavy sewage and slow flowing rivers. Robach (1974) reports that single species can occur in several zones leading to errors in classification. The major criticisms have been listed by Persoone and De Pauw (1979) as: (1) necessity to identify organisms to species with only poor taxonomic tools available; (2) paucity of basic knowledge of the ecological characteristics and requirements of individual species and communities; (3) failure to properly quantitate species relationships; (4) failure to accommodate the uniqueness of each stream, each pollutant, and each problem; and (5) restriction of the use of saprobic systems to pollution caused by municipal sewage or similar acting organic wastes. These factors limit the application of the system and indicate precautions necessary when it is used.

Somewhat similar systems have also been developed elsewhere. The "Trent biotic index" is based on the known tolerance of species of indicator invertebrates, weighted by the number of defined groups present (Hawkes, 1979). Another index is the "score system" which has been developed in Britain by Chandler (1970), who described the system as based on a point system for taxon abundance according to sensitivity to organic pollution. There are many other indices that have also been developed as described by Dickson and Cairns (1978), James and Evison (1979), Herricks and Cairns (1982), and Hellawell (1978).

Ecological Effects Related to Spatial and Temporal Patterns of Change

Reduction in dissolved oxygen content of water and the other effects associated with organic enrichment often lead to a systematic sequence of physicochemical and biological changes related to the source of discharge, time, and environmental conditions. These are best considered by discussion of related groups of situations as set out below.

Streams

With streams, there is a unidirectional flow of water which leads to the formation of the dissolved oxygen sag described previously. This gives a set of zones grading into one another but having different dissolved oxygen content, organic matter, and biota. In Chapter 4, Figures 4.14 and 4.15 show variations of species diversity as related to a discharge of organic matter and consequent changes in dissolved oxygen and the associated effects. More detailed chemical and related biological changes are shown in Figure 5.13. The stream section below the discharge point can be classified into various zones using the indices previously described (see Figure 4.5). Table 5.8 indicates the organisms associated with each zone in the saprobic system under continental European conditions.

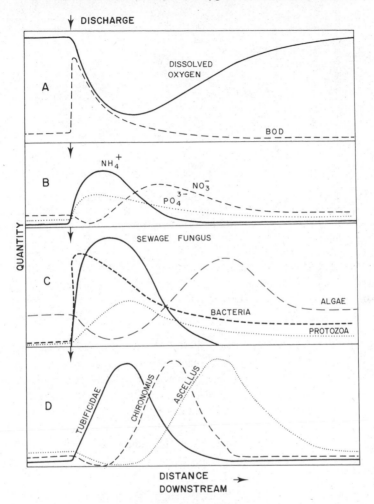

Figure 5.13.   Diagrammatic representation of the effects of an organic discharge on the downstream section of a stream: (A) and (B) chemical changes; (C) microorganisms; and (D) other organisms. [Compiled from Hynes (1960). Reprinted with permission from Liverpool University Press.]

Lakes and the Open Sea

The classification and systematic investigation of ecological changes resulting from organic discharges have been most successful in flowing streams. The unidirectional flow of water gives a clear sequence of events which simplifies study. Such events are not so clearly systematic in lakes and the sea and thus investigations of these areas have not yielded clear results. In lakes and the sea, there is a multidirectional flow of water which, in the sea, is related to tides and, to a lesser extent, ocean currents. Changes in water level in the intertidal zone as well as turbulence due to wave action and salinity gradients can also cause variations in the ecological patterns

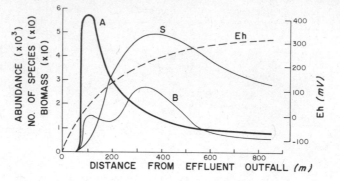

Figure 5.14.   Effects of distance from an outfall of an organic discharge in a marine area, where s = species numbers, A = abundance, and B = biomass (g cm^{-2}) with Eh measured at 40 mm depth. [From Pearson and Stanley (1979). Reprinted with permission from Springer-Verlag, New York.]

in these areas (Perkins, 1979). Also, most marine organisms have planktonic larvae, in contrast to freshwater organisms, thus factors which influence the planktonic larval stage may have an important impact on the distribution of organisms. These factors add to the complexity of interpreting changes in biota in marine areas which are more subtle than in freshwater areas.

The oceans contain a large reservoir of dissolved oxygen due to the comparatively large volume of water compared with freshwater streams. Thus, comparatively large volumes of organic matter are generally needed to cause reductions in dissolved oxygen. Many of the effects due to discharges of organic matter into the sea occur in areas of relatively shallow and confined water.

Nevertheless, somewhat similar ecological changes occur in lakes and the sea as compared to rivers. There is a reduction in the number of species and diversity in heavily polluted areas with an increase in the biomass of some tolerant organisms. With increasing distance from the discharge point, this zone leads gradually through to clean conditions with the intervening situation showing intermediate circumstances. Pearson and Stanley (1979) have related this situation to Eh values in sediment which decrease as the organic content in marine areas increases (see Figure 5.14). The macroinvertebrate species involved in the various zones of pollution resulting from discharges of organic matter are shown in Table 5.9. Similar results, involving different species, have been observed in New York Bight and the Firth of Clyde (Perkins, 1979). In addition, a decrease in the size of animals has been observed by Pearson and Rosenberg (1978) with increasing organic pollution. Increasing organic matter was also found to cause a change in the trophic structure of the affected communities with the proportion of deposit feeders increasing whereas carnivores and omnivores decrease (Pearson, 1980).

The geographical pattern of distribution of waters containing reduced dissolved oxygen and organic matter is related to the flow patterns of water in the discharge, the discharge or dumping locations, and a variety of other environmental factors. An example of the distribution patterns of areas low in dissolved oxygen, or in Eh, is shown in Figure 5.15.

Table 5.9. Changes in Dominant Macroinvertebrate Fauna with Increasing Organic Enrichment in Some Marine Areas[a]

Area/Effluent Type	Normal	Transitory	Polluted	Grossly Polluted
Lochsin, Scotland, and Fjords in Sweden/ paper mill effluent	*Nucula* *Amphiura* *Terebellides* *Rhodine* *Echinocardium* *Nephrops*	*Lapidoplax* *Corbula* *Goniada* *Thyasira* *Pholoe* *Chaetozone* *Anaitides* *Pectinaria* *Myriochele* *Ophiodromus*	*Capitella* *Scolelepis*	No fauna. Surface covered by a fiber blanket
Off Marseilles/ sewage effluent	Not described	*Nereis caudata* *Staurocephalus rudolphii* *Cirriformia tentaculata* *Corbula gibba* *Thyasira flexuosa* Rich polychaete fauna	*Capitella capitata* *Scolelepis fuliginosa*	No fauna

[a] Compiled from Bellan (1970) and Pearson and Rosenberg (1978).

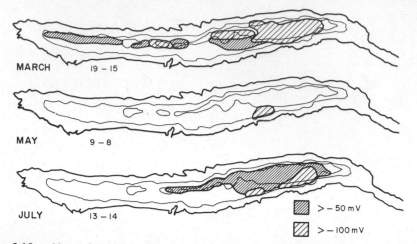

Figure 5.15.   Maps of Loch Eil (20 and 40 mm depth contours marked), showing areas of low redox potential in 1977. Values under each map are mean daily tonnage of suspended solids in pulp mill effluent discharged to the loch for the preceding 2 months. Low redox values in March followed high effluent discharge levels in the preceding months. Low input levels in March and April were followed by a marked reduction in the area of highly reduced sediment but higher inputs in May and June resulted in an increase in the area again in July. [From Pearson and Stanley (1979). Reprinted with permission from Springer-Verlag, New York.]

Vertical Profiles (see also Seasonal Variations and Vertical Profiles of Dissolved Oxygen)

Vertical stratification of water bodies leads to the isolation of bottom waters causing dissolved oxygen reduction. This can occur in water bodies ranging in size from the ocean itself to small pools. For example, Figure 5.16 shows vertical profiles of temperature and dissolved oxygen in New York Bight in 1976. The vertical

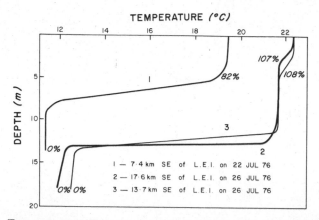

Figure 5.16.   Temperature profiles and associated dissolved oxygen saturation values at ocean stations in the vicinity of Little Egg Inlet, New Jersey. (1) 7.4 km SE of L.E.I. on 22 July 1976; (2) 17.6 km SE of L.E.I. and (3) 13.7 km SE of L.E.I. on 26 July 1976. [From Garlo et al. (1979). Reprinted with permission from Academic Press, Inc. (London) Ltd.]

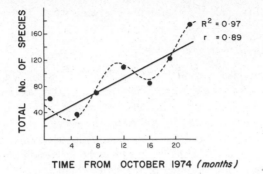

Figure 5.17.   Changes in number of species over time in an estuarine area in tropical Australia. [From Saenger et al. (1980). Reprinted with permission from the Queensland Museum.]

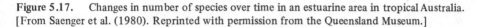

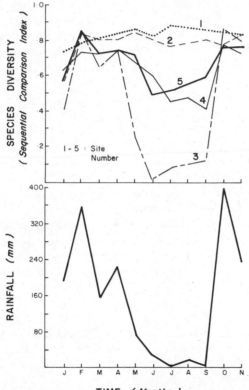

Figure 5.18.   Variation of species diversity and rainfall in a stream subject to a seasonal pattern of rainfall. (Sites 1 and 2 are above the discharge and sites 3, 4, and 5 are in numerical order below the discharge.) [From McIvor (1976). Reprinted with permission from the Australian Water and Wastewater Association.]

distribution of fish and of other mobile organisms in these areas is related to these factors.

Temporal Patterns

Temporal patterns take a wide variety of forms. Firstly, the population in many aquatic areas including those affected by organic matter, exhibit seasonal patterns of change. Figure 5.17 shows an annual cycle in the number of species of macro-benthos in an estuarine area. In addition, the species diversity shows a steady increase over the approximately 2 years of investigation. This is believed to be part of a longer time sequence of change in which the species-number increase is due to recovery from freshwater flooding of the estuary which occurs on a periodic basis. These changes provide a basic natural sequence of change in aquatic areas on which changes due to dissolved oxygen and organic matter enrichment are superimposed.

Seasonal effects are particularly marked where rainfall shows distinct seasonal variations. Reduced rainfall leads to a lower streamflow and therefore lower dilution of discharges and a greater impact on dissolved oxygen and the organic matter content of the area. This pattern of change leads to a consequent change in the community structure of biota in the area (see Figure 5.18). Diurnal changes in dissolved oxygen in some areas require biota able to tolerate the overnight minimums.

Initial introduction of organic matter into an area gives a temporal sequence of changes in dissolved oxygen, and associated factors, with consequent effects on biota. These are generally related to changes observed with distance from a source. Recovery of an area after discharge has ceased shows a somewhat similar reverse sequence which has been described by Pearson (1980).

REFERENCES

APHA (1975). Standard Methods for the Examination of Water and Waste Water. American Public Health Association, American Water Works Association and Water Pollution Control Federation, 14th Edition, Washington, D.C.

Armstrong, R. S. (1977). "Climatic Conditions Related to the Occurrence of an Anoxia in the Waters off New Jersey during the Summer of 1976." In *Oxygen Depletion and Associated Environmental Disturbance in the Middle Atlantic Bight in 1976*. Technical Series Report No. 3, Northeast Fisheries Center, National Marine Fisheries Service, National Oceanographic and Atmospheric Administration, p. 17.

Arthington, A. A., Conrick, D. L., Connell, D. W., and Outridge, P. M. (1982). The Ecology of a Polluted Urban Creek with Particular Reference to the Value of Biological Monitoring, Technical Report 68, Australian Water Resources Council, Canberra.

Bellan, G. (1970). Pollution by sewage in Marseilles. *Mar. Pollut. Bull.* 1, 59.

Cairns, J. and Dickson, K. L. (1971). A simple method for the biological assessment of the effects of waste discharges on aquatic bottom-dwelling organisms. *J. Water Pollut. Control Fed.* 43, 755.

Chandler, J. R. (1970). A biological approach to water quality management. *Water Pollut. Control* 4, 415.

Connell, D. W., Morton, H. C., and Bycroft, B. M. (1982). An oxygen budget for an urban estuary. *Aust. J. Mar. Freshwater Res.* **33**, 607.

Davis, G. E., Foster, J., and Warren, C. E. (1963). The influence of oxygen concentration on the swimming performance of juvenile Pacific salmon at various temperatures. *Trans. Am. Fish Soc.* **92**, 111.

Dawes, C. J., Moon, R. E., and Davis, M. A. (1977). The photosynthetic and respiratory rates and tolerances of benthic algae from a mangrove and salt marsh estuary: A comparative study. *Estuar. Coast. Mar. Sci.* **6**, 175.

Dickson, K. L. and Cairns, J. (1978). Biological data in water pollution assessment: Quantitative and statistical analyses, ASTM, Special Technical Publications 652, American Society for Testing and Material, Philadelphia.

Doudoroff, P. and Shumway, D. L. (1967). Dissolved oxygen criteria for the protection of fish. In *A Symposium on Water Quality Criteria to Protect Aquatic Life.* American Fish Society, Spec. Publ. No. 4, p. 13.

Edwards, R. W. (1968). Plants as oxygenators in rivers. *Water Res.* **2**, 243.

Erichsen Jones, J. R. (1964). *Fish and River Pollution.* Butterworth, London.

Fox, H. M., Simmonds, B. G., and Washbourn, R. (1935). Metabolic rates of ephemerid nymphs from swiftly flowing and still waters. *J. Exp. Biol.* **12**, 179.

Garlo, E. V., Milstein, C. B., and Jahn, A. E. (1979). Impact of hypoxic conditions in the vicinity of Little Egg Inlet, New Jersey in Summer, 1976. *Estuar. Coast. Mar. Sci.* **8**, 421.

Hawkes, H. A. (1962). "Biological Aspects of River Pollution." In L. Klein (Ed.), *River Pollution II. Causes and Effects.* Butterworths, London, p. 311.

Hawkes, H. A. (1979). "Invertebrates as Indicators of River Water Quality." In A. James and L. Evison (Eds.), *Biological Indicators of Water Quality.* John Wiley & Sons, Chichester, pp. 2-1 to 2-45.

Hellawell, J. M. (1978). *Biological Surveillance of Rivers: A biological monitoring handbook.* Water Research Centre Comp, Stevenage.

Herricks, E. E. and Cairns, J. (1982). Biological monitoring. Part III — Receiving system methodology based on community structure. *Water Res.* **16**, 141.

Hynes, H. B. N. (1960). *The Biology of Polluted Waters.* Liverpool University Press, Liverpool.

James, A. and Evison, L. (1979). *Biological Indicators of Water Quality.* John Wiley & Sons, Chichester.

Johannes, R. E. and Betzler, S. B. (1975). "Introduction . . . Marine Communities Respond Differently to Pollution in the Tropics than at Higher Latitudes." In E. J. F. Wood and R. E. Johannes (Eds.), *Tropical Marine Pollution.* Elsevier, Amsterdam.

Kolkwitz, R. and Marsson, M. (1902). Grundsatz fur die biologische Beuntielung des wassers nach seiner flora und fauna. *Mipp. Pruf. Anst. was Versong. Abwasserbeseit. Berl.* **1**, 33.

Manahan, S. E. (1975). *Environmental Chemistry*, 2nd ed. Willard Grant Press, Boston.

McIvor, C. C. (1976). The effects of organic and nutrient enrichments on the benthic macroinvertebrate community of Moggill Creek, Queensland. *Water* **3**, 16.

Nemerow, N. L. (1974). *Scientific Stream Pollution Analysis*. McGraw-Hill, New York.

Odum, H. T. (1956). Primary production in flowing waters. *Limnol. Oceanogr.* **I**, 102.

Pantle, R. and Buck, H. (1955). Die biologische uberwachung der Gewasser und die Danstelling der Ergebnisse. *Gas. u. Wasserfach.* **96**, 604.

Pearson, T. H. and Rosenberg, R. (1978). Macrobenthic succession in relation to organic enrichment and pollution of the marine environment. *Oceanogr. Mar. Biol.* **16**, 229.

Pearson, T. H. (1980). Marine pollution effects of pulp and paper industry wastes. *Helgolander Meeresunters* **33**, 340.

Pearson, T. H. and Stanley, S. O. (1979). Comparative measurement of the redox potential of marine sediments as a rapid means of assessing the effect of organic pollution. *Mar. Biol.* **53**, 371.

Perkins, E. J. (1979). "The Effects of Marine Discharges on the Ecology of Coastal Waters." In A. James and L. Evison (Eds.), *Biological Indicators of Water Quality*. John Wiley & Sons, Chichester, pp. 12-1 to 12-39.

Persoone, G. and DePauw, N. (1979). "Systems of Biological Indicators for Water Quality Assessment." In O. Ravera (Ed.), *Biological Aspects of Freshwater Pollution*. Pergamon Press, Oxford, p. 39.

Poole, N. J., Wildish, D. J., and Kristmanson, D. D. (1978). The effects of the pulp and paper industry on the aquatic environment. *Crit. Rev. Environ. Control* **8**, 153.

Roback, S. S. (1974). "Insects". In C. W. Hart and S. L. H. Fuller (Eds.), *Pollution Ecology of Freshwater Invertebrates*. Academic Press, New York.

Saenger, P., Stephenson, W., and Moverley, J. (1980). The estuarine macro benthos of the Calliope River and Auckland Creek, Queensland. *Memoirs of the Queensland Museum* **20**, 143.

Sladecek, V. (1979). "Continental Systems for the Assessment of River Water Quality." In A. James and L. Evison (Eds.), *Biological Indicators of Water Quality*. John Wiley & Sons, Chichester, pp. 3-1 to 3-32.

Sladecek, V., Hawkes, H. A., Alabaster, J. S., Daubner, I., Nothlich, I., Solbe, J. F. D. L. G., and Uhlmann, D. (1982). "Biological Examination." In M. J. Suess (Ed.), *Examination of Water for Pollution Control – A Reference Handbook*. World Health Organisation – Regional Office for Europe, Pergamon Press, Oxford, Vol. 3, p. 1.

Shumway, D. L., Warren, C. E., and Doudoroff, P. (1964). Influence of oxygen concentration and water movement on the growth of steelhead trout and coho salmon embryos. *Trans. Am. Fish Soc.* **93**, 342.

Steimle, S. W. and Sindermann, C. J. (1978). Review of oxygen depletion and associated mass mortalities of shellfish in the middle Atlantic Bight in 1976. *Mar. Fish. Rev.* **40**, 17.

Stewart, N. E., Shumway, D. L., and Doudoroff, P. (1967). Influence of oxygen concentration on the growth of juvenile largemouth bass. *J. Fish. Res. Board Can.* **24**, 475.

USEPA (1972). Water Quality Criteria (Advance Copy), U.S. Environmental Protection Agency (to be published as EPA-R3-73-033), Washington, D.C., p. 134.

USEPA (1974). Methods for Chemical Analysis for Water and Wastes. Methods Development and Quality Assurance Research Laboratories. U.S. Environmental Protection Agency, Washington, D.C.

Velz, C. J. (1970). *Applied Stream Sanitation*. Wiley-Interscience, New York.

Welch, E. B. (1980). *Ecological Effects of Waste Water*. Cambridge University Press, Cambridge, p. 337.

Westlake, D. F. (1966). A model for quantitative studies of photosynthesis by higher plants in streams. *Air Water Pollut. Int. J.* **10**, 883.

6

NUTRIENT ENRICHMENT
AND EUTROPHICATION

The enrichment of aquatic areas with plant nutrients is an important process in aquatic pollution and a significant aspect of this is eutrophication. Eutrophication was described by Weber in 1907 when he introduced the terms oligotrophic, mesotrophic, and eutrophic (Hutchinson, 1969). These terms describe the eutrophication process as a sequence from a clear lake to a bog by enrichment with plant nutrients and increased plant growth. Since that time there have been many descriptions and criteria for these terms and the introduction of new terms has proliferated. Hutchinson (1969) has suggested that the trophic state of an aquatic area should be considered in terms of the whole water system including the catchment and bottom sediments rather than the water alone. The nutrient status should be related to total nutrients present, and potentially available, rather than simply the concentration in water at any particular time. The OECD has defined eutrophication as "the nutrient enrichment of waters which results in stimulation of an array of symptomatic changes among which increased production of algae and macrophytes, deterioration of fisheries, deterioration of water quality and other symptomatic changes are found to be undesirable and interfere with water uses" (Wood, 1975).

In the natural eutrophication process, plant detritus, salts, silt, and so on from a catchment are entrained in runoff water and deposited in the water body over geological time. This leads to nutrient enrichment, sedimentation, infilling, and increased biomass. Figure 6.1 illustrates in general terms how eutrophication is related to aging. The final stage of the process results in the formation of bogs, swamps, and the extinction of the water body. It is believed the process slows with increasing time due to increased turbidity causing limited light penetration and a consequent fall in primary production (see Figure 6.1).

Water bodies with little flushing, such as lakes, dams, enclosed seas, and so on, become eutrophic through nutrient enrichment over an extended time scale as described above. This follows the generally accepted eutrophication pattern. However, nutrient enrichment also occurs in situations where infilling and increased sedimentation leading to the formation of swamps and so on is less likely due to comparatively rapid water movement and flushing. This situation arises in streams, estuaries, the continental shelf, and the open seas. Nevertheless, these water bodies

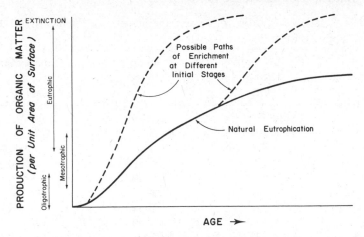

Figure 6.1. Hypothetical curve of the course of eutrophication in a water body. The broken lines show the possible course of accelerated eutrophication when enrichment from pollution occurs.

may show many of the characteristics of eutrophication and are often referred to in eutrophic terms.

Nutrient enrichment and eutrophication can be greatly accelerated by human activities. In fact, many lakes have been shown to have been rapidly enriched with nutrients over the last 100 years due to pollution (Vollenweider, 1971; Allen and Kramer, 1972; NAS, 1969). Discharges, such as domestic sewage, septic tank runoff, some industrial wastes, urban runoff, runoff from agricultural and managed forests, and animal wastes, contain plant nutrients that often lead to nutrient enrichment and accelerated eutrophication.

Eutrophication can cause quite a number of important problems in water use. An increase in the populations of plants can lead to a decrease in the dissolved oxygen content of the water on plant death and decomposition by microorganisms. This decreases the suitability of the area as a habitat for many species of fish and other organisms. The increase in turbidity and color which occurs during eutrophication renders the water unsuitable for domestic use or difficult to treat to a suitable standard for this purpose. Odors are also produced by many of the algal growths which create problems in domestic use. Blooms, pulses, and so on of aquatic plants become more frequent and, if toxic, lead to the death of fish and other aquatic organisms and also of terrestrial organisms using the water. Floating macrophytes and algal scums can render a water body unsuitable for recreation and water sports and also cause navigation problems.

NUTRIENTS AND PLANT GROWTH

If the growth of algal cells is not limited by any environmental or nutrient factor then population growth occurs according to an exponential function. Thus, the

Table 6.1.   Relative Quantities of Essential Elements in Plant Tissue (Demand) and Their Supply in River Water

Element	Demand Plants (%)	Supply Water (%)	Demand Plants/ Supply Water (approx.)
Oxygen	80.5	89	1
Hydrogen	9.7	11	1
Carbon	6.5	0.0012	5,000
Silicon	1.3	0.00065	2,000
Nitrogen	0.7	0.000023	30,000
Calcium	0.4	0.0015	< 1,000
Potassium	0.3	0.00023	1,300
Phosphorus	0.08	0.000001	80,000
Magnesium	0.07	0.0004	< 1,000
Sulfur	0.06	0.0004	< 1,000
Chlorine	0.06	0.0008	< 1,000
Sodium	0.04	0.0006	< 1,000
Iron	0.02	0.00007	< 1,000
Manganese	0.0007	0.0000015	< 1,000
Boron	0.001	0.00001	< 1,000
Zinc	0.0003	0.000001	< 1,000
Copper	0.0001	0.000001	< 1,000
Molybdenum	0.00005	0.0000003	< 1,000
Cobalt	0.000002	0.000000005	< 1,000

Source: Vallentyne (1973). Reprinted with permission from the Federation of American Societies for Experimental Biology.

number of cells at any time regulates the rate of growth:

$$\frac{dN}{dt} = \beta N$$

where N is the number of cells, t the time, and β the growth rate constant. This expression can be integrated to give

$$N_t = N_0 e^{\beta t} \tag{1}$$

where N_0 is the number of cells at time zero and N_t the number of cells at time t.

The growth rate is often expressed as the doubling time (t_d) which is constant under fixed conditions and can be derived from the Equation (1) as

$$t_d = \frac{\ln 2}{\beta}$$

Golterman (1975) has reported that the growth rate constants (β) for a wide variety of algae range from 0.12 to 3.55 ($\log_{10}$, day units) with doubling times from 2.0 to 97.7 days.

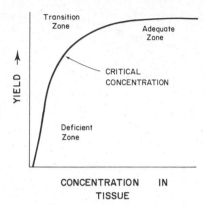

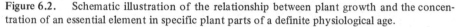

Figure 6.2.   Schematic illustration of the relationship between plant growth and the concentration of an essential element in specific plant parts of a definite physiological age.

Exponential growth of the kind outlined above cannot be maintained for very long by algae due to limitations of various kinds. For example, elements, such as carbon, nitrogen, sulfur, hydrogen, and oxygen, are needed to construct plant tissue, particularly proteins and carbohydrates. In addition to these major elements, phosphorus, iron, magnesium, sodium, and a variety of other elements are needed to construct other vital components. Table 6.1 lists the relative quantities of essential elements which occur in plant tissue. The function of all of these elements in plant processes is not clear at the present time.

In most aquatic areas, carbon and oxygen are readily available from carbon dioxide in the atmosphere and hydrogen and oxygen can also be readily obtained from water. On the other hand, the other elements mentioned above are usually obtained from dissolved salts in the water or sediments. However, these substances are not always available in the quantities required to maintain maximum growth. For example, Table 6.1 shows a comparison of the relative quantities of the elements required for plant growth with their occurrence in river water. This indicates that phosphorus and nitrogen are in comparatively short supply compared with all the other elements and that phosphorus is likely to be less available than nitrogen. This situation is generally applicable to aquatic areas; thus these elements are often growth limiting and an addition of them to a water body will stimulate plant growth. Of course, these data are generalized and, in individual cases, other elements or combinations of elements may be limiting.

Figure 6.2 shows the pattern of uptake of an essential element, when there is a deficiency of that element and addition leads to a rapid increase in plant yield. The vertical section of the curve in Figure 6.2 shows rapid growth with little change in the plant tissue concentration of the element. But, when all needs are satisfied the plant continues to take up the nutrient but this leads to no increase in yield. This is the adequate zone or the zone of luxury consumption (see Figure 6.2).

Dugdale (1967) and Caperon et al. (1971) found that the uptake of nutrient by phytoplankton follows Michaelis–Menton kinetics as illustrated in Figure 6.3.

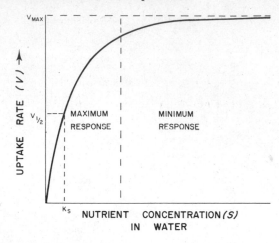

Figure 6.3.   Relationship between uptake rate of nutrient and nutrient concentration in medium for aquatic plants.

This can be expressed mathematically as

$$V = \frac{V_{max}\,S}{K_S + S}$$

where V is the rate of nutrient assimilation; V_{max} the maximum rate of nutrient assimilation; $V_{1/2}$ half the maximum assimilation rate; S the nutrient concentration and K_S the nutrient concentration at $V_{1/2}$ (see Figure 6.3).

Thus phytoplankton show a maximum growth response at low concentrations and a minimum response at somewhat higher concentrations (see Figure 6.3). Vollenweider (1971) has compiled information on the concentrations of nutrients required by various algal species and concluded that the phosphorus requirements increased from *Aserionella formosa, Tabellaria, Fragilaria, Scenedesmus*, and *Oscillatoria rubescens.*

In assessing the effects of nutrient discharges on a water body, it is necessary to establish which are the limiting nutrients and their relationships to plant growth. This can be done by bioassay of cultures of the appropriate plant, enriched with essential elements singly and in combinations. The results of such a bioassay of lake water are shown in Table 6.2. The results suggest that nitrogen is the prime factor in plant growth but phosphorus and iron are also important. There are difficulties in applying the results of bioassay experiments to actual situations since the test water may not accurately represent the water body. Water composition in natural bodies can show large variations with environmental conditions.

Ryther and Dunstan (1971) by a combination of bioassay experiments and field measurements identified nitrogen as the limiting nutrient in coastal bays on Long Island, New York, while phosphorus was present in comparatively high concentrations. This resulted from a low nitrogen to phosphorus ratio in discharges to the bays and a low regeneration rate of phosphorus compared with nitrogen from decomposition of organic matter.

Table 6.2.   Growth of *Microcystis aeruginosa* in Surface Water from Lake Mendota With and Without Additions of Essential Mineral Elements

Treatment	*Microcystis* (cells per mm^3)^a	Percent of Growth with N, P, and Fe Added
A Autoclaved lake water	300	1.1
B NO$_3$	1,750	6.5
C PO$_4$	325	1.2
D Fe	325	1.2
E PO$_4$ and Fe	175	0.7
F NO$_3$ and Fe	1,950	7.2
G NO$_3$ and PO$_4$	6,050	22.3
H NO$_3$, PO$_4$, and Fe	27,100	100.0
I All essential elements	21,025	77.6
J Synthetic nutrient solution	17,550	–

Source: Gerloff (1969). Reprinted with permission from National Academy Press.
^{a} The number of cells added as inoculum has been subtracted from the cell counts.

SOURCES AND TRANSFORMATIONS OF NITROGEN AND PHOSPHORUS IN AQUATIC SYSTEMS

The availability of nitrogen and phosphorus to growing plants depends on a complex set of biologically mediated reactions. Nitrogen occurs in the aquatic environment in a wide variety of forms and chemical combinations involving different oxidation states. Organic nitrogen is bound into cellular constituents of living organisms, for example, purines, peptides, and amino acids, while inorganic nitrogen, for example, ammonia, nitrite, nitrate, and nitrogen gas, is dissolved in the water mass. These components are linked as illustrated in Figure 6.4. The transformation in the water mass of inorganic nitrogen into organic nitrogen occurs by photosynthetic growth of aquatic plants (see Figure 6.4). The reverse of this

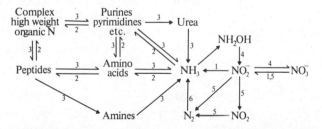

Figure 6.4.   Simplified nitrogen cycle showing main molecular transformations: (1) nitrate assimilation; (2) ammonia assimilation; (3) ammonification; (4) nitrification; (5) denitrification; (6) nitrogen fixation. [From Brezonik (1972). Reprinted with permission from John Wiley & Sons, Inc.]

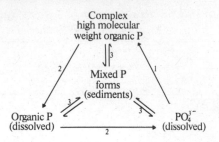

Figure 6.5. Simplified phosphorus cycle in aquatic areas illustrating the main molecular transformations: (1) photosynthesis; (2) mineralization; (3) equilibrium between sedimentary and other P forms.

process results in the formation of ammonia from organic matter by a number of mechanisms involving cell autolysis, microorganisms, and excretion from large organisms. Ammonia can be lost from water by volatilization but oxidation results in nitrification mainly by microorganisms, and produces nitrate which is non-volatile. Nitrate may undergo the denitrification reaction resulting in a loss of nitrogen gas to the atmosphere (Brezonik, 1972). Nitrogen fixation can be carried out by various organisms but is believed to be an adaptive process which is only important to the overall process when nitrogen is in limited supply (Brezonik, 1972).

Phosphorus exists in a single oxidation state as inorganic phosphorus or organic phosphorus. Inorganic forms are mainly orthophosphate (PO_4^{3-}) and polyphosphates. Organic forms are usually associated with complex cellular substances and most phosphorus in natural waters is in the organic form. The inorganic forms, particularly orthophosphate, are readily assimilated during photosynthesis. Figure 6.5 sum-marizes the main phosphorus transformations in aquatic areas.

Sediments play a major role in the availability of phosphorus in many aquatic areas. A high proportion of phosphorus is removed from the water mass by sorption onto sediment minerals. One fraction adsorbs onto anionic sites and another into the crystal lattice structure by substitution of hydroxyl ions (Golterman, 1973). The principal factors controlling this process are oxidation–reduction potential, pH values, and concentrations of other substances. A large proportion of phosphorus is adsorbed onto ferric hydroxide and oxides which dissolve, releasing the phosphorus at low redox potentials. These conditions result when the dissolved oxygen is less than 2 mg L^{-1} (Brezonik, 1972). Conversely, phosphorus is adsorbed at dissolved oxygen concentrations greater than 2 mg L^{-1}. Dissolved phosphate in natural waters is usually at a minimum at pH values of 5–7 (Brezonik, 1972). The metal content of water has also been shown to influence phosphorus dynamics since other metals apart from iron can complex with phosphorus.

In any aquatic system there is a set of nutrient sources and processes for their removal as summarized in Table 6.3 for nitrogen in a lake. Somewhat similar sources and losses would be expected for phosphorus. Major pollution sources of nutrients are surface and subsurface agricultural and urban drainage, animal waste

Table 6.3.   Sources and Sinks for the Nitrogen Budget of a Lake

Sources	Sinks
Airborne	Effluent loss
Rainwater	
Aerosols and dust	Groundwater recharge
Leaves and miscellaneous debris	
	Fish harvest
Surface	
Agricultural (cropland) and drainage	Weed harvest
Animal waste runoff	Insect emergence
Marsh drainage	
Runoff from uncultivated and forest land	Volatilization (of NH_3)
Urban storm water runoff	Evaporation (aerosol formation from surface foam)
Domestic waste effluent	
Industrial waste effluent	
Wastes from boating activities	Denitrification
Underground	
Natural groundwater	Sediment deposition of detritus
Subsurface agricultural and urban drainage	
Subsurface drainage from septic tanks near lake shore	Sorption of ammonia onto sediments
In situ	
Nitrogen fixation	
Sediment leaching	

Source: Brezonik (1972). Reprinted with permission from John Wiley & Sons.

runoff, as well as domestic and industrial waste effluents including sewage (see Table 6.4).

These wastes contain a variety of nitrogen- and phosphorus-containing substances. For example, nitrogen can occur as organic nitrogen, ammonia, nitrite, or nitrate which are derived from protein, nucleic acids, urea, and other substances. Phosphorus compounds result from degradation of compounds such as nucleic acids and phospholipids and occur as inorganic phosphates, that is, orthophosphate or polyphosphate, or organic phosphorus. In addition, phosphorus can originate from phosphate builders in detergents. These can be readily hydrolyzed to yield orthophosphate which is readily assimilated by plants.

Major sources of nitrogen and phosphorus in aquatic areas result from food production or waste in the form of sewage (e.g., see Table 6.4). This nitrogen and phosphorus is being mobilized from new sources in the global environment. Nitrogen originates from atmospheric fixation and phosphorus from the mining

Table. 6.4.   Summary of Estimated Nitrogen and Phosphorus Reaching Wisconsin Surface Waters

Source	N	P	N	P
	(kg y^{-1})		(% of total)	
Municipal treatment facilities	9,000,000	3,200,000	24.5	55.7
Private sewage systems	2,200,000	130,000	5.9	2.2
Industrial wastes[a]	680,000	45,000	1.8	0.8
Rural sources				
Manured lands	3,670,000	1,200,000	9.9	21.5
Other cropland	261,000	174,000	0.7	3.1
Forest land	197,000	19,700	0.5	0.3
Pasture, woodlot and other lands	245,000	163,000	0.7	2.9
Groundwater	15,600,000	129,000	42.0	2.3
Urban runoff	2,020,000	1,570,000	5.5	10.0
Precipitation on water areas	3,150,000	70,000	8.5	1.2
Total	35,205,000	5,700,700	100.0	100.0

Source: Hasler (1968). Reprinted with permission from D. Reidel Publishing Company.

[a] Excludes industrial wastes that discharge to municipal systems. Table does not include contributions from aquatic nitrogen fixation, waterfowl, chemical deicers, and wetland drainage.

of phosphorus-rich minerals which are converted into fertilizers and consequently food and sewage. Wastes originating from these sources enter aquatic areas and also the atmosphere.

Of the total nitrogen fixed at the earth's surface, 20–30% is used in the production of nitrogenous fertilizers (Simpson, 1977). Thus there has been a significant alteration to the earth's nitrogen cycle. Similarly, the earth's phosphorus cycle has been significantly changed. In fact, nitrogen and phosphorus are being utilized from these new sources in the ratio of about 1:7 which, with the addition of carbon to the atmosphere due to the combustion of fossil fuels, could provide the basis for increased plant production, probably mainly in aquatic areas. This could assist in partially removing carbon dioxide from the atmosphere but would result in increased eutrophication of waterways (Simpson, 1977).

PHYSICAL FACTORS AFFECTING NUTRIENT ENRICHMENT AND EUTROPHICATION

Primary production depends on solar radiation reaching the earth's surface. This radiation is attenuated on passing through water according to the Beer–Lambert Law:

$$I_Z = I_0 e^{-KZ}$$

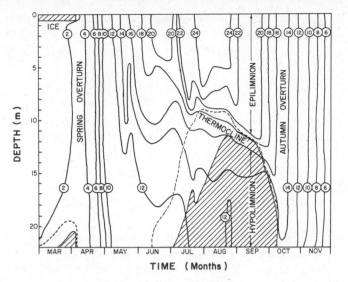

Figure 6.6. The seasonal cycle of temperature (circled) and oxygen conditions in Lake Mendota, Wisconsin. Shaded areas represent the seasonal occurrence of near-anaerobic conditions (0.2 ppm O_2 or less). The broken line represents the O_2 value of 2 ppm. [After Mortimer (1956). Reprinted with permission from the University of Wisconsin Press.]

where I_Z is the intensity at any depth; I_0 is the incident radiation; Z the depth; and K the extinction coefficient which varies with turbidity and the presence of dissolved substances. Thus, light intensity decreases exponentially with depth and when it is less than 1% of the incident radiation, I_0, causes insignificant primary production. From this depth to the surface is the photic zone where primary production occurs. However, at the immediate water surface there is inhibition of growth due to excess light. The compensation depth occurs where photosynthesis is equal to respiration but this concept is only applicable to well-mixed photic zones.

The incidence of light and related primary production varies with latitude with a minimum production at high latitudes.

Water temperature also has a strong influence on the growth of aquatic plants. Phytoplankton species have an optimal growth temperature (Eppley, 1972), but with mixed species in natural areas between 0 and 50°C, an increase in growth rate with temperature occurs for the whole community. The increase factor for each 10°C is about double in accord with the Q_{10} Law (Welch, 1980).

Thermal stratification can occur due to solar heating of the surface waters of a water body. In this situation, the epilimnion and hypolimnion can exhibit different physicochemical characteristics (see Figure 6.6). The centre of gravity of a stratified water body is lower than that of the same body in an unstratified state. Thus for mixing to occur, the center of gravity must be raised and the work required to do this provides a measure of the stability of stratification. A simple expression can be obtained by viewing the water body as having perpendicular sides and, in the stratified state, a horizontal thermocline (see Figure 6.6) at depth Z, uniform temperatures in the epilimnion and hypolimnion which lead to correspondingly uniform water densities of D_1 and D_2, respectively. In the unstratified state, the center of gravity lies at a depth h. The stability per unit area (S) can be derived as (Ruttner, 1974)

$$S = (D_2 - D_1)(2h - Z)\left(\frac{Z}{2}\right)$$

The principal agent providing the work to mix the different layers in natural water bodies is the wind. However, water inflow from a catchment may play a role in some cases. Some important generalizations regarding stratification can be made from the expression above. For instance, water density is not directly related to temperature. There is a maximum density at about 4°C, and above this temperature the density decreases at a more rapid rate than the temperature increases. Thus small differences in temperature at elevated temperatures (e.g., 25–30°C) produce larger differences in density than the same changes at lower temperatures (e.g., 1–10°C). Hence, for the same temperature differential water bodies with warmer temperatures show greater stability than those with cooler temperatures.

If all other factors are unchanged, as the thermocline descends, the stability increases until a maximum results at $Z = h$. At further depths, the stability decreases.

Seasonal patterns of vertical water movement are a common characteristic of water bodies. In temperate areas, stratification in the summer leads to "summer stagnation" and in the winter stratification does not occur due to the lack of a temperature differential. Mixing of the epilimnion and hypolimnion can occur when stability is at its lowest. Bodies which are ice-covered, although unstable, are protected from wind-generated movement but may exhibit a "turnover" in spring when the ice melts and produces water at a maximum density of 4°C before surface heating generates high stability. Similarly, mixing may occur in the autumn and winter when cooling of the surface waters leads to instability (see Figure 6.6). According to these patterns of movement, lakes can be classified into a number of different types, for example, dimictic, which means exhibiting two turnovers usually in autumn and spring; monomictic, with one turnover usually in autumn; and polymictic, with no stratification pattern (see Golterman, 1975).

Due to turbulence, thermal stratification is not common in flowing streams and estuaries. Estuaries can exhibit stratification due to the formation of a "salt wedge" by incoming seawater flowing along the estuary floor. Seasonal thermal stratification has been noted on the continental shelves, such as the New York Bight (Steimle and Sindermann, 1978). In fact, the deep oceans exhibit a high level of stable thermal stratification unaffected by the seasons. In all oceans, water below about 500 m (i.e., not on the continental shelf) has a temperature of about 4°C. Movement of bottom waters to the surface occurs in areas of upwelling, principally due to movement of ocean currents and turbulence caused by the interaction of ocean and wind-induced currents.

Stratification has a number of important implications for nutrient enrichment and eutrophication. Figure 6.7 indicates the nitrogen transformations that would be expected to occur in an idealized stratified body of water. In the epilimnion, photosynthesis produces organic matter, principally phytoplankton, containing

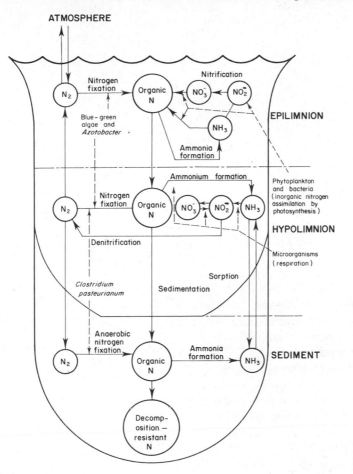

Figure 6.7.   Nitrogen cycle reactions in an idealized stratified lake. Note that both aerobic and anaerobic transformations are shown in the hypolimnion, although in a real lake they would not occur simultaneously. [From Brezonik (1972). Reprinted with permission from John Wiley & Sons, Inc.]

carbon, nitrogen, phosphorus, and a variety of other elements. Death followed by sedimentation transfers this organic matter to the hypolimnion where microbiological degradation occurs, resulting in the formation of carbon dioxide, orthophosphate, ammonia, nitrite, and nitrate (under different conditions) with consumption of dissolved oxygen. In an oligotrophic situation this may not result in any marked differences between the hypolimnion and the epilimnion since all processes occur at a low rate. However, in eutrophic conditions the hypolimnion is usually depleted in dissolved oxygen since direct reaeration from the atmosphere cannot occur. In addition, there is usually an accumulation of the inorganic materials formed during anaerobic respiration, for example, ammonia, hydrogen sulfide, and so on. Oxygen depletion of the hypolimnion occurs in many water

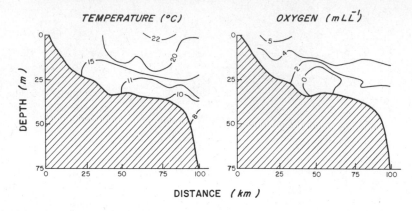

Figure 6.8.   Temperature and dissolved oxygen stratification off central New Jersey in the New York Bight area along latitude 39° 30′ N in August 1976. [From Armstrong (1977).]

bodies throughout the world due to the introduction of plant nutrients yielding excess plant growth. An interesting example occurs on the northeast continental shelf of the United States, in the New York Bight (see Figures 6.8 and 5.6).

Somewhat similar processes have been occurring in the oceans over geological time. In many areas the bottom waters of the oceans have dissolved oxygen levels of zero due to the accumulation of organic matter from primary production in the photic zone.

If dissolved oxygen is zero in the bottom waters, the nitrification reaction leading to nitrate cannot proceed and organic nitrogen yields ammonia as the principal end product. This substance is not formed in surface waters since these are well aerated by atmospheric and photosynthetic oxygen. Here the principal nitrogen form is nitrate. Eutrophic conditions produce vertical profiles of nitrate and ammonia which correspond with the above observations.

Dissolved oxygen depletion also results in the release of orthophosphate as discussed previously. Therefore, while the epilimnion may be depleted by photosynthesis and subsequent sedimentation, the hypolimnion becomes enriched. During periods of mixing of bottom and surface waters there can be an increase in plant growth in the photic zone if other environmental conditions are satisfactory. Thus surface waters may exhibit a seasonal cycle of nutrient concentrations and and phytoplankton growth related to the factors outlined above as well as the pattern of vertical mixing previously outlined (see Figure 4.8). In the oceans it is well known that primary production and fisheries are most productive in areas of upwelling where nutrient-rich bottom waters enter the photic zone. On the other hand, these processes in surface waters where upwelling does not occur, cause depletion in nutrients giving oligotrophic conditions. This depletion occurs in about 40% of the ocean surface (Bunt, 1975).

The surface area of a lake, its catchment, and depth have a strong influence on primary production. If the inputs from the catchment and precipitation have reached a steady state, nutrient input will be proportional to the total area and its

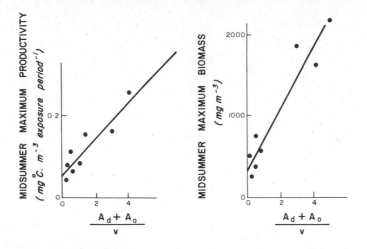

Figure 6.9.   Relationships of lake response to watershed area (A_d) plus lake (A_0) area divided by volume (v). [From Schindler (1971). Reprinted with permission from the *Journal of Fisheries Research Board of Canada*.]

influence on the nutrient concentration in the water will be related to the lake volume. This has been demonstrated by Schindler (1971), as shown in Figure 6.9.

Flushing of a water body also influences the nutrient concentration and therefore affects primary production. In a stream or estuary, flushing occurs at a relatively rapid rate and somewhat different conditions prevail to those in water bodies where water movement is at a lower level. In streams, nutrient levels decrease downstream from a discharge leading to a set of related chemical and biological changes (see Chapters 4 and 5).

NUTRIENT LOADINGS, PRIMARY PRODUCTION, AND TROPHIC STATE

The trophic state of a water body depends on a variety of physical, chemical, and biological properties. Clearly, the supply of any chemical factor which promotes plant growth is of importance. The nutrient loading approach has been most successful with lakes in which productivity is controlled by the phosphorus loading (Vollenweider and Kerekes, 1980). The nutrient loading must take into account external inputs as well as the availability of nutrients already present within the body. However, it has been found that the availability of nutrients already within the body is not of primary significance in many lake systems. The models relating nutrient loading to primary productivity consist of two submodels:

1.   The relationships between nutrient input and nutrient concentration in the water body.

2.   The relation between nutrient concentrations in the water body and primary productivity.

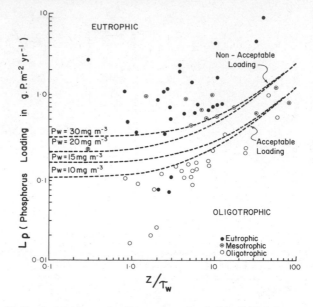

Figure 6.10.   Plots of phosphorus loading against depth (z) divided by water flushing time (τ_w) for 60 lakes classified as oliogotrophic, mesotrophic, or eutrophic and also the phosphorus loading tolerance according to the equation $L_P = P_w(z/\tau_w + 10)$. [From Vollenweider (1976). Reprinted with permission from the Istituto Italiano di Idrobiologia.]

The establishment of these relationships is only applicable if it is assumed that complete mixing of the lake occurs and that steady-state conditions operate (Golterman, 1980).

The relationship between nutrient concentration in the water (or load per unit area) and nutrient input has been reported in a series of papers by Vollenweider and co-workers (e.g., Vollenweider, 1976, Vollenweider and Kerekes, 1980). One relationship that has found wide use is

$$L_P = P_w\left(\frac{Z}{\tau_w} + Z\sigma_P\right) \tag{2}$$

where L_P is the surface loading of phosphorus per unit area per year; P_w the average total phosphorus concentration of dissolved and particulate phosphorus in water at the spring overturn; Z the mean depth; τ_w the average residence time of water; σ_P relates to the sedimentation rate of phosphorus. As a general rule, σ_P is approximately $10/Z$. Thus substituting in Equation (2),

$$L_P = P_w\left(\frac{Z}{\tau_w} + 10\right)$$

If oligotrophic lakes are considered to have $P_w < 10$ mg m^{-3}, mesotrophic 10–20 mg m^{-3}, and eutrophic > 20 mg m^{-3} with a depth to residence time ratio of 0.1, tolerance lines are obtained for the limiting positions of trophic state as shown in Figure 6.10. In addition, Figure 6.10 indicates plots for 60 lakes in North America

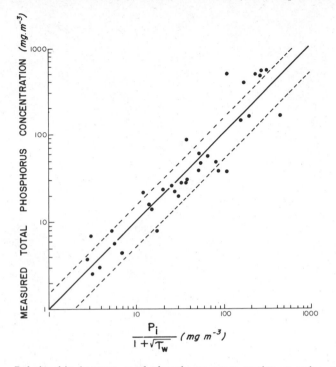

Figure 6.11. Relationship between total phosphorus concentration at spring overturn and total phosphorus predicted from Vollenweider's equation, $P_w = P_i/(1 + \sqrt{\tau_w})$. [From Vollenweider and Kerekes (1980). Reprinted with permission from Pergamon Press, Inc.]

with expert classification into trophic class. These points are in good agreement with the tolerance lines.

Vollenweider (1980) has also reported the relationship

$$P_w = \left(\frac{\tau_P}{\tau_w}\right) P_i \tag{3}$$

where τ_P/τ_w is the average residence time of phosphorus relative to the average residence time of water and P_i the average inflow concentration of total phosphorus. In addition, Vollenweider (1976) found the following approximate relationships:

$$\frac{\tau_P}{\tau_w} = \frac{1}{1 + \sqrt{\tau_w}}$$

Thus, substituting in Equation (3)

$$P_w = \frac{P_i}{1 + \sqrt{\tau_w}}$$

A plot of predicted phosphorus concentrations in water (P_w) versus actual phosphorus at spring turnover for selected OECD lakes are shown in Figure 6.11.

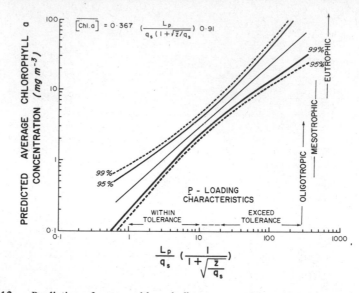

Figure 6.12.   Prediction of average chlorophyll-a concentration and trophic state as related to phosphorus loading characteristics as expressed by $P_w = (L_p/q_s)[1/(1 + \sqrt{z}/q_s)]$. [From Vollenweider (1976). Reprinted with permission from the Istituto Italiano di Idrobiologia.]

The level of accuracy is indicated by ± 50% lines. Similar results were obtained using Dillons relationship (Vollenweider, 1980):

$$P_w = P_i(1 - R_D)$$

where R_D is the phosphorus retention coefficient.

Vollenweider (1976) has also shown that his relationship above for P_w can also be expressed as

$$P_w = \frac{L_P}{q_s}\left(\frac{1}{1 + \sqrt{Z}/q_s}\right)$$

where q_s is the hydraulic load.

The second part of the loading model is concerned with the relationships between primary production and phosphorus concentrations. Primary production in terms of a biomass parameter (e.g., chlorophyll) should be related to phosphorus concentration in water (P_w) and hence to phosphorus loadings (P_i). Imbodem and Gachter (1979) have reviewed the relationships between primary production and phosphorus loadings as well as physical factors influencing this relationship. Vollenweider and Kerekes (1980) have found significant relationships of the form

$$\text{Average chlorophyll a} = 0.37\left(\frac{P_i}{1 + \sqrt{\tau_w}}\right)^{0.91}$$

This relationship has been found to hold for 60 lakes classified subjectively as oligotrophic, mesotrophic, or eutrophic.

A plot of calculated water concentration of phosphorus (P_w) against predicted chlorophyll-a concentration derived from the above relationship is shown in Figure 6.12. Using this relationship Vollenweider (1976) was able to calculate biomass in terms of chlorophyll-a from phosphorus load characteristics. The relationship is adequate for lakes of low and medium productivity but less certain for lakes of high productivity. This also allows trophic status to be determined (see Figure 6.12). It has been suggested that deviations from this relationship can occur in situations where factors other than phosphorus control primary productivity.

Tapp (1978) and Yeasted and Morel (1978) have described a number of different types of models used to evaluate trophic state. Reckhow (1979) has reviewed the success of models for phosphorus in aquatic systems. In addition, Golterman (1980) and Allan (1980) have discussed inadequacies in the modeling approach.

ECOSYSTEM CHANGES RESULTING FROM NUTRIENT ENRICHMENT AND EUTROPHICATION

1. *Changes in Community Metabolism.*   Vollenweider and Kerekes (1980) and other authors have related increases in chlorophyll-a concentrations and primary production to increasing eutrophication (see Figure 6.12). A substantial proportion of this increased primary production would be expected to be transferred to the hypolimnion where deoxygenation occurs as described previously. Thus oxygen depletion rates in the hypolimnion can be related to trophic state (see Table 6.5).

Odum (1956) has described the relationship of primary production to respiration in different water bodies which leads to a scheme of classification and comparison (see Figure 5.11). In this scheme, oligotrophic and eutrophic water bodies have approximately equal rates of primary production and photosynthesis, that is, P/R is approximately equal to 1. In contrast, organically polluted areas have a P/R ratio of much less than 1.

2. *Community and Population Changes with Nutrient Enrichment.*   Physicochemical conditions change with trophic state and stimulate changes in biological composition. With phytoplankton in lakes there is a seasonal change in community composition related to temperature, light, and other seasonal factors. Welch (1980) has reported that in temperate regions diatoms generally dominate in the spring,

Table 6.5.   Two Sets of Results Relating Hypolimnetic Oxygen Depletion Rates to Trophic State

Area	Oxygen Depletion Rates (mmol O_2 m^{-2} day^{-1})	
Oligotrophic lakes	$<-$ 5.3	$<-$ 7.7
Mesotrophic lakes	$-$ 5.3 to $-$ 10.3	$-$ 7.7 to $-$ 17.2
Eutrophic lakes	$>-$ 10.3	$>-$ 17.2

Source: Burns and Ross (1972). Reprinted with permission from John Wiley & Sons.

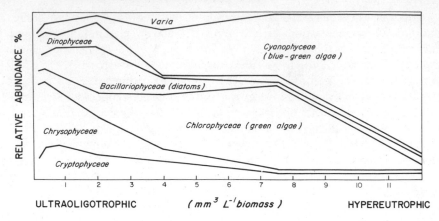

RELATIVE ABUNDANCE %

Varia

Dinophyceae

Cyanophyceae
(blue - green algae)

Bacillariophyceae (diatoms)

Chrysophyceae

Chlorophyceae (green algae)

Cryptophyceae

1   2   3   4   5   6   7   8   9   10   11

ULTRAOLIGOTROPHIC   *(mm³ L⁻¹ biomass)*   HYPEREUTROPHIC

Figure 6.13.   Composition of phytoplankton and algal biomass expressed by volume in relation to increasing lake fertility. [After Skulberg (1980).]

green algae in summer, blue-green algae in late summer, and possibly diatoms in late autumn. However, there can be considerable variation in this pattern, since different phytoplankton also have different dynamics and requirements for nitrogen phosphorus, carbon dioxide, and other factors, which produce changes in the community composition with increasing eutrophication. The most prominent change with increasing nutrient enrichment is that the blue-green algae (*Cyanophyceae*) become increasingly dominant (see Figure 6.13) (Wood, 1975).

In flowing waters the periphyton (attached biota) include the most important primary producers (Welch, 1980). The physical form and method of attachment are very important particularly with regard to the ability to withstand erosion and scouring by water movement. Under these conditions periphyton usually take the form of slick coatings on rocks or strands of filamentous formation. The most common species are filamentous blue-green algae (e.g., *Oscillortoria*), filamentous green algae (e.g., *Cladophora*), and filamentous bacteria (e.g., *Sphaerotilum*). Figure 5.13 indicates the pattern of downstream changes resulting from an organic discharge. Although some effects are due to deoxygenation, the occurrence and concentration of nutrients would be expected to influence primary production.

Rooted macrophytes may obtain nutrients from sediments or the water mass but the water is probably of secondary importance (Welch, 1980). However, growth is generally greatest in organically enriched sediments and this group of plants can display large growths in restricted and shallow waterways.

The zooplankton communities are probably not directly modified by nutrient enrichment. But indirect effects are of considerable importance. For example, the rate of primary production and hypolimnetic oxygen concentrations have a significant impact on zooplankton. Ravera (1980) has reviewed the effects of eutrophication on zooplankton and, as a general rule, if the diversity of the phytoplankton is high, the diversity of zooplankton will also be high. Zooplankton grazing also has an impact on phytoplankton community composition and biomass. Increasing eutrophication leads to decreasing efficiency of utilization of energy in the

Table 6.6.   Expected Character of Phytoplankton–Zooplankton Relationships
with Respect to Trophic State

Variable	Oligotrophic–Mesotrophic Moderately Fertilized Pelagic Marine	Eutrophic, Heavily Fertilized Ponds, Neritic Marine
Ratio of zooplankton consumption/ primary production	$\simeq 100\%$	$\simeq 30\%$
Efficiency of energy conversion	$\simeq 20\%$	$\simeq 10\%$
Size of zooplankton	Large	Small and do not control phytoplankton
Size of phytoplankton	Nanoplankton $< 50\ \mu m$	Nanoplankton not used; bacteria and detritus consumers dominate

Source: Hillbricht-Ilkowska (1972). Reprinted with permission from the International Biological Programme.

phytoplankton-based food web but a decreasing biomass of zooplankton. The
overall effects are summarized in Table 6.6.

The benthos generally reflect conditions in the water mass. Benthic animals
are influenced by material produced in the water mass or discharged into the
water body by sedimentation. Macrozoobenthos are often used as these are most
suitable for investigation (usually > 0.5–0.6 mm mesh size seive). Early spring or
autumn are commonly used as standard sampling periods to minimize seasonal
variations. The soft bottom communities are the richest and usually selected for
investigation. These communities are dominated by chironomids and polychaetes
with lesser numbers of molluscs, crustacea, and other insects.

Saether (1979; 1980) has delineated 15 chironomid communities as indicative
of trophic classes or types designated from α to Ω with increasing eutrophication.
Figure 6.14 shows the chlorophyll-a content of a variety of lakes in North America
and Europe plotted against trophic type, as designated from the chironomid communities present. A similar relationship was found when total phosphorus divided
by depth was plotted against type for the same set of lakes.

Weiderholm (1980) found that species richness decreased with increasing eutrophication, measured as chlorophyll-a, in Swedish lakes (see Figure 6.15). However,
he has suggested that species richness would be expected to reach a maximum at
some intermediate lake fertility. Also, oligochaete numbers increase in relationship
to sedentary chironomid number with increasing eutrophication and this ratio can
be used as a diagnostic characteristic.

The principal effect of increasing eutrophication on fish is due to dissolved
oxygen depletion. However, moderate increases in primary production may be

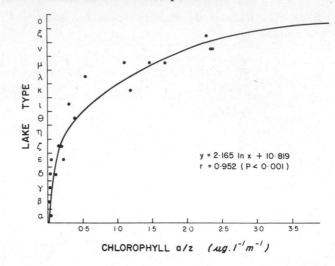

Figure 6.14.   Chlorophyll-a/z (mean lake depth) in relation to 15 lake types based on chiron-omid communities (α–o). [From Saether (1979). Reprinted with permission from *Holoartic Ecology*.]

beneficial to fish numbers and biomass (Leach et al., 1977). Summer stratification causes a similar vertical stratification of fish species according to their tolerance to temperature and dissolved oxygen reduction (Reid and Wood, 1976). But benthic areas depleted in dissolved oxygen, which expand with increasing eutrophication, decrease the habitat suitable for fish and can lead to an overall decline in fish numbers. Figure 6.16 shows the general trends in species tolerance to increasing eutrophication. The cyprinids are comparatively tolerant to environmental stress and become increasingly dominant with increasing trophic state.

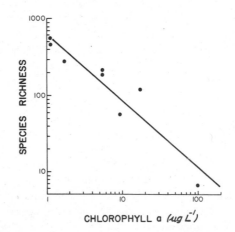

Figure 6.15.   Chlorophyll-a in relation to species richness adjusted for depth in the Swedish Great Lakes. [From Weiderholm (1980). Reprinted with permission from the Water Pollution Control Federation.]

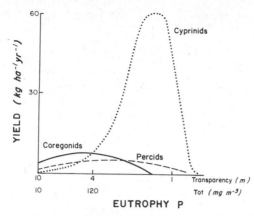

Figure 6.16. Summary of relations between increasing eutrophy and yields of coregonids, percids, and cyprinids for 17 European lakes. [From Hartmann (1978). Reprinted with permission from A. F. Z. Fishwaid Verlagesellschaft.]

Toxic algal blooms generally increase in occurrence with increasing eutrophication. This can cause mortalities of large numbers of aquatic organisms and terrestrial animals utilizing the water. Also, "red tides" can occur in enriched marine areas resulting in the occurrence of toxic components in edible shellfish.

3.   *Characteristics and Criteria for Trophic State.*   A clear assessment of the degree of eutrophication is needed for the development of procedures for water quality management. Most of the characteristics of trophic state presently available have been based on freshwater lakes as outlined in Table 6.7. A set of primary

Table 6.7.   Qualitative Characteristics of Oligotrophic and Eutrophic Lakes

	Oligotrophic	Eutrophic
Depth	Deep	Shallow
Hypolimnion: epilimnion	>1	<1
Primary productivity	Low	High
Rooted macrophytes	Few	Abundant
Density of plankton algae	Low	High
Number of plankton algal species	Many	Few
Frequency of plankton blooms	Rare	Common
Depletion of hypolimnetic oxygen	No	Yes
Fish species	Cold water, slow growth, restricted to hypolimnion	Warm water, fast growth, tolerate low O_2 in hypolimnion and high temperature of epilimnion
Nutrient supply	Low	High

Source: Welch (1980). Reprinted with permission from Cambridge University Press.

Table 6.8.   Summary of Quantitative Limits for Several Characteristics that Define Trophic State[a]

	Oligotrophy $\leqslant$	Mesotrophy	Eutrophy $\geqslant$
$\mu g\ L^{-1}$ total P (winter)	10–15		20–30
$\mu g\ L^{-1}$ chlorophyll-a (summer)	2–4		6–10
m Secchi disk (summer)	5–3		2–1.5
mg m^{-2} yr^{-1} P loading, 10–$20\ q_s[1 + (\bar{Z}/q_s)^{0.5}]$	200		400
mg m^{-2} yr^{-1} P loading, 100–$200\ (q_s)^{0.5}$	316		632
mg m^{-2} day^{-1} ODR (based on above P loading limits)	250–310		330–400
Mean daily growing rates in a growing season, mg C cm^{-2} day^{-1}	30–100		300–3000
Total annual rate of primary productivity, g C cm^{-2} yr^{-1}	7–25		75–700

Source: Welch (1980). Reprinted with permission from Cambridge University Press.

[a] For loading and ODR a 10 m deep lake with a 1 yr retention time is assumed.

quantitative characteristics for freshwater lakes is shown in Table 6.8 (Welch, 1980). These characteristics interrelate reasonably well but some water bodies fall into indeterminate classes. A significant increase in the characteristics outlined during a 5 year period is generally classified as due to cultural eutrophication (Welch, 1980).

Phosphorus has been identified as the most critical growth factor in most lakes. Vollenweider's loading technique described previously has been widely used to relate phosphorus to trophic state (see Figures 6.10, 6.11, and 6.12).

Biotic indices have also been suggested and Table 6.9 summarizes some biological characteristics related to plankton. Saether (1980) has developed the index previously discussed based on chironomid communities (see Figure 6.14) and Weiderholm (1980) a benthic quality index (BQI) with the following characteristics:

$$BQI = \sum_{i=0}^{5} \frac{k_i n_i}{N}$$

where k_i represents a value constant for each species; n_i the number of individuals in the various groups; and N the total number of indicator species. This expression combines the principal indicator species into a single index. Figure 6.17 shows the relationship between BQI and total phosphorus divided by depth. Similarly, other indices, for example, species richness and oligochaete to chironomid ratio, have been shown to be useful.

Table 6.9. Plankton of Oligotrophic and Eutrophic Lakes

Parameter	Oligotrophic	Eutrophic
Quantity	Poor	Rich
Variety	Many species	Few species
Distribution	To great depths	Trophogenic layer
Diurnal migration	Extensive	Limited
Water blooms	Very rare	Frequent
Characteristic algal groups or genera	Chlorophyceae	Cyanophyceae
	Desmids	*Anabaena*
	Staurastrum	*Aphanizomenon*
	Diatomaceae	*Microcystis*
	Tabellaria	Diatomaceae
	Cyclotella	*Melosira*
	Chrysophyceae	*Stephanodiscus*
	Dinobryon	*Asterionella*

Source: Sawyer (1966). Reprinted with permission from the Water Pollution Control Federation.

In the oceans and on the continental shelves, many of the characteristics in Table 6.7 apply. Officer and Ryther (1980) have reported that all excessive marine phytoplankton growth leading to adverse eutrophication effects have been related to flagellate blooms, which are not utilized by most grazing organisms. Nutrient elements usually stimulating the growth of marine phytoplankton have been identified principally as nitrogen (Ryther and Dunstan, 1971) and silicon (Officer and Ryther, 1980) although phosphorus has been implicated as well. Primary

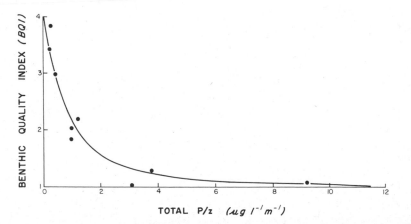

Figure 6.17. Total phosphorus/mean lake depth in relation to a benthic quality index (BQI) based on indicator species of oligochaetes. The index was calculated as $BQI = \Sigma_{i=1}^{4}[(n_i - K_i)/N]$ where $K_i = 4$ for *Stylodrilas heringianus*, *Chelmis limosella*, 3 for *Pelosolex ferox*, 2 for *Potamothrix hamneniensis*, and 1 for *Limaodrilus hoffmeisteri*; n_i = number of individuals of the various groups, respectively; N = total number of indicator species. [From Ahl and Weiderholm (1977). Reprinted with permission from the National Swedish Environment Protection Board.]

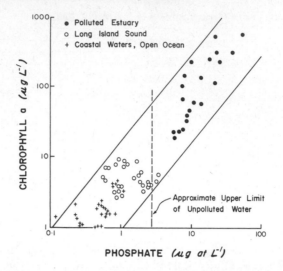

Figure 6.18.   Relationship between inorgani phosphate and chlorophyll-a content of waters ranging from unpolluted seawater to polluted estuaries. [From Ketchum (1969).]

production and nitrate levels in the ocean surface waters generally range from 70 mg C m^{-2} day^{-1} and 1 μg N L^{-1} in oligotrophic waters of subtropical halistatic areas to 1000 mg C m^{-2} day^{-1} and about 300 μg N L^{-1} in open coastal waters. Figure 6.18 indicates the concentrations of chlorophyll-a, orthophosphate, and trophic state for seawater and estuaries.

Estuaries also exhibit many of the characteristics in Table 6.7 and reported in Figure 6.18. Ketchum (1969) quotes a value of 2.8 μg P L^{-1} as the approximate upper limit of unpolluted water.

In streams, periphyton contains the major proportion of primary production resulting from nutrient enrichment. The relationships between nutrient enrichment and plant growth are not well established in this situation except that enrichment leads to increased growth. Hynes (1969) reported the enrichment characteristics by growth of *Potamogeton, Stigeoclonium, Cladophora, Ulothrax, Rhizoclonium, Oscillatoria, Phormidium, Gomphonema, Nipzschia, Navicula*, and *Surirella.*

REFERENCES

Ahl, T. and Weiderholm, T. (1977). Svenska vattenkvalitetskriterier. Eurofierande ammen. SNV PM (Swed)., 918.

Allen, R. J. (1980). The inadequacy of existing chlorophyll-a to phosphorus concentration correlations for assessing remedial measures for hypertrophic lakes. *Environ. Pollut. (Ser. B)* **1**, 217.

Allen, H. E. and Kramer, J. R. (1972). *Nutrients in Natural Waters.* John Wiley & Sons, New York, p. 457.

Armstrong, R. S. (1977). Climatic conditions related to the occurrence of anoxia in the waters off New Jersey during the summer of 1976. In Oxygen Depletion and Associated Environmental Disturbances in the Middle Atlantic Bight in 1976. pp. 17–35. Technical Series Report No. 3, North East Fisheries Centre, National Marine Fisheries Service, National Oceanic and Atmospheric Administration.

Brezonik, P. L. (1972). "Nitrogen: Sources and Transformations in Natural Waters." In H. E. Allen and J. R. Kramer (Eds.), *Nutrients in Natural Waters.* John Wiley & Sons, New York, pp. 1–50.

Bunt, J. S. (1975). "Primary Productivity of Marine Ecosystems." In H. Leith and R. H. Whittaker (Eds.), *Primary Productivity of the Biosphere.* Springer-Verlag, New York, pp. 169–183.

Burns, N. M. and Ross, C. (1972). "Oxygen–Nutrient Relationships Within the Central Basin of Lake Erie." In H. E. Allen and J. R. Kramer (Eds.), *Nutrients in Natural Waters.* John Wiley & Sons, New York, pp. 193–250.

Caperon, J., Cattell, S. A., and Krasnick, G. (1971). Phytoplankton kinetics in a subtropical estuary: Eutrophication. *Limnol. Oceanogr.* **16,** 599.

Dugdale, R. C. (1967). Nutrient limitation in the sea: Dynamics, identification and significance. *Limnol. Oceanogr.* **12,** 685.

Eppley, R. W. (1972). Temperature and phytoplankton growth in the sea. *Fish. Bull.* **70,** 1063.

Gerloff, G. C. (1969). "Evaluating Nutrient Supplies for the Growth of Aquatic Plants in Natural Waters." In *Eutrophication: Causes, Consequences, Correctives.* National Academy of Sciences, Washington, D.C., pp. 537–555.

Golterman, H. L. (1973). Natural phosphate sources in relation to phosphate budgets: A contribution to the understanding of eutrophication. *Water Res.* **7,** 3.

Golterman, H. L. (1975). *Physiological Limnology – An Approach to the Physiology of Lake Ecosystems.* Elsevier Scientific Publishing, Amsterdam, p. 489.

Golterman, H. (1980). Quantifying the eutrophication process: Difficulties caused for example by sediments. *Progr. Water Technol.* **12,** 63.

Hartmann, J. (1978). Sukzession der Fischertrage in kulturbedingt eutrophierenden Seen. *Fischwirt* **28,** 70.

Hasler, A. D. (1968). "Man-Induced Eutrophication of Lakes." In S. F. Sanger (Ed.), *Global Effects of Environmental Pollution.* D. Reidel, Dordrecht, pp. 111–125.

Hillbricht-Ilkowska, A. I. (1972). Interlevel energy transfer efficiency in planktonic food chains. International Biological Programme – Section PH, December 13, 1972. Reading, England.

Hutchinson, G. E. (1969). "Eutrophication, Past and Present." In *Eutrophication: Causes, Consequences, Correctives.* National Academy of Sciences, Washington, D.C., pp. 17–28.

Hynes, H. B. N. (1969). "The Enrichment of Streams." In *Eutrophication: Causes, Consequences, Correctives.* National Academy of Sciences, Washington, D.C., pp. 188–196.

Imbodem, D. M. and Gachter, R. (1979). "The Impact of Physical Processes on the Trophic State of a Lake." In O. Ravero (Ed.), *Biological Aspects of Freshwater Pollution.* Pergamon Press, Oxford, pp. 93–110.

Ketchum, D. H. (1969). "Eutrophication of Estuaries." In *Eutrophication: Causes, Consequences, Correctives*. National Academy of Sciences, Washington, D.C., pp. 197–209.

Leach, J. H., Johnson, M. G., Kelso, J. R. M., Hartmann, J., Numann, W., and Entz, B. (1977). Response of persid fishes and their habitats to eutrophication. *J. Fish. Res. Board Can.* **33**, 1964.

Mortimer, C. H. (1956). "An Explorer of Lakes." In G. C. Sellery and E. A. Birge (Eds.), University of Wisconsin Press, Madison, pp. 165–211.

National Academy of Sciences (1969). *Eutrophication: Causes, Consequences, Correctives*, Proceedings of a Symposium. National Academy of Sciences, Washington, D.C., p. 661.

Odum, H. T. (1956). Primary production in flowing waters. *Limnol. Oceanog.* **1**, 102.

Officer, C. B. and Ryther, J. H. (1980). The possible importance of silicon in marine eutrophication. *Marine Ecol. – Progr. Ser.* **3**, 83.

Ravera, O. (1980). Effects of eutrophication on zooplankton. *Prog. Water Technol.* **12**, 141.

Reckhow, K. H. (1979). "Empirical Lake Models for Phosphorus: Development, Applications, Limitations and Uncertainty." In D. Savaia and A. Robertson (Eds.), *Perspectives on Lake Ecosystem Modelling*. Ann Arbor Science Publishers, Ann Arbor, Michigan, pp. 193–222.

Reid, G. K. and Wood, R. D. (1976). *Ecology of Inland Waters and Estuaries*, 2nd ed. D. Van Nostrand, New York, p. 485.

Ruttner, F. (1974). *Fundamentals of Limnology*, 3rd ed. University of Toronto Press, Toronto, p. 307.

Ryther, J. H. and Dunstan, W. M. (1971). Nitrogen, phosphorus, and eutrophication in the coastal marine environment. *Science* **171**, 1008.

Saether, O. A. (1980). The influence of eutrophication on deep lake benthic invertebrate communities. *Progr. Water Technol.* **12**, 161.

Saether, O. A. (1979). Chironomid communities as water quality indicators. *Holarctic Ecol.* **2**, 65.

Sawyer, C. N. (1966). Basic concepts of eutrophication. *J. Water Pollut. Control Fed.* **38**, 737.

Schindler, D. W. (1971). A hypothesis to explain the differences and similarities among lakes in the experimental lakes area, North Western Ontario. *J. Fish. Res. Board Can.* **28**, 295.

Simpson, H. J. (1977). "Man and the Global Nitrogen Cycle-Group Report." In W. Stumm (Ed.), *Global Chemical Cycles and their Alterations by Man*. Abakon, Verlagsgesellschaft, Berlin, pp. 253–274.

Skulberg, O. M. (1980). Blue-green algae in Lake Mojasa and other Norwegian lakes. *Progr. Water Technol.* **12**, 121.

Steimle, F. W. and Sindermann, C. W. (1978). Review of oxygen depletion and associated mass mortalities of shellfish in the middle Atlantic Bight in 1976. *Marine Fisheries Rev.* **40**, 17.

Tapp, J. S. (1978). Eutrophication analysis with simple and complex models. *J. Water Pollut. Control Fed.* **50**, 484.

Vallentyne, J. R. (1973). The algae bowl — A Faustian view of eutrophication. *Proc. Fed. Am. Soc. Exp. Biol.* **32**(7), 1754.

Vollenweider, R. A. (1971). Scientific fundamentals of the eutrophication of lakes and flowing waters, with particular reference to phosphorus and nitrogen as factors in eutrophication. OECD Technical Report DAS/CSI/68.27, Revised 1971, p. 159.

Vollenweider, R. A. (1976). Advances in defining critical loading levels for phosphorus in lake eutrophication. *Mem. Ist. Ital. Idrobiol.* **33**, 53.

Vollenweider, R. A. and Kerekes, J. (1980). The loading concept as basis for controlling eutrophication philosophy and preliminary results of the OECD programme on eutrophication. *Progr. Water Technol.* **12**, 5.

Welch, E. B. (1980). *Ecological Effects of Waste Water.* Cambridge University Press, Cambridge, p. 337.

Wiederholm, T. (1980). Use of benthos in lake monitoring. *J. Water Pollut. Control Fed.* **52**, 537.

Wood, G. (1975). An Assessment of Eutrophication in Australian Inland Waters. Australian Water Resources Council Technical Paper No. 15. Australian Government Publishing Service, Canberra, pp. 238.

Yeasted, J. G. and Morel, F. M. M. (1978). Empirical insights into lake response to nutrient loadings with application to models of phosphorus in lakes. *Environ. Sci. Technol.* **12**, 195.

7

PESTICIDES

Natural pesticides, such as derris dust, sulfur, nicotine, and pyrethrins, have been in use for a considerable period of time but have not been highly successful due to a lack of potency, lack of specificity, and high cost. In the last 50 years, there has been a steady growth in the use of synthetic organic chemicals as pesticides. These have considerable advantages over the natural products in that they are relatively potent, selective, and comparatively cheap.

The use of synthetic chemicals to control pests, principally insects, weeds, fungi, and nematodes, is now an integral part of primary production and disease control (see O'Brien, 1967; Metcalf, 1971; Farm Chemicals Handbook, 1976; Audus, 1976). Estimates of economic and health benefits from the use of these substances have received considerable attention (e.g., Khan, 1980). It is clear that food production and the control of insect-borne disease have improved enormously and usage is expected to increase substantially in the next 15 years. This applies particularly with herbicides where the quantity used is expected to increase several fold (see Khan, 1980). However, these substances have an ecological impact which must be taken into account in pest management strategies. Substantial areas of the earth's surface are treated annually with quantities which amount to 4.53×10^8 kg in the United States (Pimentel and Goodman, 1974).

The environmental impact of pesticide use is related to several fundamental properties essential to their effectiveness as pesticides. Firstly, pesticides are toxicants capable of affecting all taxonomic groups of biota, including nontarget organisms, to varying degrees dependent on physiological and ecological factors. Secondly, many pesticides need to be resistant to environmental degradation so that they persist in treated areas and thus their effectiveness is enhanced. This property also promotes long-term effects in natural ecosystems.

Broadcast methods of application are usually used which result in substantial quantities being transferred to nontarget, natural ecosystems. Dispersal in the atmosphere results in treatment of natural terrestrial areas while water runoff transfers quantities to freshwater areas and, ultimately, the oceans.

CHEMICAL NATURE AND PROPERTIES

Pesticides include a vast array of natural and synthetic substances of widely different chemical nature (see Figures 7.1–7.5). Their only common property is their

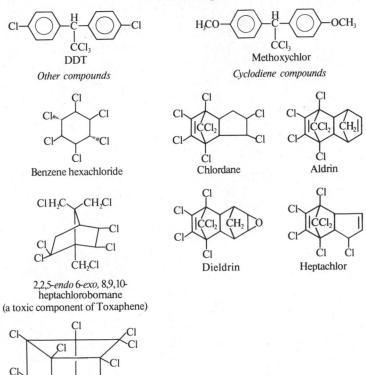

DDT and Analogues

DDT

Methoxychlor

Other compounds

Cyclodiene compounds

Benzene hexachloride

Chlordane

Aldrin

Dieldrin

Heptachlor

2,2,5-*endo* 6-*exo*, 8,9,10-
heptachlorobornane
(a toxic component of Toxaphene)

Mirex

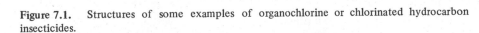

Figure 7.1.   Structures of some examples of organochlorine or chlorinated hydrocarbon insecticides.

ability to destroy biotic pests of some kind. Pollution ecology and ecotoxicology are principally concerned with the synthetic organic chemicals, for reasons previously outlined, although others can have a significant environmental effect in some situations. According to function the pesticides of major ecological significance are the insecticides, herbicides, and fungicides.

Insecticides

Commonly used insecticides are the chlorinated hydrocarbons, or organochlorine compounds, and the organophosphorus compounds. The chemical structures of some widely used chlorinated hydrocarbons are shown in Figure 7.1. This group includes DDT and related substances which have been extensively used in agriculture, until recent years, due to their effectiveness and low cost. Usage reached a

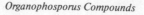

Organophosporus Compounds

General structure: $\begin{array}{c} R \\ R_1 \end{array} \!\!\!> \!\! \overset{\overset{\displaystyle S(O)}{\|}}{P} \!\! - X$

Phosphates

$$(CH_3O)_2\overset{\overset{\displaystyle O}{\|}}{P}-O-\overset{\overset{\displaystyle H}{|}}{C}=CCl_2$$

Dichlorvos

Phosphorothioates

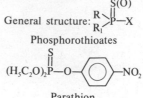

$$(H_5C_2O)_2\overset{\overset{\displaystyle S}{\|}}{P}-O-\!\!\langle\bigcirc\rangle\!\!-NO_2$$

Parathion

Phosphorothiolothionates

$$(CH_3O)_2\overset{\overset{\displaystyle S}{\|}}{P}-S-\overset{\overset{\displaystyle}{\underset{\overset{\displaystyle |}{CH_2-\overset{\overset{\displaystyle O}{\|}}{C}-OC_2H_5}}{CH}}}{CH}-\overset{\overset{\displaystyle O}{\|}}{C}-OC_2H_5$$

Malathion

Carbamates

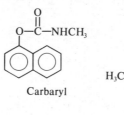

Carbaryl

Methiocarb

Synthetic Pyrethroids

Permethrin

Figure 7.2.   Structures of some examples of organophosphorus and other insecticides.

maximum in the early 1970s but has been restricted recently due to concern regarding their ecological impact. The cyclodiene chlorinated hydrocarbons (see Figure 7.1) also include well-known substances which have a somewhat similar usage and history.

The chlorinated hydrocarbon insecticides show some general similarities in properties. They have low water solubility, high lipophilicity, and are comparatively persistent in the natural environment. In addition, they bioaccumulate in individual organisms and may biomagnify in food chains.

The organophosphorus compounds have the general structure shown in Figure 7.2. About 40 compounds are available commercially as insecticides and the chemical structures of some of the most common are shown in Figure 7.2. Their general properties differ markedly from the chlorinated hydrocarbons in that many

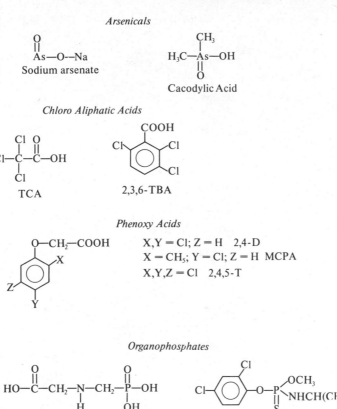

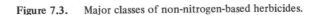

Figure 7.3. Major classes of non-nitrogen-based herbicides.

members have limited persistence in the natural environment, are water soluble, are nonbioaccumulative, and do not biomagnify in food chains.

Herbicides

A wide variety of substances have been used as herbicides. The chemical structures of some of the most common are shown in Figures 7.3, 7.4, and 7.5. But the development of the selective herbicides, with plant growth control or auxin-like properties, has resulted in extensive use of these substances for weed control in agriculture and forestry. Plants are destroyed by a gross dislocation of their growth processes.

The most effective and widely used herbicides in this group are the phenoxy acids, including such substances as 2,4-dichlorophenoxy acetic acid (2,4-D) and 2,4,5-trichlorophenoxy acetic acid (2,4,5-T) (see Figure 7.3). As a general rule the phenoxy acids have limited persistence in the natural environment, are

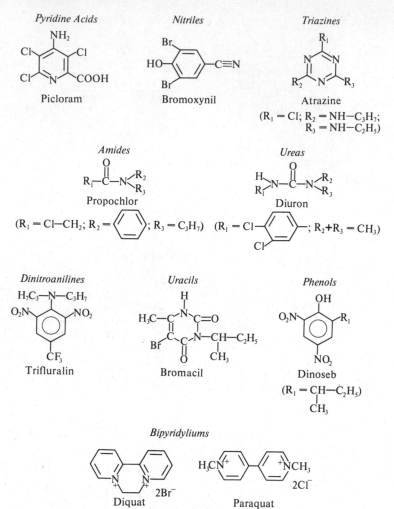

Figure 7.4. Chemical structures of some examples of major classes of nitrogen-based herbicides.

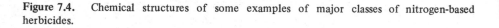

moderately water soluble, are nonbioaccumulative, and do not biomagnify. Usage is expected to increase substantially in the future.

Fungicides

Lesser quantities of the fungicides are used than either the insecticides or herbicides. However, usage is more varied than the other two groups, with these substances being used in such operations as inhibition of fungal growth in paper manufacture, mildew-proofing of fabrics, control of post-harvest rots in fruit, vegetables, stored

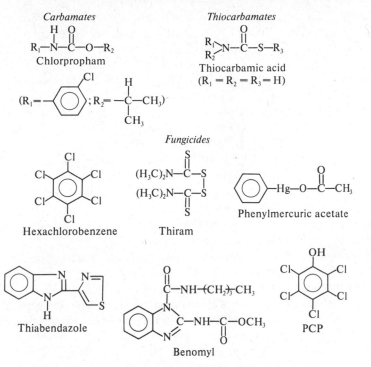

Figure 7.5. Chemical structure of some nitrogen-based herbicides and some fungicides.

grains, and so on. Discharge to the environment and impact on natural ecosystems is generally low but may be important in some cases. Chemically, fungicides are quite diverse, as shown in Figure 7.5, and it is not possible to generalize on their properties.

BEHAVIOR AND FATE IN THE ABIOTIC ENVIRONMENT

The Atmosphere

Worldwide use of pesticides has resulted in the distribution of the more-persistent pesticides throughout the earth's ecosphere. The atmosphere has been postulated as a major route for this widespread distribution (Risebrough et al., 1968; Woodwell et al., 1971).

Pesticides escape into the atmosphere either in particulate form or as a vapor from usage zones. In general, atmospheric pesticides are partitioned between both vapor and particulate forms (Seiber et al., 1975). Particulates may be either liquids (aerosols) or solids and may range in composition from active pesticide molecules to mostly an inert carrier with absorbed or adsorbed pesticide molecules. Combustion particulates may act as carrier substrates (Wheatley, 1973).

Pesticides can enter the atmosphere in an intermittent pattern from point sources, for example, physical drift during agricultural application, or by more continuous processes (Wheatley, 1973). For example, continuous exchange and redistribution of pesticides occurs between deposits on substrates, for example, soil and plant surfaces, and the atmosphere as a part of dynamic equilibration processes.

Robinson (1973) has concluded that the rate of entry and the distances over which pesticides move are dependent upon their vapor pressure and the meteorological conditions. Most of the pesticides entering the atmosphere in a particular area are deposited locally. This is attributed to scavenging processes such as deposition of dust, photochemical decomposition, and washout by rain. Aeolian transport over long distances appears to be a minor phenomenon. Overall, the dominant processes determining the fate of pesticides in the atmosphere can be identified as (1) vapor phase and particulate behavior, (2) photochemical reactions, and (3) dry deposition and rain washout.

Evaporation and soil erosion losses from application sites appear to be more significant sources of atmospheric pesticides than spray drift since this involves particulate matter which settles out in a relatively short time (Seiber et al., 1975). However, pesticide molecules will interchange between vapor and particulate states as the system approaches equilibrium. Wheatley (1973) suggests that the highly dynamic natural systems equilibrate rapidly tending to damp-out perturbations in pesticide concentrations caused by the intermittent escape of relatively large quantities of pesticide as air moves from a pesticide source to a sink. For example, the oceans are a large sink since the net exchange is from air to water.

Vaporization and vapor phase movements are important in the dissipation of pesticides from plant, water, and soil surfaces (Spencer and Cliath, 1973). Vaporization of chemicals from surface deposits is considered to be dependent on vapor pressure of the chemical and its rate of movement away from the evaporating surface. However, many variables are involved under field conditions, complicating the development of predictive approaches.

Concentrations of pesticides likely to occur in the atmosphere and in air around zones of usage are indicated in Table 7.1. In application areas, concentrations may range up to $10 \, \mu g \, m^{-3}$. In contrast, background concentrations of organochlorine compounds, DDT plus metabolites and dieldrin, in dust collected in Barbados ranged from 10^{-9} to $10^{-8} \, \mu g \, m^{-3}$ (Risebrough et al., 1968). This dust was believed to originate from the European and African continents although Robinson (1973) suggests the existence of an efficient scavenging process during airborne trans-Atlantic transport. The Barbados air levels were smaller than those found in London air, by a factor of 10^5–10^6. Thus the removal process can be represented by a mass conservation equation (Robinson, 1973):

$$Q_K = Q_{EA} + Q_{in}^q - Q_{out}^q$$

where Q_K is the amount of a pesticide in unit mass of dust deposited at Kitridge Point, Barbados: Q_{EA} the amount present in unit mass of dust at the western edge of the European–African continents; Q_{in}^q and Q_{out}^q are the amounts absorbed and

Table 7.1.   Concentrations of Pesticides in Air ($\mu g\ m^{-3}$) Associated with Different Usage Zones and Atmospheric Transport

Insecticides	Operation/Site	Concentrations in Air	
		Range ($\mu g\ m^{-3}$)	Mean
	Vicinity of Usage[a]		
Parathion	Orchards	40–15,000	—
Azinphos-methyl	Apple orchard	50–2,550	670
Malathion	Apple orchard, vegetable fields	410–760	590
	Mosquito control	70–90	—
Carbaryl	Apple orchards	180–810	600
Endrin	Vegetable fields, aerial spraying	20–90	50
BHC	Forests, ground-treated	2,600–12,500	—
	Forests, aerially-treated	540–4,100	—
DDT	Forests, ground-treated	2,600–4,600	—
	Forest, aerially-treated	1,900–171,000	—
	Nearby Usage Zones[a]		
Azinphos-methyl ⎫ Carbaryl ⎪ Malathion ⎬ Parathion ⎭	300–600 m from application Rougemont Quebec	up to 500	—
DDT	Six small communities in	0.0001–0.022	—
Chlordane	usage areas,	0.0001–0.006	—
Aldrin	Florida	0.0001–0.004	—
DDT	One rural area during usage	0.003–8.5	—
Chlordane		0.001–0.031	—
Malathion	Mosquito control usage	0.0001–0.031	—

(continued on next page)

Table 7.1. (continued)

Insecticides	Operation/Site	Concentrations in Air Range (μg m^{-3})	Mean
DDT	During and after fogging, Atlantic coast resort	During 0.100–8.0 After 0.002–0.011	— —
Malathion		During 0.001–0.03 After < 0.0001–< 0.001	— —
Urban Air			
pp^1DDT	London[b]	—	0.0038
pp^1DDE		—	0.0071
pp^1DDD		—	0.0038
Dieldrin		—	0.026
DDT	United States[c]	0.0001–1.56	—
Background			
pp^1DDT	Barbados[d]	—	4.5×10^{-8}
pp^1DDT		—	5×10^{-9}
pp^1DDE		—	1.4×10^{-8}
pp^1DDD		—	4.3×10^{-9}
Dieldrin		—	5.4×10^{-9}

[a] Compiled from Wheatley (1973).
[b] Compiled from Abbott et al. (1966).
[c] Compiled from Stanley et al. (1971).
[d] Compiled from Risebrough et al. (1968).

lost during aeolian transport over the Atlantic ocean. The removal processes were presumed to include dust deposition, washout by rain, photochemical decomposition, and partition of pesticides (in the vapor form) between the atmosphere and the surface of the ocean. Input processes include spray from the oceanic surface which may contain traces of pesticides. Quantitative approaches at such low concentrations are inhibited by quantitative variability in methods and uneven distribution of pesticides in air masses.

Wheatley (1973) suggests that the atmosphere serves as a global transport route whereby atmospheric circulation effectively redistributes stable pesticides ubiquitously. Thus, significant amounts of some pesticides are in transit at any given time (see Woodwell et al., 1971).

Photochemical transformations occur readily in the atmosphere with the rate and nature of the transformation products dependent primarily upon the chemical and spectral absorption characteristics of the parent pesticides. The wavelength range of solar radiation at the earth's surface is generally greater than 290 nm which yields sufficient energy on absorption to result in the rupture of bonds in organic molecules (Tinsley, 1979). However, photochemical reactions are mediated by the presence of naturally occurring sensitizers, oxidants, and catalytic surfaces, resulting in photoisomerization and intramolecular condensation.

Significant photochemical transformations under atmospheric conditions (e.g., Figure 7.6) can be summarized as (1) dechlorination reactions which often result in the loss of pesticide toxicity; (2) photoisomerization reactions which may form relatively stable and more toxic photoisomers; (3) photooxidation reactions in which oxidants such as O, O_2, O_3, and NO_2 may react strongly with olefinic and aromatic groups; and (4) photomineralization processes where chlorinated hydrocarbon pesticides can be converted into CO_2 and HCl (Korte, 1978). However, parathion exposed to UV light does not produce photoproducts in the vapor phase in contrast to its photooxidation when adsorbed on fine dust particulates. This suggests that airborne particulate surfaces may play an important role in the environmental photochemistry of pesticides (Moilanen et al., 1975). In certain cases, photochemical transformations produce more toxic and stable compounds, for example, the epoxidation of aldrin to dieldrin (see Figure 7.6) and heptachlor to heptachlor epoxide.

Quantitative assessment of the rates of removal of pesticides from the atmosphere by deposition and rainout are limited to several chlorinated insecticides, principally DDT (see Woodwell et al., 1971; Robinson, 1973; and Wheatley, 1973). Woodwell et al. (1971) calculated that DDT removal from the atmosphere by rain would have a time constant of about 3.3 yr and that its mean residence time was probably not longer than 4 yr.

In general, conclusions that the atmosphere acts as a major transport route for pesticides are based on estimates of the atmospheric levels and deposition patterns of DDT and, to a lesser degree, dieldrin. Most pesticides are clearly deposited within, or nearby, usage zones but substantial amounts of DDT (and dieldrin) are resident in the atmosphere. This amouts to about 4×10^5 tonnes of DDT or about one-sixth of estimated world usage up to 1974. Estimates of worldwide surface

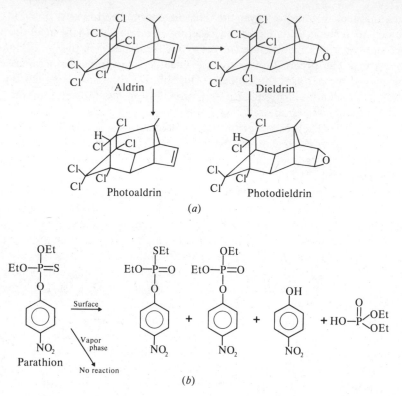

Figure 7.6.   Photochemical reactions of selected insecticides in air. (*a*) Vapor phase photochemical transformations of aldrin, dieldrin, and photoaldrin. (From Crosby and Moilanen, 1974. Reprinted with permission from Springer-Verlag). (*b*) Photoproducts from parathion adsorbed on dust particles. (From Moilanen et al., 1975. Reprinted with permission from Plenum Press.)

distributions of DDT in oceans and land surfaces, for example, up to 2.4×10^3 tonnes of DDT in Antarctic snow, suggest considerable atmospheric circulation of stable pesticides (see Wheatley, 1973).

The Lithosphere

The behavior of pesticides in the lithosphere is principally related to the thin outer layer, several meters deep, on the earth's surface (Robinson, 1973). The soil environment contains a relatively immobile matrix of solid, liquid, and gaseous phases. The solid phase contains clay minerals, organic matter, oxides, and hydroxides of aluminum and silicon, whereas the liquid phase consists of water and dissolved salts termed the soil solution. Air forms most of the gaseous phase.

Pesticides enter the soil environment via direct application, dust deposition, rainout or precipitation, flooding of land by rivers, and waste disposal. Interactions between pesticides and the soil environment are controlled by three important factors: (1) sorption/desorption processes; (2) leaching–diffusion; and (3) degradation (Haque, 1975). Movement and loss of pesticides in soil can occur by leaching, runoff, and volatilization as well as transformations such as chemical degradation, photodecomposition, and microbial action.

Adsorption

Major factors influencing adsorption/desorption on soil colloids include soil characteristics and composition (e.g., water, clay and organic matter content, pH, and eh) and pesticide characteristics (e.g., chemical character, shape and configuration of the pesticides, pK_a or pK_b, water solubility, charge distribution on cations, polarity, and ionization). Organic matter content of soils exerts an important influence on adsorption. In soils with an organic matter content of up to about 6%, adsorption involves both mineral and organic surfaces but, at high organic contents, adsorption will occur mostly on organic surfaces (Khan, 1980).

Bailey and White (1970) identified four structural factors in a pesticide molecule that relate to soil adsorption potentials:

1. Nature of functional groups present, particularly acidic, carboxyl, alcoholic hydroxyl, and amine groups.
2. Nature of substituent groups that may alter the behavior of functional groups.
3. Position of substituent groups with respect to functional groups that may enhance or hinder intramolecular bonding or permit coordination with transition metal ions.
4. Presence and magnitude of unsaturation in the molecule which may affect the lipophilic–lipophobic balance.

Khan (1980) suggests that the charge characteristics of a pesticide are probably the most important properties governing its adsorption. This may range from weakly polar to a relatively strong ionic charge. It follows that the pH of the soil system will also be an important factor.

Some pesticides exhibit an inverse relationship between water solubility and adsorption. This has been shown for some acidic herbicides on muck soil and for some nonionic pesticides on organic matter and several substituted ureas. But in other cases, the relationship is either direct or nonexistent (Khan, 1980).

Haque (1975) found that the highly water soluble diquat and paraquat readily adsorbed on clay surfaces, the moderately soluble 2,4-D physically adsorbed to most soil surfaces, and the hydrophobic chlorinated hydrocarbons, DDT and PCBs, strongly adsorbed to organic surfaces. Khan (1980) has reviewed the adsorption of ionic and nonionic types of pesticides on soil surfaces, and the mechanisms are set out in Table 7.2.

The adsorption process on solid surfaces can be represented by adsorption isotherms (see Chapter 2). In general, the Freundlich and Langmuir isotherms have been used to develop mathematical relationships for pesticide adsorption

Table 7.2. Proposed Mechanisms of Pesticide Adsorption on Soil Colloids[a]

Type of Mechanism	Nature of Mechanism	Examples
van der Waals attraction	Adsorption of nonionic, nonpolar, and parts of pesticide molecules; short-range dipole–dipole interactions.	Adsorption of isocil on montmorillonite and kaolinite, carbaryl and parathion on soil organic matter, picloram by humic materials.
Hydrophobic bonding	Nonpolar pesticides on molecules with significantly high proportions of nonpolar regions. Association of nonpolar pesticides with lipid fraction of soil organic matter and humus; partitioning process.	Adsorption of nonpolar chlorinated hydrocarbon pesticides by lipids in organic matter.
Hydrogen bonding	Polar nonionic organic molecules, anionic pesticides and organic matter involving interactions of oxygen-containing functional groups as well as amino groups.	s-Triazines and organic matter interactions (humic compounds). Adsorption of weak anionic pesticides; 2,4-D adsorption on montmorillonite may involve hydrogen bonding of the C=O group to the hydroxyls of the clay surface.
Charge transfer	Formation of charge transfer complexes by electrostatic attraction, over short distances, between interacting species.	Postulated mechanism: adsorption of s-triazines onto soil organic matter and clay minerals; paraquat and diquat interactions with clay minerals.
Ion exchange	Adsorption of cationic pesticides, e.g., paraquat and diquat via cation exchange functions through — COOH and phenolic — OH groups associated with organic matter.	Paraquat and diquat cationic exchange with organic matter and clay minerals. Considered to be dominant mechanism in comparison to hydrogen bonding; also s-triazines on organic matter and clay minerals.
Ligand exchange	Formation of coordination complexes with various metals on clay minerals; replacement of one or more ligands by adsorbent molecule.	Coordination of aminotriazole to Ni^{2+} and Cu^{2+} cations on montmorillonite; also the coordination of trillate and linuron to exchangeable cations on clay minerals through the oxygen on the carbonyl group.

[a] Compiled from Khan (1980).

on soil materials. These isotherms are expressed as

$$\text{Freundlich:} \quad \frac{X}{m} = KC^{1/n} \tag{1}$$

$$\log \frac{X}{m} = \log K + \frac{1}{n} \log C \tag{2}$$

$$\text{Langmuir:} \quad \frac{X}{m} = \frac{K_1 K_2 C}{1 + K_1 C} \tag{3}$$

where X/m is the ratio of pesticide (X) to the adsorbent mass (m), C the pesticide concentration in solution at equilibrium, and K and n constants. In Equation (3), K_1 is a constant for the system dependent on temperature and K_2 is the mono-layer capacity. Nonionic pesticide–soil adsorption isotherms are generally near-linear for the relationship between amount adsorbed and residual concentration in solution. Most of the data can be represented by the Freundlich equation (1). Ideally, the relationship between $\log (X/m)$ and $\log C$ is linear and the slope and intercept of such a plot gives the value of the constants n and K.

For ideal situations, where the isotherm slope is unity (i.e., $n = 1$), there would be unlimited adsorption of pesticides as the equilibrium concentration continually increases. In practice, however, this is rarely observed and slopes are character-istically variable for different pesticide–soil systems and environmental conditions. In certain cases, the K value may be considered a useful index for comparing the degree of adsorption between pesticides and soil surfaces (Khan, 1980). Audus (1976) and others have pointed out the empirical nature of the Freundlich equation and suggested that it should not be used for extrapolation outside the experimental range.

Freundlich and Langmuir isotherms are not applicable if the adsorption is pre-dominantly due to an ion exchange mechanism. In this case the Rothmund–Kornfield equation has been used to distinguish between ionic and other mecha-nisms of adsorption (Khan, 1980):

$$\frac{[\bar{P}^{2+}]}{[\bar{H}^+]^2} = \frac{K[(P^{2+})]^{1/n}}{[H^+]^2} \tag{4}$$

where $\bar{P}^{2+}$ and $\bar{H}^+$ refer to ions in the adsorbent and

$$A = \log K + \left(\frac{1}{n}\right) S \tag{5}$$

where $A = \log [\bar{P}^{2+}] - 2 \log [\bar{H}^+]$ and $S = \log [P^{2+}] - 2 \log [\bar{H}^+]$.

Leaching and Diffusion

The movement of pesticides in the soil environment involves the interaction of several processes: diffusion and dispersion, mass transfer, leaching, and volatilization. Diffusion, leaching, and vapor loss of pesticides in soil are inhibited as adsorption increases (Haque, 1975). Physical transport processes are discussed in more detail in Chapter 2. Khan (1980) has discussed the movement of pesticides in soil and Hartley (1976) has reviewed physical transfer mechanisms in soil as applied to herbicides.

Diffusion of pesticides in the soil environment can occur in solution, at the air–water or air–solid interfaces, and in air. In general, diffusion coefficients (D) of pesticides, as derived from Fick's Law of Diffusion, are 1×10^4 to 3×10^4 times greater in air than in water (Khan, 1980). Thus pesticides with a water to air ratio under 1×10^4 should diffuse primarily through air whereas those with ratios over 3×10^4 should diffuse principally through water.

The overall diffusion of pesticides in soil, however, is complex and influenced by environmental variables such as solubility, vapor density, adsorption, bulk density, soil water content and porosity, and uptake by plants. Increasing temperature tends to increase diffusion which may be significant in summer and in tropical regions.

The water leaching behavior of pesticides through soil is dependent upon soil adsorption principally (Khan, 1980) although diffusion and hydrodynamic dispersion, adsorption dynamics, and evapotranspiration play a role (Hamaker, 1975). Techniques used in predicting the distribution of pesticides through soil profiles do not reliably predict leaching behavior under field conditions, particularly the slow moving or trailing fraction. There is less movement of chemical than indicated by laboratory leaching tests (Hamaker, 1975).

Evaporation of pesticides can be a major route of loss to the atmosphere. The rate of volatilization is related to the vapor pressure of the pesticide within the soil and its rate of movement to and away from the evaporating surface. Adsorption on soils and other surfaces, as well as solution in soil water, significantly alters the vapor pressure of the pesticide and its rate of movement within soils. The magnitude of these effects is dependent mainly on the nature of the chemical, soil chemical concentration, soil water content, and soil characteristics, for example, organic matter and clay content (Spencer and Cliath, 1975). In general, pesticides vaporize faster from wet soils than dry soils because of mechanisms such as water competition for adsorption sites and upward capillary movement of the soil solution as water evaporates from the soil surface.

For many pesticides (e.g., dieldrin and lindane) an inverse relationship between the rate of pesticide volatilization and soil organic matter content has been widely reported. Temperature also affects the volatilization of pesticides from soils primarily by a direct influence on vapor pressure and the physical and chemical properties of the soil (Khan, 1980).

Pesticide Degradation Processes

The degradation of pesticide residues is a major route for the loss of pesticides from soils. The chemical nature of the pesticide is particularly important in determining the pathways and rates of degradation (see Figure 7.7). However, the transformation products may also exhibit significant biological activity. Also, many conversion products persist in soil and aqueous media and may have ecological effects in the same manner as the original compound — for example, the formation of photoproducts from chlorinated cyclodienes by the action of sunlight and microbiological organisms in soils (see Figure 7.6).

The degradation pathways of pesticides in soils have been extensively reviewed indicating three main types of pesticide transformations in soils; photochemical, chemical, and microbiological. These are considered below.

1. *Photodecomposition.* Although photodecomposition of pesticides in air and water occurs widely, Goring et al. (1975) suggest that photodecomposition of pesticides in soil is of doubtful significance. Radiant energy is strongly absorbed by soil and consequently little photodecomposition of pesticides would be expected, except on or very near to the surface.

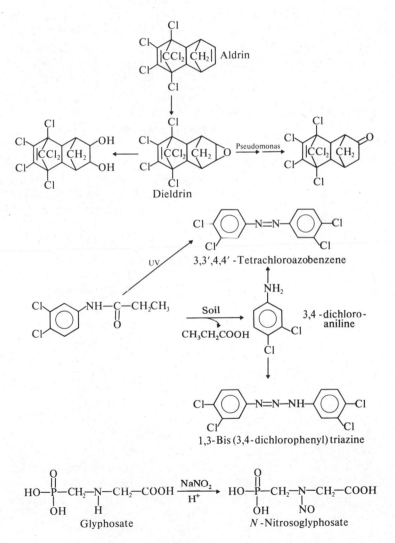

Figure 7.7. Some pathways for the formation of pesticide degradation products.

Khan (1980) points out that experiments on photochemical degradation of pesticides in soil have been carried out under conditions involving exposure to comparatively high intensity light and frequently in nonaqueous solvents. Several insecticides and herbicides have been shown to decompose when applied experimentally on dry soil in thin layers exposed to sunlight (Khan, 1980). Natural photosensitizers, for example, humic substances, may also play a role in facilitating degradation of soil pesticides (Matsumura, 1973). Under field conditions, assessment of photochemical breakdown is complicated by competing soil and microbiological processes.

2. *Chemical Transformations.*   Chemical transformations of pesticides are important processes for their removal from soils. Khan (1980) has concluded that reactions are mediated by water functioning as a reaction medium, a reactant, or both. Common reactions involve hydrolysis — for example, many organophosphorus compounds and s-triazines — and oxidation — for example, many sulfur containing pesticides — but reduction and isomerization are important with certain compounds.

N-Nitrosation of certain nitrogenous pesticides, for example, atrazine in soils, has received recent attention because of the formation of nitrosoamines. Nucleophilic substitution reactions, other than hydrolysis, take place with reactants dissolved in water or with groups in soil organic matter. Soil reactions with free radicals is also a distinct possibility. Degradation reactions in soil may be catalyzed by clay surfaces, metal oxides, metal ions, and organic matter. Extracellular soil enzymes play a significant role in the degradation of many pesticides and represent the transition between chemical and intracellular microbiological breakdown (Goring et al., 1975).

3. *Microbiological Degradation.*   The major groups of soil microorganisms (actinomycetes, fungi, and bacteria) can readily adapt to and degrade pesticides through oxidation, ether-cleavage, ester and amide hydrolysis, oxidation of alcohols and aldehydes, dealkylation, hydroxylation, dehydrohalogenation, epoxidation, reductive dehalogenation, and *N*-dealkylation (Matsumura, 1973). Dehydrohalogenation is a major process since a large proportion of pesticides contains halogens. Several reviews on microbiological processes affecting pesticides in soil are available, including specific structural characteristics of pesticides associated with microbial degradation (e.g., Kaufman and Plimmer, 1972; Matsumura, 1973; Audus, 1976; and Khan 1980). Goring et al. (1975) concluded that processes of pesticide degradation in soil include chemical reaction, microbial enrichment, and cometabolism. Simultaneous chemical and microbiological transformations in soil are difficult to distinguish. The complex major pathways of degradation that have been established include dehydrohalogenation and isomerization reactions. The end products of pesticide transformations in soil are carbon dioxide, water, mineral salts, metabolites naturally occurring in soil, and humic substances. However, the fate of many major pesticide metabolites in soil is relatively unknown. As Kaufman and Kearney (1976) point out, of approximately 150 chemicals currently used as herbicides throughout the world, complete metabolic degradation pathways are known for only three or four.

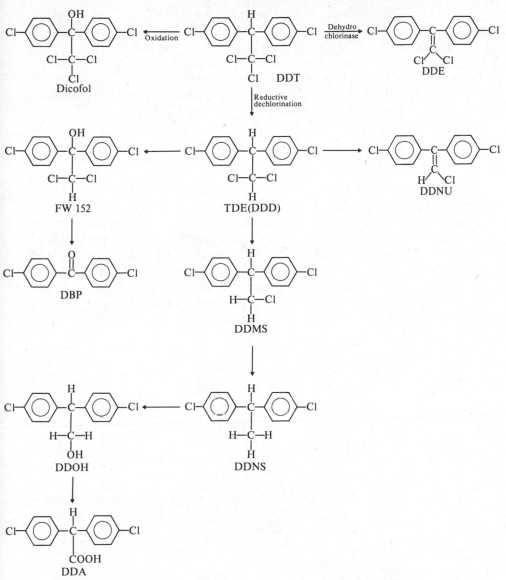

Figure 7.8.   Some pathways for the formation of pesticide degradation products of DDT.

Pesticide Degradation Products

The environmental significance of pesticide degradation products has received relatively little attention, despite the hundreds of potential metabolites, free and conjugated, that can be formed from parent compounds (e.g., see Figure 7.7). The most thoroughly studied degradation products are DDE and DDD derived from the biological dehydrochlorination and reductive dechlorination of DDT (see Figure 7.8). These are stable and biologically active. Photooxidation and microbial epoxidation of cyclodienes to toxic products occur in soil resulting in the conversion of aldrin to dieldrin, heptachlor to heptachlor epoxide, and isodrin to endrin (Matsumura, 1973) (see Figure 7.7).

Many sulfur-containing pesticides are also modified in soils by oxidation. For example, parathion can be oxidized to toxic paraoxon but this reaction is considered unimportant in soil (Lichtenstein and Schulz, 1964). Similarly, malathion can form malaoxon as an intermediate degradation product in soil.

Khan (1980) has reviewed the production of N-nitrosoamines from the oxidation of amine groups and reduction of nitro groups in nitrogen-containing pesticides in soils. Generally, soil pH conditions of about 3–4 and excess nitrite favor the N-nitrosation reaction and yield only trace quantities of nitrosamines. Also the high levels of pesticides and sodium nitrite employed in certain studies to demonstrate the formation of N-nitroso compounds in soil are not likely to be encountered in practical agriculture. On the other hand, the persistence of N-nitroso derivatives, such as N-nitrosoglyphosate, has been established in various soils. In addition, N-nitroso compounds are found in technical and formulated pesticide products used in agricultural and domestic applications (see Fine et al., 1980). The possible ecotoxicological significance of N-nitroso compounds in soils is yet to be established.

The environmental and metabolic transformations and interconversions of primary aromatic amines and related compounds appear to occur more rapidly

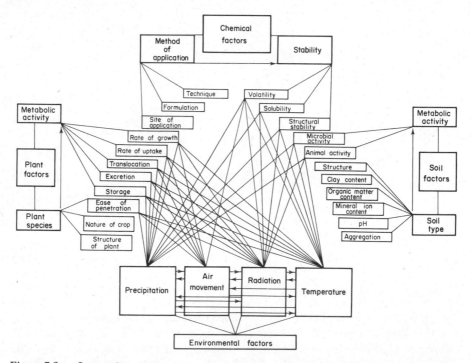

Figure 7.9. Interactions between factors that influence the persistence of pesticides in plants and soils. [From Edwards (1975). Reprinted with permission from the International Union of Pure and Applied Chemistry.]

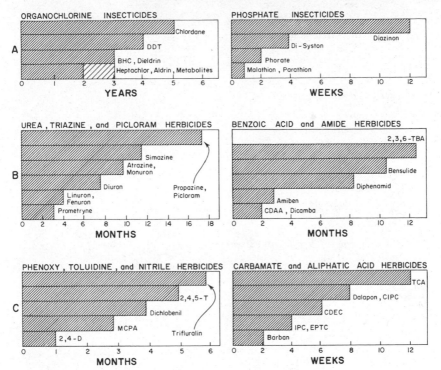

Figure 7.10. Persistence of pesticides in soils: time taken for a compound applied to soil at the normal dosage to decrease by more than 75%. [A, B, and C from Kearney et al. (1969). Part A reprinted with permission from Charles C. Thomas, Publisher; B and C reprinted with permission from Springer-Verlag.]

than mineralization reactions (Parris, 1980). Consequently, it appears the related compounds can be regarded as latent forms of aromatic amines in the environment. One interesting set of condensation products of aromatic amines is the chlorinated azobenzenes. For example, 3,3′,4,4′-tetrachloroazobenzene has received attention because of its steric similarity to 2,3,7,8-tetrachlorodibenzodioxin (TCDD) (see Chapter 9).

Persistence

The persistence of a pesticide in a soil is dependent on the effect of the many interacting factors as shown in Figure 7.9 (Goring et al., 1975; Khan, 1980). In general, persistence may be interpreted as the residence time of a pesticide in the soil environment, or as a half-life (see Chapter 2) assuming first-order kinetics (see Figure 7.10 and Table 7.3). The most persistent group of pesticides are the chlorinated hydrocarbon insecticides. The herbicides show a wide spectrum of persistence ranging from a few days or weeks for the carbamates and aliphatic acids, to over a year for certain of the *s*-triazines and picloram. Organophosphorus and carbamate insecticides are generally short lived in soils from a few days to several months. However, Lichtenstein (1980) cautions that the apparent loss of residue

Table 7.3.   Half-Lives of Some Relatively Persistent
Insecticides in Soils[a]

Insecticide	Approximate Half-Life (yr)
Organochlorines	
DDT	3–10
Heptachlor	7–12
Isodrin/Endrin	4–8
Toxaphene	10
Aldrin	1–4
Dieldrin	1–7
Chlordane	2–4
BHC	2
Organophosphorus	
Dyfonate	0.2
Chlorfenvinphos	0.2
Carbophenothion	0.5
Carbamates	
Carbofuran	0.05–1

[a] Compiled from Khan (1980).

from conventional soil studies may be due to unextractable or "bound" soil residues.

Kearney and Plimmer (1970, 1972) have compared structure–activity–degradability relationships for several major classes of herbicides. In some chemical classes those structural features necessary for herbicidal activity were coincident with those necessary for degradability, whereas in other chemical classes an opposite relationship existed (Kaufman and Kearney, 1976). The relative persistence and initial degradation reactions of various herbicide classes are shown in Table 7.4. Herbicides that initially undergo ester hydrolysis are relatively short lived in comparison to those that experience an initial dealkylation reaction. In this latter case persistence is variable but tends to increase with complexity of the basic molecule.

Considerable differences in herbicide persistence may exist within closely related classes. For example, some methoxy-s-triazines and their metabolites are reported to be persistent in soils for up to several years but the chloro- and methylthio-s-triazines are much less persistent. Soil conditions have a strong influence on persistence as shown by the half-lives of 2,4,5-T which varied from 4 days at 35°C and 34% soil moisture to 60 days at 10°C and 20% soil moisture (Walker and Smith, 1979). The persistence of the bipyridylium herbicides, paraquat and diquat, appears to depend on the organic matter and clay contents of soil. In a mineral soil, about 8% of the total paraquat applied accumulated as a residue over a 9-yr period (Khan, 1980). Overall, certain herbicide residues, for

Table 7.4.   Relative Persistence and Initial Degradative Reactions of Nine Major Herbicide Classes[a]

Chemical Class	Persistence	Initial Degradative Process
Carbamates	2–8 weeks	Ester hydrolysis
Aliphatic acids	3–10 weeks	Dehalogenation
Nitriles	4 months	Reduction
Phenoxyalkanoates	1–5 months	Dealkylation, ring hydroxylation, or β-oxidation
Toluidine	6 months	Dealkylation (aerobic) or reduction (anaerobic)
Amides	2–10 months	Dealkylation
Benzoic acids	3–12 months	Dehalogenation or decarboxylation
Ureas	4–10 months	Dealkylation
Triazines	3–18 months	Dealkylation or dehalogenation

[a] Adapted from Kaufman and Plimmer (1972). Reprinted with permission from Academic Press Inc. (London) Ltd.

example, phenoxy acids, uracils, substituted ureas, and s-triazines, may persist in agricultural soils for periods of over 1 yr, especially soils receiving repeated applications.

The Hydrosphere

The hydrosphere acts as a major reservoir for persistent pesticide residues. Pesticides enter into the hydrosphere via many pathways including: (1) direct application for pest and disease vector control; (2) urban and industrial waste-water discharges; (3) runoff from nonpoint sources including agricultural soils; (4) leaching through soil; (5) aerosol and particulate deposition, rainfall; and (6) absorption from the vapor phase at the air–water interface. The relative inputs from these sources are difficult to assess. But generally, water bodies associated with urban regions receive substantial pesticide inputs from industrial and domestic effluents. Overall, the major input probably originates from agricultural and forestry practices in conjunction with control programs for human disease vectors (see Kerr and Vass, 1973).

The major routes of pesticide translocation into the hydrosphere are generally accepted as surface runoff and aerial transport and deposition. Goldberg et al. (1971) estimated that insecticide washed out of the atmosphere into the oceans amounted to 2.4×10^7 kg of DDT-type compounds corresponding to approximately one-quarter of the world's annual production at that time. Kerr and Vass (1973) consider that this estimate is likely to be excessive since it did not provide for potential recycling across the air–sea interface. But they concluded that aerial fallout greatly exceeds river discharge as the principal source of input for pesticides to the oceans.

Pesticide runoff varies according to the rate of surface water flow and soil types, while leaching depends primarily on adsorption/desorption between the soil constituents and the water percolating through it (Robinson, 1973). Generally, surface run off from agricultural lands and leaching do not appear to be significant sources of pesticide contamination of water bodies whereas significant amounts may originate from industrial effluents, waste disposal, for example, sheep dips and washing of spraying equipment, and so on (Edwards, 1973).

Many environmentally significant pesticides are highly lipophilic and therefore relatively insoluble in water. For example, the organochlorine insecticides generally have solubilities $< 10 \text{ g L}^{-1}$ (Edwards, 1973). In natural waters, the major proportion of the lipophilic pesticides are either adsorbed onto suspended and settled particles or partitioned into organic substrates. These pesticides show strong affinity for the lipoid components of living and dead organic matter. The quantities involved depend on the chemical characteristics and solubility of the pesticide as well as characteristics of the sediment — for example, organic content, clay content, and pH. In addition, pesticide residues tend to be associated with the small particulates in the water column and sediments.

pH has a marked effect on the sorption of pesticides containing acidic functional groups. For example, at any pH the amount of 4-amine-3,5,6-trichloropicolinic acid adsorbed is correlated with the amount of unionized acid present (Hamaker et al., 1966). In addition, temperature affects solubility and volatility of pesticides in solution.

Removal of pesticides from the hydrosphere may occur by (1) volatilization, (2) absorption by aquatic organisms, and (3) settling of particles to which pesticides are adsorbed (Robinson, 1973). Removal by degradation processes occurs by transformation and ultimately mineralization.

Recycling of pesticides into the atmosphere through diffusion and codistillation at the air–water interface is possibly a major source of pesticide loss for the oceans given the vast surface areas involved. However, the importance of pesticide interchange between air and water is uncertain (Robinson, 1973; Edwards, 1973), although Goldberg et al. (1971) have provided some basic calculations for the exchange of DDT-type compounds from the upper mixed zone of the ocean.

Available evidence suggests absorption of pesticides by biota is only a very minor route for removal from the hydrosphere. Woodwell et al. (1971) concluded that the world's biota probably contains about 5.4×10^9 g, less than 1/30 of 1 year's production of DDT during the mid-1960s and most of this was estimated to be in marine organisms.

Pesticides in the water column may be removed by (1) sedimentary sorption, (2) degradation by microorganisms, (3) uptake by organisms, or (4) diluted further, especially upon translocation in the oceans (Hassett and Lee, 1975). Photochemical decomposition of pesticides may also proceed via a series of photolysis reactions — for example, photooxidation, photonucleophilic hydrolysis, and reductive dechlorination — influenced by factors such as natural photosensitizers, pH conditions, and availability of dissolved oxygen. For example, photodecomposition of the aquatic herbicide trifluralin (2,6-dinitro-N,N-dipropyl-α,α,α-trifluoro-p-toluidine) proceeds rapidly under acidic and aerobic conditions in comparison to basic anaerobic conditions.

In the oceans, DDT residues, principally in association with particulate matter (Hartung, 1975) circulate initially in the upper mixed layer which frequently extends to a depth of 75–100 m. Sedimentation slowly transfers the DDT below the thermocline into the much larger volume of the abyss with a transfer time of about 4 yr (Woodwell et al., 1971). Within the abyss, transfer rates to sediments may extend from hundreds to thousands of years. Once in the sediment, a pesticide may be (1) re-released in the water, (2) absorbed by organisms, (3) transformed or degraded by microbiological organisms, or (4) permanently buried. In certain cases, degradation may proceed more favorably under anaerobic conditions (e.g., DDT and lindane).

TRANSPORT AND TRANSFORMATION IN BIOTA

Uptake Processes

Uptake and distribution of pesticides and other chemicals in organisms involves the interactions of several phases represented diagrammatically in Figure 7.11 (see Kenaga, 1975). Pesticides are distributed in the environment by flow and transport systems (air, water) to biotic surfaces where they are adsorbed and from which they may be absorbed by living organisms. They are then partitioned through surfaces, organs, and cell barriers, via aqueous flow systems (e.g., water, blood, and sap), and returned to external flow systems either unchanged or transformed by metabolic processes. General mechanisms of pollutant transfer and transformation in organisms are described in Chapter 2. Factors influencing the uptake and distribution of pesticides in biological systems are related to: (1) inherent physical and chemical properties of the pesticide (e.g., volatility, water and fat solubility, and sorption characteristics); (2) physiological characteristics of various species (e.g., feeding behavior, routes of uptake, and habitat); and (3) ecosystem specific properties (e.g., types of flow systems, temperature, pH, organic matter, food web structure, etc.).

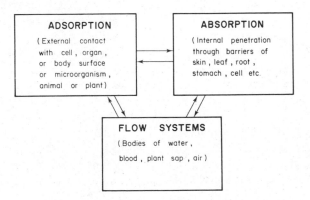

Figure 7.11. Interactions of uptake phases for pesticides.

Animals

Pesticide uptake by animals may occur either directly from the physical environment or from gastrointestinal absorption. For aquatic organisms, intake can result from (1) ingestion of contaminated food, (2) uptake from water passing over gill membranes, (3) cuticular diffusion, and (4) direct absorption from sediments (Livingston, 1977). Terrestrial species may absorb pesticides by the following processes: (1) via the gut through contaminated food and water; (2) directly via percutaneous absorption; and (3) inhalation of airborne pesticides. For example, insects absorb pesticides via the cuticle and trachea whereas with soil fauna it occurs through ingestion and also direct contact with pesticides adsorbed to soil minerals or dissolved in soil water (Walker, 1975).

Following absorption, pesticides are distributed to the organs and tissues of the body by the circulatory systems (e.g., blood and lymph in vertebrates and the hemolymph in insects) and by movement across membranous barriers. The distribution pattern is dependent upon the nature of the pesticide, the route of intake, and its metabolism, as well as characteristic species (Walker, 1975). Lipophilic compounds, such as organochlorine insecticides, tend to bind reversibly to plasma proteins as part of an internal transport mechanism and are readily stored in depot fat. Rapid mobilization, however, of depot fat may redistribute stored substances.

Plants

Pesticides penetrate the outer layers of plants through foliage, the epidermis of stems, bark, and the roots. The most common pathways are considered to be (1) the walls of root hairs or epidermal cells of roots, (2) stomata and the cuticle of cells in the spongy mesophyll, and (3) lenticles or cracks in the cuticle and periderm (Finlayson and MacCarthy, 1973). The degree and routes of pesticide absorption through plant surfaces are strongly influenced by factors such as properties of the pesticides, type of formulation, environmental conditions, and biological properties of the plant. In penetrating internal structures, pesticides have to pass through the cuticle which covers leaf surfaces exposed to air. The two possible routes are referred to as the lipoid and aqueous pathways (Wain and Smith, 1976). Lipophilic pesticides enter comparatively rapidly via the lipoid components of the cuticle. The cuticle is also somewhat permeable to polar molecules as demonstrated by the penetration of the ions of a variety of herbicidal compounds. Diffusion through this aqueous route is slower than the lipoid route but can be enhanced through use of appropriate emulsifications (Crafts, 1964). Stomatal penetration of herbicides may also be an effective route if open at time of application. Residues of insecticides in plants have also been attributed to penetration of leaves by the vapor phase of compounds with comparatively low vapor pressures, for example, DDT (Finlayson and MacCarthy, 1973).

Uptake of pesticides by the roots is considered the most important site of absorption, particularly in the zone of root hairs (Finlayson and MacCarthy, 1973). Crafts (1964) has concluded that herbicides are absorbed at greatly different rates by roots as a result of discrimination in uptake between different organic molecules in a manner similar to inorganic ions, for example, potassium and sodium. But

Wain and Smith (1976) considered the uptake of herbicides via the roots to be a less specific process than uptake by leaves. However, root absorption occurs more readily with water-soluble pesticides rather than lipid-soluble ones (Audus, 1964, 1976). There is also evidence that some applied pesticides may reach the roots as vapor (e.g., trifluralin).

Translocation or transport of pesticides in plants may be upward (acropetally), downward (basipetally), or laterally (Finlayson and MacCarthy, 1973). Crafts (1964) states there are two systems by which herbicide molecules may move rapidly in plants, the phloem and xylem. Pesticides absorbed through the leaves, for example, 2,4-D, tend to move in the assimilate stream, that is, via the phloem, and accumulate in developing parts of the plant such as the root and shoot tips. Translocation of root-absorbed herbicides involves movement via the sap stream of the xylem in the direction of the transpiration stream. This is the normal movement pattern for the systemic pesticides. Many of the herbicides which enter the phloem can, under certain conditions, move into the xylem and thus circulate in the plant (e.g., dalapon) (Sargent, 1976). Considerable variations in uptake and movement occur between species. Selectivity results from differential rates of penetration through the plant surfaces and different rates of translocation to sites of biological action (Sargent, 1976).

Many mechanisms and factors influence the local storage, binding, and persistence of pesticides in plants. In the case of herbicides (see Audus, 1976) immobilization of active molecules may result from adsorption during uptake or translocation. Evidence of pesticide persistence in plants is generally confined to field and experimental studies with agricultural crops (see Finlayson and MacCarthy, 1973).

Bioaccumulation

Bioaccumulation of organochlorine insecticides by biota (see, e.g., Edwards, 1973; Moriarty, 1975) has been investigated revealing significant factors influencing the process (see, e.g., Kenaga, 1972, 1975; Robinson and Roberts, 1968; Robinson, 1973; Brooks, 1975; Moriarty, 1975). In addition, mathematical models have been developed (e.g., Robinson, 1973; Moriarty, 1975, 1978) and attempts made to predict bioaccumulation potential of pesticides and other organic chemicals from experimental tests (e.g., Cairns et al., 1978; OECD, also see Chapter 2). These tests are designed to measure the physicochemical characteristics of these substances as well as uptake and elimination kinetics for selected species. However, some reservations have been expressed concerning the environmental significance of laboratory screening tests (Klein and Scheunert, 1978). The chemical structure of a pesticide exerts a major influence on bioaccumulation processes and ultimately its elimination from an organism. Recently, structural features have received increased attention as a predictive measure for bioaccumulative potential. For example, some significant differences in environmental behavior of cyclodienes, particularly in metabolism, have been related to substituents (epoxy groups) and stereochemistry. BHC and PCB isomers also exhibit significant differences in environmental behavior (Klein and Scheunert, 1978; Shaw and Connell, 1983).

The bioaccumulation of pesticides from aqueous solution can be related to a simple partitioning or sorption process between animal and water. If we assume steady-state conditions, the bioaccumulation factor (BF) for pesticide (P) from aqueous solution (w) by biota (b) is given by

$$BF = \frac{[P]_b}{[P]_w} \qquad (6)$$

This is equivalent to the thermodynamic equilibrium relationship for concentrations of a compound distributed between two environmental compartments.

In the case of sorption, the Freundlich sorption equation [see Equation (2)] is

$$\log \frac{X}{m} = \frac{1}{n} \log C_w + \log K_s \qquad (7)$$

where X/m is the amount of compound sorbed per unit mass (m) of sorbent, C_w the compound concentration in water at equilibrium, $1/n$ represents the slope of the isotherm and K_s is the sorption constant. If the slope of the isotherm is assumed to be unity, which appears to be a reasonable assumption in most cases of nonionic or undissociated organic pesticides, Equation (6) becomes

$$K_s = \frac{X/m}{C_w}$$

This is equivalent to the bioaccumulation factor (BF) in Equation (7) above. Empirical relationships have been observed between distribution phenomena, as represented by BF, and sorption constants, K_s, and n-octanol/water partition coefficients for various organisms (Kenaga, 1975; Chiou et al., 1977; Ernst, 1977; Korte, 1978; also see Chapter 2).

Ellgehausen et al. (1980) measured the water to animal bioaccumulation factor for ten pesticides using algae, daphnids, and catfish. The bioaccumulation of stable nonionic pesticides can be described as sorption and was correlated with the n-octanol/water partition coefficient. Linear relationships are also observed between the partition coefficients and water solubilities of various nonionic chemicals (Chiou et al., 1977; Ellgehausen et al., 1980). Generally, as water solubility increases the partition coefficient and bioaccumulation potential in organisms decreases (see Table 7.5).

With unstable pesticides, predicted bioaccumulation factors and depuration rates may vary greatly from experimental estimates. For example, Ellgehausen et al. (1980) found that the pesticide profenofos, that is, O-(4-bromo-2-chlorophenyl)-O-ethyl-S-n-propylphos-phorothioate, was readily degraded in catfish and the calculated bioaccumulation after 24 hr exposure was less than half the value that would be expected from the n-octanol/water partition coefficient.

Kenaga (1975) suggests that initial adsorption of pesticides is often related to the surface area to mass ratio of the animal. Although Ellgehausen et al. (1980) concluded from their studies with three different organisms that the final degree

Table 7.5. Inverse Relationship of Pesticide Water Solubility to Accumulation in Fish

Pesticide	Solubility in Water (ppm)	Maximum Accumulation (Whole Fish)
Lindane	10	100 ×
Toxaphene	3	10,000 ×
Dieldrin	0.25	10,000 ×
DDT	0.0012–0.037	(100,000–1 million) ×
2,4-D	725	150 ×

Source: Hamelink et al. (1971). Reprinted with permission from the American Fisheries Society.

of bioaccumulation was related to the total biomass and not the surface area to mass ratio. Bioaccumulation factors decreased with increasing biomass at the given pesticide concentration.

Models for Bioaccumulation

The theoretical basis and limitations of bioaccumulation models used to estimate the uptake and elimination kinetics of pollutants such as pesticides are discussed in Chapter 2. Compartmental models (see Atkins, 1969) have been applied to the kinetics of organochlorine insecticide uptake and elimination by Robinson and his co-workers (e.g., Robinson and Roberts, 1968; Robinson et al., 1969; Robinson, 1973) and further developed by Moriarty (1975, 1978). The biodynamics of organochlorine insecticides in living organisms have been reviewed by Brooks (1975). Moriarty (1975) has used the compartmental model to analyze selected experimental data from the literature (see Tables 7.6 and 7.7).

Much of these data indicate that a single-compartment model, following first-order kinetics, can provide a reasonable approximation to an aquatic organism. But in many cases a better model is the two-compartment model, both compartments following first-order kinetics. Moriarty (1975) has used this latter model to explain the kinetics of uptake and depuration with several mammals. Alternatively, Ellgehausen et al. (1980) found that depuration of pesticides from aquatic organisms followed second-order reaction kinetics.

The half-lives of the pesticides noted in Tables 7.6 and 7.7, as well as the reciprocal of the rate constants of depuration, were found to be highly correlated with the lipophilicity of the pesticides. However, Moriarty (1975) concluded that bioaccumulation of organochlorines does not depend solely on passive diffusion, but many active processes of transport, metabolism, and excretion. Therefore, the rates of transfer between compartments are considered more relevant than partition coefficients.

Table 7.6. Loss of Residues from Aquatic Species After Exposure When Kept in Continuous Flow Systems[a]

Species	Pesticide	Tissue Analyzed	Initial Concentration (ppm)	Period Covered by Analyses (days)	λ (per day)	$t_{1/2}$ (days)
Cottus perplexus (reticulate sculpin)	Dieldrin	Whole fish	2.3	89	0.020	34.7
Lebistes reticulatus (guppy)	Dieldrin	Whole fish	11	30	0.0047	148
Tubifex sp. (tubificid worm)	Dieldrin	Whole worm	149	16	0.057	12.2
Amblema plicata (bivalve mollusc)	Dieldrin	Gills	0.056	84	0.022	31.6
Lampsilis siliquoidea (bivalve mollusc)	Dieldrin	Whole body except shell	16.8	84	0.039	17.6
Anodonta grandis (bivalve mollusc)	DDT	Whole body except shell	1.4	14	0.20	3.5
			1.5	28	0.053	13.1

[a] Adapted from Moriarty (1975). The number of exponential terms in all cases was one.

Table 7.7. Intake of Pesticides by Aquatic Species

Intake from Continuous-Flow Systems		Concentration in Water (pp 10^9)	Rate of intake[a] ÷ Concentration in Water (per day)	Estimated Steady-State Concentration (ppm)
Species	Pesticide			
Lepomis macrochirus (bluegill sunfish)	Endrin	0.2 2.0	756 385	
Cottus perplexus (reticulate sculpin)	Dieldrin	0.5	1,121	28
Lebistes reticulatus (guppy)	Dieldrin	0.8–1.1	1,064	230
Salmo gairdnerii (rainbow trout)	Toxaphene	0.84	1,123	
Tubifex sp. (tubificid worm)	Dieldrin	80	250	352
Daphnia magna	Dieldrin	2.1 4.5 12.8	8,790 13,200 10,350	
Amblema plicata (bivalve mollusc)	Dieldrin	0.02 20	165 75	0.055 16.9
Tetragoneuria spp. (dragonfly nymph)	^{14}C-DDT	3.6 4.0 6.1 10.8 13.3 20.0	888 850 688 611 616 790	

(continued on next page)

Table 7.7. (continued)

Species	Pesticide	Concentration in Food (mg per kg body weight per week)	Rate of Intake per day (pp 10⁹)
Salmo clarki lewisi (cut throat trout)	DDT	1	34
Leptodius floridanus (crab larvae)	Dieldrin	3	193
		0.213	10
		(ppm)	(dry weight)

Source: Moriarty (1975). Reprinted with permission from Academic Press Inc. (London) Ltd.

[a] Residues usually refer to whole organisms and rates of intake were therefore estimated by linear regressions or by one-compartment models.

Food Chain Transfer and Biomagnification

The biomagnification process, as discussed in Chapter 2, has been widely used to explain the occurrence of increases in certain pesticide residue concentrations with trophic level. The persistent lipophilic organochlorine insecticides are usually cited as substances which exhibit this phenomenon (e.g., DDT, DDD) and, in fact, few clearcut examples are found outside this group (Ware, 1980). Several classic cases include behavior of DDD residues in Clear Lake, California, 1958 (Hunt and Bischoff, 1960; Rudd, 1964), DDT in Green Bay and Lake Michigan, Wisconsin, 1964 (Hickey et al., 1966), and DDT in a Long Island salt marsh area (Woodwell et al., 1967).

In the Clear Lake investigation, it was concluded that DDD accumulated through the aquatic food chain and was magnified approximately 80,000 times in predaceous fish from a theoretical water concentration of 0.02 ppm (see Figure 7.12). For aquatic species, Moriarty (1975) has suggested that the evidence indicates that the direct intake of insecticides from the physical environment is more important than dietary intake. Residue levels in organisms appear to depend on the rates of intake and loss of pollutants rather than position in the food web.

A number of studies have demonstrated that uptake of pesticides from water is far more important than ingestion from food. For example, Reinert (1972) found that more dieldrin was absorbed directly from water than food by daphnids and guppies. Also Canton et al. (1975) concluded from their food chain studies (*Chlorella → Daphnia → Lebistes*) that the uptake of α-hexachlorocyclohexane (α-HCH) directly from water was the major route. Similarly, Macek et al. (1979) found that exposure of bluegill sunfish to di(2-ethylhexyl)phthalate, 1,2,4-trichlorobenzene, and lepthophos indicated that the steady-state body burden due to dietary exposure was not significantly different from that due to aqueous exposure. They also concluded that biomagnification via the food chain was insignificant compared to bioconcentration directly from water. Also, Ellgehausen et al. (1980) have demonstrated that only part of the residue present in the lower trophic levels was transferred to the higher levels of the food chain (Table 7.8). Direct uptake of pesticides from water was more significant for the final concentrations in aquatic organisms. In comparison, Macek et al. (1979) observed that DDT may have some measurable potential for biomagnification but added that experimental

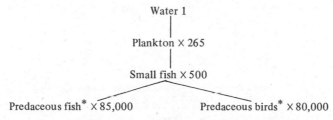

Figure 7.12.   Biomagnification of DDD from the treated water of Clear Lake, California. * Residues in visceral fat; whole-body residues for plankton and small fish. (From Rudd, 1964. Reprinted with permission from University of Wisconsin Press.)

Table 7.8.   Transfer of Residues Within the
Food Chains Algae/Daphnids (I) and Daphnids/
Catfish (II)

Compound	Transfer (%)	
	I	II
pp'-DDT	23.0	44.0
Fluorodifen	–	14.1
Terbutryn	11.9	9.1
Atrazine	3.9	9.1

Source: Ellgehausen et al. (1980). Reprinted
with permission from Academic Press Inc.

errors alone would probably exceed the incremental change in body burden from
food transfer.

Thus, current evidence indicates that biomagnification of pesticides in aquatic
food chains is not a significant phenomenon for most organisms in natural systems.
However, aquatic birds must bioaccumulate insecticides from prey organisms,
rather than their physical environment, and since concentrations are comparatively
high, biomagnification is a plausible mechanism. For terrestrial food chains, higher
concentrations of organochlorine insecticides in organisms at higher trophic levels
are widely recorded, particularly for top predators. Presumably, dietary intake is
the dominant route for pesticide accumulation although Moriarty (1975, 1978) has
pointed out that the increase may result from the slower rate of elimination in the
higher levels of the food chain.

Biotransformations

The transformations of pesticides by living organisms include biodegradation, detoxification, and metabolism and have been the subject of extensive investigation.
Most pesticides along with other pollutants are transformed in organisms by several
major reaction pathways in which the pesticide is oxidized, hydrolyzed, or reduced
(Phase I) or conjugated (Phase II), or both Phase I and Phase II reactions occur
(see Figure 7.13). In general, lipophilic pesticides are metabolized into more-water-
soluble and less-toxic metabolites.

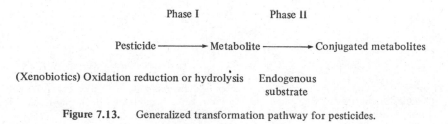

Figure 7.13.   Generalized transformation pathway for pesticides.

Table 7.9. Enzymic Systems Which Metabolize Pesticides

Enzymic System	Location	Compounds Metabolized
	Phase I Reactions	
Mixed function oxidases	Microsomes, notably from vertebrate liver and insect fat body	Many liposoluble pesticides
Phosphatases	Present in nearly all tissues and subcellular fractions of species investigated	Organophosphorus insecticides and "nerve gases"
Carboxyesterases	In most tissues of insects and vertebrates	Malathion and malaoxon
Epoxide hydrase	Microsomes – particularly mammalian liver microsomes	Epoxide – including dieldrin and heptachlor epoxide and various arene epoxides
DDT dehydrochlorinase	Found in virtually all insects and vertebrates so far investigated	pp'-DDT and pp'-DDD
	Phase II Reactions	
Glucuronyl transferases	Mainly in microsomes, notably those from liver, adrenal cortex, and alimentary tract. Widespread in vertebrates other than fish but not in insects	Many compounds with labile hydrogen including hydroxylated metabolites of dieldrin, benz-pyrene, chlorfenvinphos and biphenyl
Glutathione-S-transferases	70,000 g supernatants of vertebrate liver homogenates. Also occur in insects	Many chlorinated compounds including γ-BHC and tetrachloronitrobenzene (TCNB). Some epoxides

Source: Walker (1975). Reprinted with permission from Academic Press Inc. (London) Ltd.

Table 7.10. Some Examples of Detoxification Reactions of Pesticides by Biota[a]

Reaction	Microorganisms	Plants	Animals
Oxidation			
Aryl hydroxylation	Carbaryl	Carbaryl; 2,4-D	Carbaryl, biphenyls; Baygon
Aliphatic hydroxylation	Aldehydes, ketones; β oxidation of 2,4-dichlorophenoxyalkanoic acids	Aldrin; alkyl groups; β oxidation of MCPB	Cyclodienes; alkyl groups; DDT; p-nitrotoluene and other alkyl-benzenes; rotenone, pyrethrins
Dealkylation	–	Malathion, Abate	Malathion; parathion; Baygon; methoxychlor
N-dealkylation	Alkylamines, carbaryl; zectran, simazine	Phenylureas, triazines; diphenamid; trifluralin	N-methyl- and N,N'-di-methyl carbamates, Bidrin
N-methylhydroxylation	–	Bidrin, azodrin	N-methyl carbamates; Bidrin
O-desulfuration	–	Phosphorothioates and dithioates	Phosphorothionates
Sulfoxidation	Temik; Migamoto; fenitrothion, sulfothion	Temik, Abate, Mesurol, Phorate	Temik; Disyston
Epoxidation	Cyclodienes	Cyclodienes	Cyclodienes
Epoxidation hydration	Dieldrin	Dieldrin	Cyclodienes
Ether cleavage	–	2,4-D	Benzodioxoles: piperonyl butoxide, sesamex
Hydrolysis			
Ester hydrolysis	Dichlorvos, diazinon, parathion, DFP, phorate Hinosan, malathion Trichlorafon, Carbaryl	Organophosphate insecticides and fungicides; 2,4-D; carbamate ester	Organophosphate and carbamate insecticides
Amide hydrolysis	–	Propanil; CMPT	Dimethoate
Reduction			
Dechlorination			
Reductive	DDT: s-triazine; 2,4-D	Not common; γBHC; 2,4-D; 2,4,5-T	DDT: γBHC
Oxidative			Cyclodienes
Dehydrochloration			
Anaerobic	DDT	Not common	DDT
Oxidative	–	–	Photodieldrin
Reductive	EPN; denitrothion	C-6989; trifluralin	Parathion, EPN
Ring Cleavage	2,4-D; DDT, simazine		

[a] Adapted from Khan et al. (1975).

Reactions of both phases are catalyzed by enzymes occurring in the plasma or various organs, such as the liver. These enzymes may be distributed in the microsomal subcellular fraction which is derived from the endoplasmic reticulum, the mitochondria, or are present in cell sap (Adamson, 1974). Enzymic systems which metabolize pesticides are summarized in Table 7.9.

Major detoxification pathways mediated by these enzyme systems include initial metabolic alterations such as oxidations and hydrolyses (Khan et al., 1975). Table 7.10 sets out examples of major detoxification reactions for pesticides in microorganisms, plants, and animals but in many cases, detoxification may involve more than one reaction pathway.

The polar metabolite resulting from the initial detoxification of a pesticide may be conjugated before excretion (Walker, 1975). Several pathways for the conjugation and elimination of pesticides and their metabolites from biota exist. Microorganisms do not perform conjugation reactions because excretion takes place through the cell surface. However, plants do not excrete pesticides and their metabolites. Instead, they conjugate these substances with endogenous compounds and deposit them in metabolically inactive sites in the cell (e.g., vacuoles) (Khan et al., 1974). Animals excrete conjugates or pesticides primarily in urine and bile but other means include eggs, milk, and sweat (Walker, 1975).

Rates of these various reactions determine selective toxicity of a pesticide to different organisms and are also related to differences in the activities of detoxifying enzyme systems (Khan et al., 1974; 1975). The presence and activity of these enzyme systems in major phylogenetic groups of nontarget organisms is a critical factor in the susceptibility or resistance of these organisms to xenobiotic exposure.

The formation of biologically active metabolites of certain broad-spectrum pesticides, for example, DDT, aldrin, heptachlor, parathion, and malathion, in various biota may be of ecological significance. This needs to be kept in mind in the evaluation of the effects of residues (see, e.g., O'Brien, 1967; Khan et al., 1974, 1975).

TOXIC EFFECTS ON ORGANISMS

Toxicity Mechanisms

For many pesticides the primary site of toxicological action is well defined although, generally, the detailed modes of action are poorly understood (Corbett, 1974). Figure 7.14 presents a summary of the major modes of pesticide action and interrelationships between these actions. It should be noted that pesticides with a nonspecific mode of action, for example, those acting as indiscriminate enzyme inhibitors, are not included in this summary.

The economically important insecticides substantially act by interfering with the passage of impulses in the nervous system. Primary sites of action for different types of neuroactive insecticides are believed to involve either (1) axonal transmission, (2) the acetylcholine receptor, or (3) acetylcholinesterase.

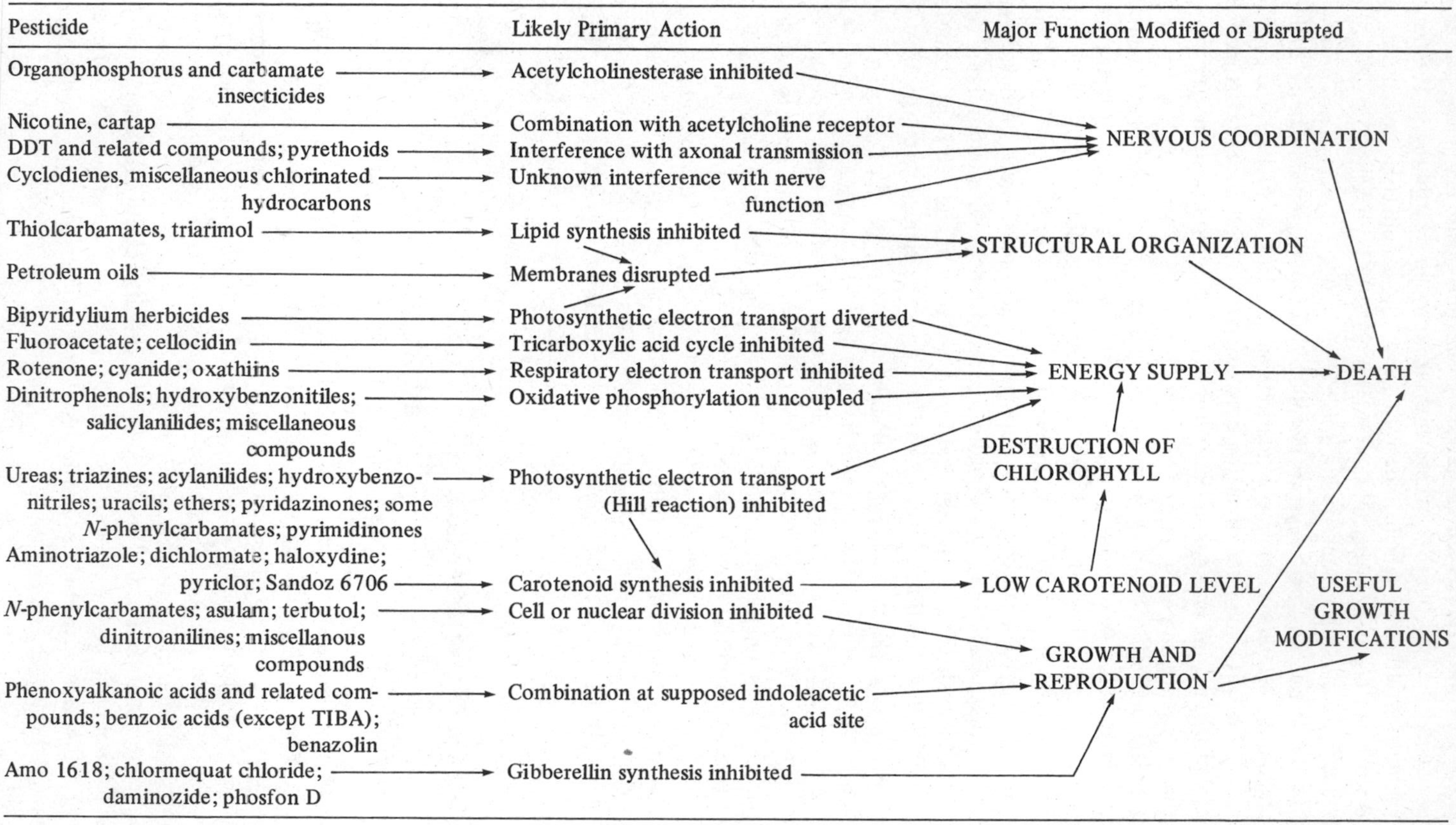

Pesticide	Likely Primary Action	Major Function Modified or Disrupted
Organophosphorus and carbamate insecticides	Acetylcholinesterase inhibited	NERVOUS COORDINATION
Nicotine, cartap	Combination with acetylcholine receptor	
DDT and related compounds; pyrethoids	Interference with axonal transmission	
Cyclodienes, miscellaneous chlorinated hydrocarbons	Unknown interference with nerve function	
Thiolcarbamates, triarimol	Lipid synthesis inhibited	STRUCTURAL ORGANIZATION
Petroleum oils	Membranes disrupted	
Bipyridylium herbicides	Photosynthetic electron transport diverted	ENERGY SUPPLY
Fluoroacetate; cellocidin	Tricarboxylic acid cycle inhibited	
Rotenone; cyanide; oxathiins	Respiratory electron transport inhibited	
Dinitrophenols; hydroxybenzonitiles; salicylanilides; miscellaneous compounds	Oxidative phosphorylation uncoupled	
Ureas; triazines; acylanilides; hydroxybenzo-nitriles; uracils; ethers; pyridazinones; some N-phenylcarbamates; pyrimidinones	Photosynthetic electron transport (Hill reaction) inhibited	DESTRUCTION OF CHLOROPHYLL
Aminotriazole; dichlormate; haloxydine; pyriclor; Sandoz 6706	Carotenoid synthesis inhibited	LOW CAROTENOID LEVEL
N-phenylcarbamates; asulam; terbutol; dinitroanilines; miscellanous compounds	Cell or nuclear division inhibited	GROWTH AND REPRODUCTION
Phenoxyalkanoic acids and related compounds; benzoic acids (except TIBA); benazolin	Combination at supposed indoleacetic acid site	
Amo 1618; chlormequat chloride; daminozide; phosfon D	Gibberellin synthesis inhibited	USEFUL GROWTH MODIFICATIONS

DEATH

Figure 7.14. General modes of action of pesticides. (Adapted from Corbett, 1974. Reprinted with permission from Academic Press, Inc. (London) Ltd.)

Chlorinated Insecticides

Several mechanistic theories have been postulated to explain primary mechanisms of action of chlorinated insecticides on insects, fish, birds, and mammals. Structural and steric interactions with receptor sites in nerve membranes and biomolecules are primarily involved. For example, the primary mode of insecticidal action exerted by DDT is related to its binding to the nerve membrane resulting in interference with ion movement into and out of the axon. Structure/activity investigations suggest that the base of the DDT wedge, the CCl_3, moiety or a similar-sized substituent group, binds to the lipoprotein of an axonal membrane. This allows the DDT apex to hold open a gate for sodium ions and thus interfere with ion movement (Corbett, 1974). Similarly, other chlorinated insecticides, for example, cyclodienes and HCH compounds, act on the nervous system in a manner related to well-defined structural forms and arrangements (see Brooks, 1975). However, the mechanisms of toxic action for this class of compounds remain inadequately understood.

Organophosphates and Carbamates

The potency and selective toxicity of organophosphorus insecticides are fundamentally attributed to their inactivation of the enzyme, cholinesterase (ChE) (see De Bruin, 1976). This enzyme breaks down the neurotransmitter, acetylcholine (ACh), at the synapse immediately its function is completed.

Essentially, organophosphorus insecticides, $R_1R_2P(O, S)X$, combine with ChE by attachment through the electrophilic P atom (see Figure 7.2) to the esteratic site of the enzyme-active center, to form an irreversible enzyme–inhibitor complex. The inhibitor–enzyme complex subsequently dissociates, releasing group X as HX and leaving the phosphorylated enzyme. This enzyme may eventually be hydrolyzed, eliminating the phosphoryl moiety and thus regaining normal enzyme activity. This hydrolysis generally proceeds at such an extremely slow rate that blockage of ChE by efficient inhibitors is comparatively irreversible (De Bruin, 1976).

Organophosphorus insecticides are ganglionic rather than axonic poisons (Brown, 1978), causing an initial stimulation of cholinergic transmission followed by depression which terminates in paralysis of all nerve synapses and motor nerve endings. The potency of organophosphorus insecticides is expressed in terms of the molar concentration of inhibitor yielding 50% loss of the ChE activity (ID_{50}). This measure is significantly correlated with mammalian toxicity (De Bruin, 1976). Brown (1978) points out that the inhibiting potency of an OP compound is decided not only by its two alkyl substituents (R_1 and R_2, usually methyl or ethyl), but also by the much larger third substituent (X usually phenyl, heterocyclic or aliphatic).

Carbamate insecticides have an analogous toxic action to the organophosphorus insecticides. Carbamates may initiate a blocking action by covalent binding of the electrophilic carbamyl groups (carbamylating, cf. phosphorylating) to the esteratic site of the ChE. In contrast to the organophosphorus insecticides, however, carbamates are reversible inhibitors of ChE since the active enzyme can be regenerated from the inhibitor–enzyme complex.

Herbicides

Mechanisms of herbicidal action in plants can be broadly described by two fundamental physiological effects: (1) action on growth processes and (2) action on photosynthesis. For effects on growth processes, successful herbicidal properties include stability and mobility of the parent chemical or its toxic reaction products within a plant. Mechanisms of action are summarized in Figure 7.15.

The phenoxyacid herbicides are the most widely used class of herbicides which cause growth reactions. These compounds, such as 2,4-D, 2,4,5-T, and MCPA, chemically simulate natural auxins (growth hormones), except that they are far more active and persistent, being translocated into all cells and upsetting normal hormone balances. Cell proliferation is uncontrolled, axis growth is abnormal, and special growth is inhibited (Brown, 1978). Tumor development can occur as a result of stimulatory action on the synthesis of DNA and proteins. Roots lose some of their ability to take up salt and water and softening of the root cortex occurs. Photosynthesis is reduced and phloem transport of food material is inhibited (Crafts, 1964). Inactivation of auxin herbicides may occur through decarboxylation and certain plant species may develop tolerance to auxin herbicides via these detoxification mechanisms (Crafts, 1964).

The other basic mode of physiological action involves inhibition of photosynthesis. In particular, the substituted urea herbicides interfere with the participation of water in the noncyclic photosynthetic electron flow. The basic mode of action is the inhibition of Photosystem II in the chloroplast, in which electrons are removed during the oxidation of water to molecular oxygen (Hill reaction). This prevents the synthesis of ATP and, ultimately, NADH, and results in the development of chlorosis followed by collapse of the young leaves. Similar effects are induced by the triazines, another class of Hill reaction inhibitors. The bipyridylium (quarternary ammonium) compounds such as diquat and paraquat also act on photosynthetic pathways.

Synergism and Antagonism

Multiple exposure to pesticides, within or between classes, may induce synergistic or antagonistic biological effects. In addition, interaction with other pollutants and environmental parameters may further alter biological activities (Ware, 1980). Much of the information on pesticide interactions, however, is focused on terrestrial invertebrates (notably insects) and mammals, although some work has been done with aquatic animals (Livingston, 1977).

Substantial potentiation in mammalian toxicity has been observed during combined treatment using EPN with malathion and its oxygen analog, malaoxon. EPN competes with malathion for its carboxyester-cleaving enzyme, malathionase, thus impeding hydrolysis of malathion dicarbethoxy ethyl ester linkages and resulting in increased persistence of malathion within the organism (De Bruin, 1976). Many other examples of potentiation with organophosphate insecticides are reported.

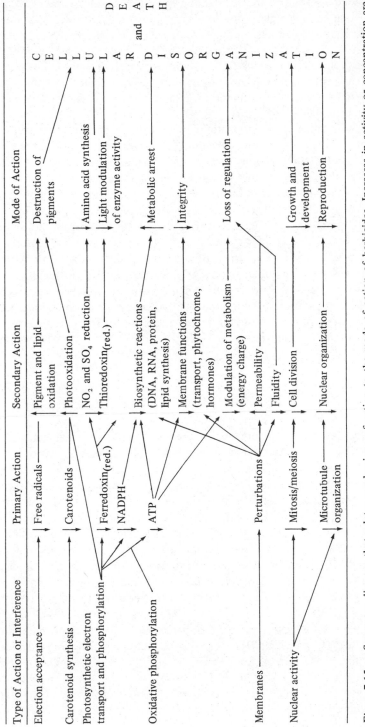

Figure 7.15. Summary diagram that relates mechanisms of action to the mode of action of herbicides. Increases in activity or concentration are indicated by arrows pointing up and decreases are shown with arrows pointing down. (From Moreland, 1980. Reprinted with permission from Annual Review, Inc.)

Some pesticide formulations, containing pyrethrins, carbamates, DDT, and cyclodienes, contain synergists, for example, piperonylbutoxide, sesamin, iso-thiocyanates, chloroethylethers, and dicarboximides, to potentiate their activity toward target insects, although this induces increased mammalian susceptibility. Methylenedioxyphenyl synergists (e.g., piperonylbutoxide) act by blocking the enzyme system responsible for detoxification of the insecticide (De Bruin, 1976). Thus, the basic mechanism of potentiation relates to the inhibition of detoxifying enzymes (Seume and O'Brien, 1960).

In comparison, antagonistic interactions involve the inhibition by one pesticide of the enzyme responsible for the activation of another. Antagonistic interactions of pesticides include combinations such as parathion and malathion and dipterex plus EPN (De Bruin, 1976). Microsomal enzyme inducers (e.g., organochlorine insecticides) usually interact with organophosphorus insecticides to exhibit acceler-ated deactivation. Substituted urea herbicides also confer protection against para-thion toxicity, probably due to increased MFO detoxication of parathion (De Bruin, 1976).

For nontarget organisms an array of synergistic and antagonistic pesticide combinations has been described particularly for fish, birds, and mammals. The majority of these results are derived from laboratory bioassay studies which are limited in their application to potential effects on natural populations. Global evidence of multiple pesticide residues in tissues and organs from nontarget species indicates the need to understand the biological implications of pesticide inter-actions.

Liver Enzyme Induction

Some pesticides, especially the organochlorine insecticides, are capable of inducing increased activity of hepatic enzyme systems, for example, mixed-function oxidase induction in mammals, birds, and fish at extremely low levels of intake. This phenomenon has the potential for influencing several environmental effects in-cluding (1) synergistic or antagonistic effects through stimulation of enzyme systems responsible for metabolizing other pesticides; (2) antagonistic storage of lipophilic insecticides in animal tissues due to accelerated metabolic detoxifi-cation which may result in lowered chronic toxicity for certain insecticide com-binations; and (3) increased hormone turnover by induced enzymes, causing disturbances to endrocrine relationships which may lead to physiological aber-rations in certain bird species, for example, thin-eggshell phenomenon (Ware, 1980).

Cyclodiene epoxides have the lowest apparent threshold (1 ppm) for inducing hepatic enzymes. The value for DDT is slightly higher, while other organochlorines (heptachlor, toxaphene, aldrin, methoxychlor) are not active below about 5 ppm in the diet (Ware, 1980). Duggan and Weatherwax (1967) have reported that organochlorine insecticide residues in diets of wild birds can range up to 10 ppm although current exposure levels suggest much lower intake values. Nevertheless, existing environmental residue levels for certain organochlorine insecticides support the suggestion (Ware, 1980) that significant enzyme induction probably occurs in some wildlife species.

Lethal and Sublethal Effects of Pesticides on Organisms

Pesticides, by virtue of their design and application, induce a broad spectrum of selective and nonselective biocidal effects influencing all taxonomic groups of organisms (Cope, 1971; Pimentel, 1971; Matsumura et al., 1972; Stickel, 1973; Pimental and Goodman, 1974; Brown, 1978; Ware, 1980). There is a considerable variability in species sensitivity to a particular pesticide as well as variation in the toxicity of different pesticides to a particular species. Other factors such as sex, age, nutritional state, stress, and habitat or microenvironment vary individual sensitivity.

For most aquatic and terrestrial species, the tolerances and sensitivities to individual pesticides have been investigated by bioassay procedures. Numerous methodological limitations exist for the application of results to natural systems. Recent interest in comparative methodologies for test organisms has developed because of the considerable variability in experimental conditions. For example, it is usual to find that continuous-flow or intermittent-flow tests on fish give lower LC_{50} values than static tests (Brown, 1978). Other critical factors include duration of exposure (e.g., proportion of life span of organism exposed or multi-generational), as well as route of pesticide uptake (e.g., oral or surface absorption). In addition, there is the difficulty of establishing quantitative relationships between short-term tolerance studies and longer-term effects resulting from sublethal exposures. Also, few studies relate tissue levels with the development of toxic effects.

Lethal Toxicities

Estimates or indices of the lethal toxicities of pesticides to organisms are derived from field observation and bioassay data. Field studies usually relate dosage rates (e.g., lb acre^{-1}) to differences in the number of species and individuals between treated and control areas. Most of these studies involve plant communities, soil organisms, and arthropod fauna of plant communities.

A vast array of experimental data is available, obtained under various non-standardized test conditions using many different test species. But comparable field or bioassay data exist for few species and pesticides. Some selected examples of comparable acute toxicity data are shown in Tables 7.11 and 7.12 for terrestrial species and in Tables 7.13 and 7.14 for aquatic species. The comparative suscepti-bility of twelve freshwater fish from five families to selected organochlorine, organophosphorus, and carbamate insecticides as measured by average rank order, has been given by Hurlbert (1975) as: brown trout (most sensitive), rainbow trout, largemouth bass, coho salmon, yellow perch, redear sunfish, bluegill, carp, black bullhead, channel catfish, fathead minnow, and goldfish.

Table 7.11. Acute Oral Toxicities as LD$_{50}$'s[a] (mg per kg body weight) of Pesticides to Birds[b]

Pesticide Group	Ring-Necked Pheasant	Coturnix Quail	Chukar Partridge	Mallard Duck	Rock Dove (Pigeon)
Organochlorines	1.8–1300	70–840	23	33–>2400	27–>4000
Organophosphorus	1.4–75	3.7–84	6.5–270	0.6–1190	2.0–50
Carbamate	4.2–>2000	3.2–2290	5.2–237	0.4–2180	60–2000
Herbicides	472–>2000	–	–	564–>2000	–
Fungicides	–	55–>2000	–	56–>2800	–

[a] 14-day observation period, oral dose in capsules.
[b] Compiled from Tucker and Crabtree (1970), Tucker and Haegele (1971), Heath et al. (1972), and Pimentel (1971).

Table 7.12. Acute Oral LD_{50} (mg kg^{-1}) of Pesticides (Organochlorine, Organophosphorus, and Carbamate Compounds) to Large Quadrupeds[a]

	Mule Deer	Domestic Goats
OC Compound		
Endrin	–	25–50
Dieldrin	75–100	100–200
Toxaphene	139–240	> 160
OP Compound		
Demeton	–	13
Parathion	33	42
Monocrotophos	38	35
Dimethoate	> 200	–
Chlorpyrifos	–	> 500
Fenitrothion	727	–
Carbamate		
Aminocarb	11	–
Methomyl	16	–
Mexacarbate	25	22
Propoxur	225	> 800
Carbaryl	300	–

[a] Adapted from Brown (1978).

Comparisons between major taxonomic groups or classes lack an unequivocal quantitative base but a semiquantitative assessment of relative acute toxicities of insecticides and herbicides to groupings of organisms is useful from an ecological view (Table 7.15). From the overall pattern of pesticide toxicities the following order of lethal toxicities to nontarget organisms is apparent: insecticides > herbicides > fungicides. But there are considerable variations in species tolerances and sensitivities within pesticide classifications.

Aquatic Species. The most obvious and acute effects of pesticides on nontarget organisms, particularly fish, have been caused by broad-scale applications of insecticides resulting in large mortalities or "kills." Organochlorine insecticides are highly toxic to many aquatic organisms. For example, with fish species, most 96 hr LC_{50} values range from 1 to 200 ppb while other pesticides tend to exhibit 96 hr LC_{50} values above this range. Organophosphorus and carbamate insecticides, however, are toxic to many aquatic invertebrates with 96 hr LC_{50} values often in the low parts per billion range (see Table 7.13). Acute toxicities of herbicides to aquatic species are generally in the parts per million range (see Table 7.14) with some exceptions including diuron, dinitrocresol, and endothal. Butler (1963) has demonstrated that organochlorine insecticides suppress photosynthesis in natural estuarine phytoplankton assemblages at the relatively high concentration

Table 7.13.   Acute Toxicities of Organochlorine and Organophosphorus Insecticides to Various Organisms (Parts Per Billion)

Insecticide	Estuarine Phytoplankton (Photosynthesis Reduction %)[a]	Daphnia (48 hr EC$_{50}$)[b]	Freshwater Amphipod[c] (96 hr LC$_{50}$)	Three Marine Decapods[d] (96 hr LC$_{50}$)
Organochlorine Insecticides				
Aldrin	85	28	9,800	8–33
BHC	–	–	–	–
Chlordane	94	29	26	–
DDT	77	0.36	1.0	0.6–6
Dieldrin	85	250	460	7–50
Endrin	46	20	3.0	1.7–12
Heptachlor	94	42	29	8–440
Lindane (γBHC)	28	460	48	5–10
Methoxychlor	81	0.78	0.8	4–12
Mirex	42	–	–	–
Thiodan	87	–	5.8	–
Toxaphene	91	15	26	–
Organophosphorus Insecticides				
Abate	–	–	82	–
Baytex	7	0.80	8.4	–
Chlorthion	–	–	–	–
Diazimon	7	0.90	200	–
Dibrom	56	0.35	110	–
Dichlorvos	–	0.066	0.50	4–45
Dursban	–	–	0.11	–
EPN	–	–	15	–
Fenitrothion	–	–	–	–
Guthion	0	–	–	–
Malathion	7	1.8	1.8	33–88
Methyl parathion	–	–	–	2–7
Parathion	–	0.60	0.60	–
Phosdrin	–	0.16	0.16	11–69
Phosphamidon	–	8.8	8.8	–
Systox	7	–	–	–
TEPP	–	–	–	–

Source: Hurlbert (1975). Reprinted with permission from Springer-Verlag.

[a] Percentage decrease in productivity of a natural estuarine phytoplankton assemblage during a 4 hr exposure to 1000 ppb.
[b] EC = concentration required to immobilize 50% of test organisms.
[c] *Gammarus lacustris*.
[d] Sand shrimp (*Crangon septemispinosa*) (most susceptible), grass shrimp (*Palaemonetes vulgaris*), and hermit crab (*Pagurus longicarpus*).

of 1000 ppb (4 hr exposure). Amphibia such as frogs, toads, and their tadpoles are considerably less susceptible to insecticide poisoning than fish (Brown, 1978). Within pesticide classes a wide range of tolerances is usually found for individual species.

Terrestrial Species.   Generally, insecticides are more toxic to warm-blooded animals than herbicides or fungicides as suggested by LD$_{50}$ values for the laboratory

Mosquito Larvae[e] (24 hr LC$_{50}$)		Stonefly Naiad[f] (96 hr LC$_{50}$)	Freshwater Minnow[g] (96 hr LC$_{50}$)	Twelve Freshwater Fish (96 hr LC$_{50}$)	Seven Estuarine Fish (96 hr LC$_{50}$)	Tadpoles[h] (96 hr LC$_{50}$)
5.3	—	1.3	—	—	5–100	150
—	—	—	—	—	—	3,200
—	—	15	69	—	—	—
70	10	7.0	34	2–21	0.4–89	1,000, 800*
7.9	3	0.50	16	—	0.9–34	150, 100*
15	—	0.25	1.3	—	0.05–3.1	120, 180*
5.4	—	1.1	56	—	0.8–194	440
27	—	4.5	56	2–131	9–66	4,400, 2,700*
67	—	1.4	35	—	12–150	330
—	—	—	—	—	—	—
—	—	—	—	—	—	—
—	—	2.3	5.1	3–18	—	140, 500*
1.6	5	10	—	—	—	—
4.2	2.3	4.5	—	980–3,404	—	—
25	—	—	3,200	—	—	—
83	—	25	—	—	—	—
—	—	8.0	—	—	—	1,700*
75	110	0.10	—	—	200–2,680	—
2.8	6.3.	10	—	—	—	—
4.4	—	—	250	—	—	—
5.6	25	—	—	—	—	—
—	—	1.5	—	4–4, 270	—	130
80	320	10	12,500	101–12,900	27–3, 250	420, 200*
18	6.5	—	7,500	2,750–9,000	5,200–75,800	—
3.2	3.1	5.4	1,600	—	—	1,000*
—	—	5.0	—	—	65–800	—
—	—	150	—	—	—	—
—	—	—	4,200	—	—	—
—	—	—	1,000	—	—	—

[e] First column: *Culex fatigans*; second column: *Anophetes albimanus*.
[f] *Pteronarcys california*.
[g] Fathead minnow (*Pina phales promelas*).
[h] *Bufo woodhousii* (no asterisk) and *Pseudacris trisciata* (asterisk).

rat, *Rattus norvegicus* (Pimental, 1971) (see also Table 7.12). Brown (1978) suggests that nearly all the compounds developed as herbicides have a low toxicity to warm-blooded animals and are unlikely to have a direct toxic effect on animals in natural systems. With insecticides the cyclodienes are particularly toxic (LD$_{50}$ values < 100 mg kg^{-1}) to mammals, along with organophosphorus insecticides, such as parathion, azinphosmethyl, phosphonidon, and methyl parathion and the carbamate, mexacarbate.

Feeding studies using game birds indicate the sensitivity of avian species to many organochlorine and organophosphorus insecticides often having LD$_{50}$ values

Table 7.14. Acute Toxicities of Herbicides to Various Aquatic Organisms (Parts Per Million)[a]

Herbicide	Daphnia[b] (EC$_{50}$)	Freshwater Amphipod (96 hr LC$_{50}$)	Stonefly Naiad (96 hr LC$_{50}$)	Bluegill[c] (48 hr LC$_{50}$)	Tadpole[d] (96 hr LC$_{50}$)
Amitrole	9.8, 0.014	—	—	—	—
Dalapon	23	—	>100	115	—
Dicamba	11*	3.9	—	130	—
Dichlobenil	—	11	7.0	20	—
Dichlone	9.8, 3.7*	1.1	—	—	—
Dinitrocresol	0.014	—	0.32	—	—
Diquat	—	—	—	91*	—
Diuron	7.1	0.16	1.2	17*	—
2,4-D acid	47, 1.4*	—	15	—	—
2,4-D butoxyethanol ester	>100	0.44	1.6	2.1	—
2,4-D dimethylamine salt	—	>100	—	166	100
2,4-D isooctyl ester	—	2.4	—	8.8	—
Endothal	—	>100	—	0.257	—
Fenac	—	4.5	12	55	—
Fenuron	—	—	—	58**	—
IPC	10*	10	—	82*	—
Monuron	106	—	—	1.8**	—
Naphtha	3.7*	0.84	2.8	—	—
Paraquat	11, 3.7*	11	>100	400	28, 26*
Picloram	—	27	48	26.5*	—
Silvex	100, 2*	—	0.34	83*	10, 10*
Simazine	—	13	—	118	—
Sodium arsenite	6.5, 1.8*	—	38	58*	—
Trifluralin	0.24*	2.2	3.0	8.4	0.10*
2,4,5-T	—	—	—	0.50	—

[a] Adapted from Hurlbert (1975).

[b] 48 hr immobilization concentrations at 15.6°C (*) and 26-hr immobilization concentrations at 21.1°C.

[c] 24 hr (*) and 96 hr (**).

[d] Fowler's toad (*Bufo woodhousii*) and Western shovel frog (*Pseudacris triseriata*)

Table 7.15.   Relative Toxicities[a] of Pesticides to Aquatic Animals

Pesticide Type	Plankton	Shrimp	Crab	Oyster	Fish
Herbicide	1	1	1	1	1
Organochlorine compounds	× 3	× 300	× 100	× 100	× 500
Organophosphorus compounds	× 0.5	× 1000	× 800	× 1	× 2

Source: Thomson (1970).
[a] Measured as induced mortalities.

under $100 \, mg \, kg^{-1}$, as shown in Table 7.11. Less sensitivity is exhibited with herbicides and fungicides. Wild populations of birds can suffer considerable mortalities from insecticide exposure.

Sublethal Toxicities

Sublethal effects of pesticides may indirectly lead to reduced chances of survival or reproduction in natural populations, at concentrations most likely to be experienced by nontarget organisms. However, it can be difficult to distinguish between an adverse effect induced by the presence of a sublethal xenobiotic chemical and an adaptive response by an organism or system. This becomes more apparent when effects are measured within natural populations. Hence there is a need to establish criteria which permit evaluation of significant adverse changes in populations with sublethal exposure.

Specific sublethal effects are numerous and diverse, and relate to a broad spectrum of physiological and behavioral responses, such as alterations in enzyme production, growth rate, reproduction, behavior and activity, production of tumors, and teratogenic effects (Hurlbert, 1975; McEwen and Stephenson, 1979). Much of the available information refers to the organochlorine pesticides.

Jefferies (1975) has suggested that most of the sublethal effects of organochlorine insecticides in mammals and birds in laboratory and field situations may be due to lesions on the thyroid, producing hyper- and hypothyroidism. Of course, sublethal effects on the transmission of nervous impulses are well established.

The widespread use of organochlorine insecticides has resulted in population declines in some bird populations due primarily to behavioral and physiological changes induced by chronic sublethal exposure to these substances in their diet (McEwen and Stephenson, 1979). In general, the most important effect has been the death of embryos due primarily to premature breakage of thin eggshells and secondly, to lethal pesticide levels in the embryos (see Table 7.16) (Ware, 1980). The relationship between DDE concentration and eggshell thickness is shown for several species in Figure 7.16. McEwen and Stephenson (1979) suggest that the following range of factors may be involved: (1) laying of eggs with thin eggshells (most species); (2) reduced clutch size; (3) high egg loss during incubation; (4) egg breakage; (5) high death rate of embryos; (6) high death rate in unhatched

Table 7.16.   Eggshell Thickness Changes (pre-DDT vs. DDT Era) and Mean Organo-chlorine and PCB Residues in Eggs of Great Lakes Birds[a]

Bird Species	% Change in Thickness Index[b]	DDE (ppm)	DDT + TDE (ppm)	Dieldrin (ppm)	PCB (ppm)
Great blue heron	− 25*	22	2	0.5	8
Red-breasted merganser	− 23*	44	3	0.9	84
Common merganser	− 15*	24	1	1.6	56
Double-crested cormorant	− 15*	8	0.2	0.3	5
Red-necked grebe	− 13*	54	3	0.7	62
Herring gull	− 14*	166	2.5	1.0	171
Black tern	− 10*	5	0.3	0.3	9
Black-crowned night heron	− 10*	5	0.4	0.6	5
Common egret	− 7*	19	1.5	0.3	7
White pelican	− 6	2	0.1	0.1	1
Hooded merganser	− 4	2	0.1	0.1	3
American bittern	− 3	0.5	0	.01	1
Pied-billed grebe	− 1	7	0.5	0.2	7

Source: McEwen and Stephenson (1979).

[a] Converted from lipid weight basis and rounded to significant figures.

[b] Rounded to nearest whole number. Asterisk indicates significant thinning.

chicks at pipping; (7) high mortality of chicks; (8) late nesting and unusual nesting behavior including aberrant parental behavior and destruction of eggs; and (9) large proportion of adults in the population.

Physiological factors cited to explain this phenomenon (Ware, 1980) are: (1) reduced levels of carbonic anhydrase; (2) inhibition of calcium metabolism; (3) induction of hepatic microsomal enzymes that metabolize steroid hormones;

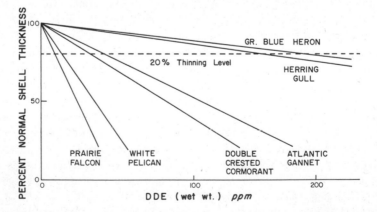

Figure 7.16.   Relationship between DDE content and eggshell thickness in six bird species. [From Brown (1978). Reprinted with permission from John Wiley & Sons, Inc.]

(4) hypothyroidism; and (5) ATP-ase inhibition. DDT, DDE, and dieldrin can interfere with some of the major functions involved in eggshell production. McEwen and Stephenson (1979) have summarized this evidence according to physiological sites where malfunctions may be critical due to interference with: (1) ATP-ases essential for the transport of calcium across cell membranes; (2) vitamin D needed for the absorption of calcium from the intestinal tract; (3) the hormones controlling the deposition of calcium in medullary bone; (4) the parathyroid and/or ultimobranchial glands controlling mobilization of calcium from medullary bone; and (5) interferences with carbonic anhydrase that controls the supply of carbonate ions in the shell gland.

Jefferies (1975) has attributed effects on eggshell thickness to changes in thyroid activity with accompanying effects on the carbonic anhydrase system and the activity of the pituitary. Correlations between effects on these parameters in various bird species and treatments with DDT, DDE, and dieldrin have been observed. Two species of birds, the chicken and the bengalese finch, which do not show thin-shelled eggs on dosing with DDT, exhibit marked hyperthyroidism following DDT treatment. Dieldrin treatment also produces little or no eggshell thinning in experimental studies with various species. McEwen and Stephenson (1979) concluded that although there is some evidence that the thyroid or parathyroid may be involved, current data are inconclusive and the physiological mechanism responsible for this phenomenon remains to be established.

Sublethal effects in animals may be more pronounced during sensitive stages of their life cycles. For example, Jefferies (1975) has predicted that marked changes in thyroid activity due to organochlorines can lead to adverse effects in: (1) active mammals during winter when thyroid activity is necessarily increased; (2) hibernating mammals after they emerge from hibernation in spring during which thyroid activity reaches a peak; and (3) birds where changes in thyroid activity are known to affect both the onset of moulting and feather structure. Adverse effects on the thyroid and pituitary may persist after organochlorines are removed from the diet. In addition, they can affect the thyroid of the embryo by passage through the placental barrier (Jefferies, 1975).

Among the herbicides, the chlorinated phenoxyacid 2,4,5-T is capable of long-term behavioral teratological effects in rats following a single prenatal exposure toward the end of the first week of pregnancy. Oral doses ($15-40$ mg kg^{-1}) produce a long-standing taste aversion, and acute doses (100 mg kg^{-1}) affect food and water intake and cause widespread changes in electrolyte distribution, with strongest effects on brain and kidney. As with organochlorine insecticides, thyroid lesions are implicated as offspring of 2,4,5-T-treated rats showed changes in weight development, thyroid activity, and brain serotonin, a transmitter hormone (Sjoden and Soderberg, 1978).

Fish exposed to sublethal concentrations of many different types of pesticides exhibit changes in physiological actions, failures in reproduction, and other effects (Holden, 1973; Brown, 1978; McEwen and Stephenson, 1979). In the early stages of pesticide poisoning, fish usually show increased activity but this subsequently decreases until death occurs (Holden, 1973). The probable mechanism of action is an initial period where the peripheral nervous system is affected and the blood–brain barrier prevents a significant penetration of pesticides into the brain (Holden, 1973). Entry into the central nervous system resulting in death follows later. Where sublethal concentrations are maintained over long periods, many subtle changes in fish behavior, resistance, susceptibilities, biochemistry, morphology, and reproduction become apparent. Reported effects in fish include: (1) alterations in learning ability or responses to natural stimuli (e.g., response to salinity and temperature regimes; reduced swimming performance and stamina under physical stress; changes in growth rates); and (2) physiological and biochemical changes (e.g., modified brain ChE and electrolyte concentrations; histopathological abnormalities in gills, liver, kidneys, and blood vessels; accumulation of pesticides, notably organochlorines, in body tissues and organs and transfer into fish eggs; marked reductions in egg hatchings and high mortalities in sac-fry). In contrast, several species of fish are reported to have developed resistance to pesticides. The mosquito fish, in particular, is tolerant to most organochlorine insecticides and some of the organophosphorus compounds. Several species of fish have also demonstrated avoidance reactions to lethal or sublethal concentrations of some insecticides and herbicides but not others (see Holden, 1973; Brown, 1978).

Chronic exposure to cholinesterase inhibitors produces neurotoxic symptoms in vertebrates, associated with reduction of brain or serum cholinesterase. However, attempts to correlate apparent behavioral and physiological effects in mammals, fish, and birds with cholinesterase levels have been largely unsuccessful due to variable and inconsistent responses. For example, bluegills becoming moribund in 750 ppb of parathion showed only 25% inhibition, while those moribund in 20 ppb showed 57% inhibition, and even 90% inhibition (Holden, 1973).

ECOSYSTEM RESPONSE

Populations and Communities

Individual species and organisms in the natural environment differ widely in sensitivity to any pesticide. This variation in response means that a pesticide can eliminate susceptible individuals from a population or an entire susceptible species from a community of organisms (Pimentel and Goodman, 1974). A considerable literature exists describing the effects of pesticides on populations and communities of organisms under field conditions. Major effects of pesticides on animal and insect populations result primarily in significant changes in species abundance and associated shifts in population dynamics. These can be summarized (Dempster, 1975) as follows:

1.   *Reductions in Populations Caused by Direct Toxic Effects, Secondary Poisoning, and Elimination of Prey Organisms.*   The magnitude and duration of direct toxic effects may vary. Relatively insignificant changes in density levels may occur due to factors such as only a localized effect on the total population or a capacity for rapid recolonization and reproduction in response to a density change.

Table 7.17.   Changes in Pest Status in Central American Cotton

	Pesticide Application/Season	
0–a few (1950)	8–10 (1955)	28 (1960)
Boll weevil	Boll weevil	Boll weevil
Leafworm	Leafworm	Leafworm
	Bollworm	Bollworm
	Cotton aphid	Armyworm (2 species)
	False pink bollworm	Whitefly
		Cabbage looper
		Plant bug

Source: Ware (1980). Reprinted with permission from Springer-Verlag.

On the other hand, apparent elimination of particularly sensitive species can result. In animal populations, secondary poisoning, through consumption of contaminated prey, has adversely affected some populations of vertebrate predators and scavengers, for example, the sparrow hawk and peregrine falcon in England and the western grebe at Clear Lake in California. Severe reductions in prey organisms may also result in a reduction of specific parasites or predators who are unable to adapt to alternative feeding strategies following loss of food.

2.   *Increases in Populations Caused by Resurgence of Pests, Increases in Non-target Species, and Replacement of One Species by Another.*   The phenomenon of pest resurgence is well documented and invariably results from the elimination of the pest's natural enemies or competitors but survival of the pest. In some cases, pest populations may actually increase significantly after spraying, or new pests may emerge. Loss of one species from a habitat may also allow another pest or nontarget species to invade. Table 7.17 illustrates these effects.

3.   *Sublethal Effects on Survival and Reproduction of Animals.*   Sublethal physiological and behavioral effects on animals can affect the survival and reproduction in populations and communities by causing: (1) accumulation of sublethal doses and lethal mobilization of lipid-stored insecticides; (2) shifts in prey–predator relationships and differential survival; and (3) changes in reproduction success, for example, alterations in fecundity in arthropods, sac-fry mortalities in fish, delayed egglaying, reduced fertility, reduced hatchability, and thin eggshells in birds.

4.   *Changes in Genetic Viability and Resistance.*   The development of resistance to pesticide toxicity in animal populations emerges due to differential survival of individuals within an exposed population as a result of genetic variability. The survivors are thus genotypes selected for toxicity resistance. With continued exposure the proportion of resistant individuals may significantly increase in succeeding generations.

The application of pesticides to agricultural, forest, grassland, and other ecosystems to control populations of specific animal and plant species may also markedly affect populations and communities of nontarget organisms, either directly or indirectly. The significant responses and effects within populations and communities or major classes of organisms subjected to different types and rates of pesticide application are set out below.

Terrestrial Invertebrates

Insecticides, nematicides, molluscicides, fungicides, and herbicides are primarily selected for their direct toxicity toward soil and plant organisms. Thus the greatest environmental effects occur with nontarget organisms in these groups. Nontarget arthropods are the most immediately susceptible because of their similar physiological and habitat relationships to target organisms. For example, soil invertebrates such as insects, earthworms, slugs, and gastropods are directly and selectively affected by a wide variety of pesticides (see Edwards, 1973; Croft and Brown, 1975; and Brown, 1978). The toxic behavior of persistent organochlorines or organophosphate insecticides on soil organism populations has received the most attention although other research, for example, indicates that systemic fungicides can be lethal to earthworms (Brown, 1978). In addition, herbicides may strongly influence invertebrate populations indirectly by their effects on vegetation which provides food and habitat (Ware, 1980).

Application of insecticides in agriculture and forestry may result, firstly, in the resurgence of pest species and, secondly, the emergence of minor pests as major or secondary pests. For example, the introduction of DDT to control the codling moth has resulted in increased numbers and finally outbreaks of aphids, scale insects and mealy bugs, tortricid leafrollers, and tetramychid mites (Brown, 1978). In Central American cotton crops, economic pest species increased from two to eight in the mid-1960s since the introduction of organochlorine insecticides in 1950. Although cotton yields increased, the number of insecticide applications needed per growing season increased from 5 in 1950 to 28 in the mid-1960s (see Table 7.17).

Resistance is common in populations of terrestrial invertebrates. Hundreds of species of arthropods are known to be resistant to insecticides (Georghiou and Taylor, 1976). For example, by 1976 insecticide resistance in arthropods had developed to DDT and related compounds in 203 species, cyclodiene resistance in 225 species, organophosphorus resistance in 147 species, carbamate resistance in 36 species, pyrethrin resistance in 6 species, and formamidine resistance in 2 species. Resistance in arthropod populations appears to develop readily with the rate of the development dependent on (1) the intensity of selection of resistant genotypes in successive populations, (2) the number of generations per year, and (3) the degree of isolation of the population from dilution by immigration from untreated populations. Development of resistance is thus species dependent, and variations in its onset between interacting species can cause disruption within an ecosystem (Brown, 1978).

Insecticides can cause reduction of species diversity within the invertebrate fauna of natural plant and soil communities. In agroecosystems, in contrast to more stable forest ecosystems, maximum species diversity may not necessarily be such a critical parameter in maintaining biological control (Way and Chancellor, 1976; van Emden and Williams, 1974; Brown, 1978).

Terrestrial Vertebrates

Field applications of organochlorine, organophosphorus, and carbamate insecticides have caused episodes of direct mortalities among wild mammals and birds. Such incidents are not common with reptiles and amphibia, however, some degree of insecticide tolerance has been observed (see Brown, 1978). Rodenticides, such as the cyclodiene insecticides, heptachlor, dieldrin, and endrin, are responsible for considerable mortalities among nontarget wild mammals such as raccoon, rabbits, ground and fox squirrels, cottontails, woodchucks, voles, shrews, and moles, after spraying programs in the United States (Scott et al., 1959; Rudd, 1964), and foxes and badgers in England following secondary poisoning after the use of treated seed dressings (Blackmore, 1963; Jefferies, 1975). In comparison, DDT spraying in forest and agricultural regions appears to have had little adverse affects on mammal populations at similar application rates to the cyclodienes, about 1–3 lb acre^{-1}. Much higher rates of DDT application are required to produce significant symptoms of poisoning in mammals (see Brown, 1978). Two possible exceptions are, firstly, strong circumstantial evidence of declines in bat populations because of high DDE residues in their insect diets and, secondly, direct mortalities and possible reproduction failure in ranch mink ingesting DDT residues, primarily in fish diets, for example, coho salmon (McEwen and Stephenson, 1979).

Certain cholinesterase inhibitors are acutely toxic to mammals (e.g., parathion and methomyl), but reported effects on mammal populations are few and generally refer to indirect mortalities in small mammal populations where food supplies have been reduced (Brown, 1978; McEwen and Stephenson, 1979). Generally, there is little indication that significant adverse effects have occurred with mammal populations.

For bird populations, the widespread use of insecticides, mainly the organochlorines, has produced significant reductions in populations of some species through lethal and sublethal effects (Brown, 1978). Many instances of acute poisonings by DDT, lindane, cyclodienes, organophosphorus, and carbamate insecticides are reported which show a wide range of species susceptibilities. DDT-exposed populations in northwestern United States forests (0.84 kg DDT per hectare) exhibited adverse effects among upper canopy feeders, particularly wood warblers and vireos, while species in the lower zones were less seriously affected (Herman and Bulger, 1979).

Experimental feeding studies with DDT and dieldrin residues, in particular, have shown the mobilization of pesticide body burdens with fat reserves and the potential transfer of lethal doses to the brains of birds during or following periods of stress (e.g., starvation and excessive activity) (Van Velzen et al., 1972; McEwen and Stephenson, 1979). Serious population declines in some species are more associated with sublethal effects primarily due to excessive eggshell thinning (see Figure 7.16 and Table 7.16) and reproductive failure in carnivorous birds. Excessive eggshell thinning has been recorded in at least some populations of 39 species since the introduction of organochlorine insecticides. This effect is considered to adversely impair reproduction in about 27 species but documented evidence of serious population declines exist for only a few species. These include the bald eagle

(*Haliacetus leucophalus*), the osprey (*Pondion haliaetus*), the peregrine falcon (*Falco peregrinus*), the European sparrow hawk (*Accipiter nisus*), and the brown pelican (*Pelicanus occidentalis*). It is apparent falconiformes are much more sensitive to eggshell thinning than gallinaceous species (McEwen and Stephenson, 1979).

Aquatic Invertebrates

Numerous laboratory and field observations on target and nontarget populations and communities of aquatic invertebrates confirm marked and differential changes in species abundance, composition, and productivity as a result of lethal and sublethal concentrations of pesticides in aquatic ecosystems (Kerr and Vass, 1973). The most obvious impact on nontarget populations of aquatic invertebrates has resulted from the direct use of larvacides in disease vector control programs against the larvae of the mosquito and the blackfly. For example, heavy mortalities of crabs and shrimps in many salt marsh areas have been recorded where DDT and other organochlorines have been extensively used. Indirect impacts have resulted from large-scale forest spraying, for example, DDT and malathion applications and runoff from agricultural and urban areas. Susceptible groups have consistently been the arthropods, microcrustacea, and larval stages of decapods.

By comparison, mature molluscs and annelids are relatively tolerant to pesticide contamination although several pesticides are used specifically as molluscicides for the control of aquatic snails that are alternative hosts of Schistosoma parasites of humans. These include copper compounds, for example, cuprous oxide and cuprous chloride, niclosamine, and *N*-tritylmorpholine. Niclosamine is toxic to both fish species and snails yet *N*-tritylmorpholine, cuprous oxide, and cuprous chloride are relatively nontoxic to fish species (Brown, 1978).

Many resistant strains of aquatic invertebrates, primarily insects and crustacea, have recently developed. Examples include increased DDT resistance by blackfly, *Simulium damnosum*; *Cyclops* copepods to cyclodiene insecticides (Brown, 1978); and several populations of freshwater shrimp, *Palaemonotes kadiakenesis*, with a variety of pesticides, for example, carbaryl, toxaphene, and cyclodienes (Naqvi and Ferguson, 1970). Kerr and Vass (1973) point out that among aquatic invertebrates the capacity for genetic response is quite variable. Thus, any development of resistance in these species will occur at differential rates leading to modification in the species composition of exposed communities.

Aquatic Vertebrates

The most significant effects on wildlife of pesticide use have been the impact of organochlorine insecticides on fish and birds (Ware, 1980). Effects on fish populations have been mainly as a result of large-scale forest and agriculture spraying programs (Brown, 1978):

1.   DDT application rates from 0.25 lb acre^{-1} to various temperate North American forest ecosystems have resulted in episodes of severe fish mortalities. Susceptible species such as trout and salmon exhibited greatest mortalities in the younger age groups. For example, DDT spraying against spruce budworm, at a dosage rate of 0.15 lb acre^{-1}, caused mortalities with adult salmon parr (40%)

and young-of-the-year (98%). At 0.25 lb acre^{-1} no parr mortalities were observed but losses occurred to fingerlings (6%) and salmon fry (78%) (Kerswill and Edwards, 1967).

2.   Cyclodiene insecticides are exceptionally toxic to fish and moderate to severe mortalities among natural freshwater and marine fish populations have been observed at application rates from 0.1 to 3 lb acre^{-1}. BHC and lindane (γBHC) are similarly toxic to fish populations at concentrations in the low parts per billion range.

3.   Organophosphorus and carbamate insecticides are much less toxic to fish populations although significant mortalities may occur with some compounds such as parathion, methyl parathion, diazinon, malathion, and chlorpyrifos at larvicidal rates below 1 lb acre^{-1}.

4.   Little detailed information has emerged from the many laboratory and field investigations on the long-term effects of fish kills on affected populations and communities. Although reductions in numbers and changes to the age structure of populations as well as species diversity are apparent, recovery periods are not well established. But with the speckled sea trout, in the Laguna Madre estuary, Texas, the declining population showed a recovery from 0.2 per acre in 1969, to a normal level of 30 per acre in 1971, which coincided with the suspension of DDT use on agricultural lands within the catchment.

5.   Deleterious effects of pesticides such as DDT on reproduction have been observed. Massive mortalities of sac-fry of brook, rainbow, and cutthroat trout in hatcheries have resulted from using DDT-contaminated feed. Rainbow trout and coho salmon have been similarly affected in DDT-contaminated lakes. Organochlorines accumulate in eggs and can lead to the death of fry as the yolk sac is absorbed (Edwards, 1973).

6.   Pesticide resistance has developed in several fish species with high reproduction rates. Fish species from the heavily sprayed cotton growing regions in Mississippi have developed resistance to toxaphene, aldrin, dieldrin, endrin, and DDT at concentrations several hundred times the normal toxic level. Available evidence suggests this level of resistance is confined to warm-water species.

Plants

Herbicides exhibit high biological activity with all plants (Way and Chancellor, 1976) with effects varying from stimulation of growth at low application rates to lethality at higher rates. They can be selective or nonselective in their activities because of their varied modes of biological action. The capacity of these compounds to selectively modify, revert, and suppress plant growth has led to many applications particularly in agriculture and forestry. For example, certain compounds are effective against grassy weeds while others are effective against broadleaf weeds in cereal crops and grasses. Selective activity is dependent on (1) the time of application (e.g., pre-planting, pre-emergence, or post-emergence), (2) the method and rate of application, and (3) the actual amount that is transported to the site of action.

Most of the information on the effects of herbicides on plant communities refers to the weed flora of cultivated or agricultural land where there is a constant recycling of the initial stages of a plant succession that rarely exceeds 1 yr and generally exists for less than 6 months. Ecological effects in these systems are outlined below (Way and Chancellor, 1976).

1.   *Pre-existing Natural Tolerances and Differences in Susceptibilities Between Plant Species.*   Pre-existing natural tolerance of weed species (Way and Chancellor, 1976) and crops (Shimabukuro et al., 1971) has been observed due to detoxification mechanisms within the plants. Corn, *Zea mays*, for example, has resistance to simazine and triazine, primarily through detoxification processes that: (a) conjugate the triazine to glutathione by means of an S-transferase enzyme located in the corn leaves (Shimabukuro et al., 1971) and (b) nonenzymic catalysis of the hydroxylation of simazine in the roots. Atrazine detoxification occurs by nucleophilic attack on the 2-Cl substituent (Hamilton and Moreland, 1962).

2.   *Changes in the Density of Plant Populations.*   Although herbicides affect the species composition of weed populations (Way and Chancellor, 1976), changes in the overall density of weed infestations are difficult to assess. Weed populations are dynamic and influenced by a number of constantly changing environmental factors as well as cultural practices in the case of arable land. However, marked reductions in viable seeds in horticultural soil and seedling density in sprayed cereals have resulted over periods of 6–8 yr (see Way and Chancellor, 1976). In general, higher original densities of weeds tend to be reduced more readily than low densities. Even under efficient conditions of weed control, residual low-density populations, rich in species, will probably persist (Way and Chancellor, 1976).

3.   *The Development of Resistant Populations.*   Herbicide-resistant plant ecotypes have been observed for several herbicides together with the emergence of herbicide-resistant strains in field populations of weed species (Brown, 1978) (see Table 7.18). Other strains of perennials e.g. *Cynodon dactylon* and *Cardaria chalepensis*, and annual weeds e.g. *Polygonum lapathifolum, Setaria viridis* and *Hordeum jubatum*, have also shown variable response to herbicide treatments (Way and Chancellor, 1976).

4.   *Changes in the Number and Composition of Plant Species.*   Changes in the composition of weed populations are readily observed in treated field crops although it is often uncertain what proportion of this can be attributed to herbicide use when there are other possible factors such as climatic conditions, cultural practices, and growing conditions (Way and Chancellor, 1976). However, there are certain distinctive characteristics as set out below:

(a)   Single applications of herbicides in arable land show little, if any, long-term changes in weed flora.

(b)   Repeated applications, season after season, especially of one herbicide, may promote the incidence of certain species of weeds rather than others. Brown (1978) cites several examples of changes in weed infestations. For example, infestations of foxtail grasses, *Setaria*, and crab grasses, *Digitaria*, developed in sugar cane in Hawaii following the use of 2,4-D to control broadleaf weeds.

Table 7.18.   Examples of Weed Populations Resistant to Herbicides

Resistant Populations/Ecotypes	Herbicide	Place/Time
Erechitites hieracifolia (Hawaiian fireweed)	2,4-D	Hawaii 1955
Cirsium arvense (Canada thistle)	MCPA	Norway 1973
Senesio vulgaris (Common groundsel)	Atrazine	Washington state 1968
Poa annua (Annual bluegrass)	Metoxuron	France 1974
Sorghum halepense (Johnson grass)	Dalapon	Arizona 1963
Convolvulus arvensis (Field bindweed)	2,4-D	New Mexico 1963
Avena fatua (Wild oats)	Propham	North Dakota 1963
Setaria lutescens (Yellow foxtail)	Dalapon	Maryland 1960
Cirsium arvense (Canada thistle)	Amitrole	Idaho 1970

Source: Brown (1978). Reprinted with permission from John Wiley & Sons.

(c)   Species tolerant of herbicides, but less competitive, tend to replace susceptible species.

Herbicides and other pesticides can enter waterways directly from use in aquatic weed control or in spray drift and water runoff. Table 7.19 lists the toxicity of a variety of pesticides to algae.

OVERALL ECOSYSTEM EFFECTS

A diverse range of ecological effects may be generated following a sequence of lethal or sublethal effects on the growth, reproduction, or survival of individual organisms. Because the dynamics of species populations with each other and their abiotic environment are extremely complex and poorly understood, relatively less emphasis has been placed on investigating the effects of pesticides on natural ecosystems than estimates of direct toxicities to individual organisms. It is the interrelations of species and abiotic factors within an ecosystem that must also be considered in any assessment of pesticide impact on individuals or groups of species. The complete or partial removal of a species from an ecosystem by use of a pesticide will be followed by changes in the prey, predator, competitor, and other species populations with which that species interacts. Again, each of these species interact with others (Hurlbert, 1975). Thus, the actual extent of pesticide effects on nontarget organisms can be seen as ecosystem dependent.

Table 7.19.   Summary Table of Pesticide Toxicity to Algae

Toxicity Level	Pesticide
Highly toxic to at least some algae (< 0.1 ppm)	Bromacil Diuron EPTC[b]
Toxic to some algae at field application levels (~ 0.1–5 ppm)	Atrazine[b] BHC Chlorpropham Dalapon[b] Dicamba[b] Dichlobanil[b] Dinaseb Dinquat MCPA[b] Paraquat[b] Pentachlorophenol Picloram Propanil Propazine Sodium pentachlorophenate Simazine TCA 2,4-D[b] Trifluralin
Toxic to most or all algae at high concentration (> 5 ppm)	Atrazine[b] BHC[a] Ceresan[a] DDT[a] Diquat[b] Methoxychlor[a] Mirex[b] Paraquat[b] 2,4,5-T
No toxicity demonstrated	Barban[b] Chloronil[b] Dicamba[b] EPTC[b] MCPA[b] Malathion[a] 2,4-D[b] 2,4-DP

Source: Kohn (1980). Reprinted with permission from Springer-Verlag.

[a] Fungicide or insecticide.

[b] Different studies have shown widely divergent toxicity levels for the pesticide listed.

Ecosystem response to extensive pesticide exposure is primarily measured in terms of changes in species composition and population numbers. Typically, these changes follow a sequence of dynamic events as outlined below (Pimentel, 1971; Pimentel and Goodman, 1974; Brown, 1978).

1. If lethal or sublethal concentrations of pesticides are dispersed in an ecosystem, the number of species in the ecosystem becomes reduced.
2. If the reduction in number of species is sufficient, this may lead to instability with the ecosystem and subsequently to population outbreaks in some nontarget species. Outbreaks result from a breakdown in the normal check–balance structure of the system.
3. After a pesticide disappears from the affected ecosystem, species in the lower trophic levels (e.g., herbivores) usually increase to outbreak levels.
4. Predators and parasites existing at the higher trophic levels become susceptible to loss of a species or large scale fluctuations in numbers of species in the lower parts of the food chain upon which they are dependent (Pimental and Goodman, 1974).

REFERENCES

Abbott, D. C., Harrison, R. B., Tatton, J. O. G., and Thomson, J. (1966). Organochlorine pesticides in the atmosphere. *Nature (London)* **211**, 259.

Atkins, G. L. (1969). *Multicompartment Models for Biological Systems*. Methuen, London.

Adamson, R. H. (1974). "An Introduction to Detoxification as a Mechanism of Survival." In M. A. Q. Khan and J. P. Bederka, Jr. (Eds.), *Survival in Toxic Environments*. Academic Press, New York, p. 125.

Audus, L. J. (Ed.) (1964). *The Physiology and Biochemistry of Herbicides*. Academic Press, London.

Audus, L. J. (Ed.) (1976). *Herbicides. Physiology, Biochemistry, Ecology*, Vol. 2, 2nd ed. Academic Press, London.

Bailey, G. W. and White, J. L. (1970). Factors influencing the adsorption, desorption, and movement of pesticides in soil. *Residue Rev.* **32**, 29.

Blackmore, D. K. (1963). The toxicity of some chlorinated hydrocarbon insecticides to British wild foxes (*Vulpes vulpes*). *J. Comp. Pathol. Ther.* **73**, 391.

Brooks, G. T. (1975). *Chlorinated Insecticides Vol. II. Biological and Environmental Aspects*. CRC Press, Cleveland, Ohio.

Brown, A. W. A. (1978). *Ecology of Pesticides*. John Wiley & Sons, New York.

Butler, P. A. (1963). "Commercial Fisheries Investigations." In Pesticide-Wildlife Studies: A Review of Fisheries and Wildlife Service Investigations during 1961 and 1962. *Fish Wildl. Serv. Circ.* **167**, 11.

Cairns, Jr., J., Dickson, K. L., and Maki, A. W. (1978). *Estimating the Hazard of Chemical Substances to Aquatic Life*. ASTM, Philadelphia.

Canton, J. W., Greve, P. A., Slooff, W., and van Esch, G. J. (1975). Toxicity, accumulation and elimination studies of γ-hexachlorocyclohexane (γ-HCH) with fresh-water organisms of different trophic levels. *Water Res.* **9**, 1163.

Chiou, C. T., Freed, V. H., Schmedding, D. W., and Kohnert, R. L. (1977). Partition coefficient and bioaccumulation of selected organic chemicals. *Environ. Sci. Technol.* **11**, 475.

Cope, O. B. (1971). Interactions between pesticides and wildlife. *Ann. Rev. Entomol.* **16**, 325.

Corbett, J. R. (1974). *The Biochemical Mode of Action of Pesticides*, Academic Press, London.

Crafts, A. S. (1964). "Herbicide Behavior in the Plant." In L. J. Audus (Ed.), *The Physiology and Biochemistry of Herbicides*. Academic Press, London, p. 75.

Croft, B. A. and Brown, A. W. A. (1975). Responses of arthropod natural enemies to insecticides. *Ann. Rev. Entomol.* **20**, 285.

Crosby, D. G. and Moilanen, K. W. (1974). Vapor-phase photodecomposition of aldrin and dieldrin. *Arch. Environ. Contam. Toxicol.* **2**, 62.

De Bruin, A. (1976). *Biochemical Toxicology of Environmental Agents*. Elsevier, Amsterdam.

Dempster, J. P. (1975). "Effects of Organochlorine Insecticides on Animal Populations." In F. Moriarty (Ed.), *Organochlorine Insecticides: Persistent Organic Pollutants*. Academic Press, London, p. 231.

Duggan, R. E. and Weatherwax, J. R. (1967). Dietary intake of pesticide chemicals. *Science* **157**, 1006.

Edwards, C. A. (1973). *Environmental Pollution by Pesticides*. Plenum Press, London.

Ellgehausen, H., Guth, J. A., and Esser, H. O. (1980). Factors determining the bioaccumulation potential of pesticides in the individual compartments of aquatic food chains. *Ecotoxicol. Environ. Safety* **4**, 134.

Ernst, W. (1977). Determination of the bioconcentration potential of marine organisms. A steady state approach I. Bioconcentration data for seven chlorinated pesticides in mussels (*Mytilus edulis*) and their relation to solubility data. *Chemosphere* **6**, 731.

Farm Chemicals Handbook (1976). Meister Publishing, Willoughby, Ohio.

Fine, D. H., Krull, I. S., Rounbehler, D. P., and Edwards, G. S. (1980). "*N*-nitroso Compound Impurities in Consumer and Commercial Products." In R. Haque (Ed.), *Dynamics, Exposure and Hazard Assessment of Toxic Chemicals*. Ann Arbor Science, Ann Arbor, Michigan, p. 417.

Finlayson, D. G. and MacCarthy, H. R. (1973). "Pesticide Residues in Plants." In C. A. Edwards (Ed.), *Environmental Pollution by Pesticides*. Plenum Press, London, p. 57.

Georghiou, G. P. and Taylor, C. E. (1976). Pesticide resistance as an evolutionary phenomenon. *Proc. XV Int. Congr. Entomol.*, 759.

Goldberg et al. (1971). Cited in Kerr and Vass (1973).

Goring, C. A. I., Laskowski, D. A., Hamaker, J. W., and Meikle, R. W. (1975). "Principles of Pesticide Degradation in Soil." In R. Haque and V. H. Freed (Eds.), *Environmental Dynamics of Pesticides*. Plenum Press, New York p. 135.

Hamaker, J. W. (1975). "The Interpretation of Soil Leaching Experiments." In R. Haque and V. H. Freed (Eds.), *Environmental Dynamics of Pesticides.* Plenum Press, New York, p. 115.

Hamaker, J. W., Goring, C. A. I., and Youngson, C. R. (1966). "Sorption and Leaching of 4-Amino-3,5,6-Trichloropicolinic Acid in Soils." In *Organic Pesticides in the Environment*, Advances in Chemistry Series 60. American Chemical Society, Washington, D.C. p. 23.

Hamelink, J. L., Waybrant, R. C., and Ball, R. C. (1971). A proposal: Exchange equilibrium control the degree cholorinated hydrocarbons are magnified in benthic environments. *Transactions Am. Fisheries Soc.* **100**, 207.

Hamilton, R. H. and Moreland, D. E. (1962). Simazine: Degradation by corn seedlings. *Science* **135**, 373.

Haque, R. (1975). "Role of Adsorption in Studying the Dynamics of Pesticides in a Soil Environment." In R. Haque and V. H. Freed (Ed.), *Environmental Dynamics of Pesticides.* Plenum Press, New York, p. 97.

Hartley, G. S. (1976). "Physical Behaviour in the Soil." In L. J. Audus (Ed.), *Herbicides. Physiology, Biochemistry, Ecology*, Vol. 2, 2nd ed. Academic Press, London, p. 1.

Hartung, R. (1975). "Accumulation of Chemicals in the Hydrosphere." In R. Haque and V. H. Freed (Eds.), *Environmental Dynamics of Pesticides.* Plenum Press, New York, p. 185.

Hassett, J. P. and Lee, G. F. (1975). "Modeling of Pesticides in the Aqueous Environment." In R. Haque and V. H. Freed (Eds.), *Environmental Dynamics of Pesticides.* Plenum Press, New York, p. 173.

Heath, R. G., Spann, J. W., Hill, E. F., and Kreitzer, J. F. (1972). Comparative dietary toxicities of pesticides to birds, U.S. Fish and Wildlife Service, Bureau of Sport Fish and Wildlife, Special Scientific Report — Wildlife No. 152, p. 57.

Herman, S. G. and Bulger, J. B. (1979). Effects of a forest application of DDT on non-target organisms. *Wildlife Monographs, Suppl. J. Wildl. Manage.* **43**, No. 69.

Hickey, J. J., Keith, J. A., and Coon, F. B. (1966). An exploration of pesticides in a Lake Michigan estuary. *J. Appl. Ecol.* **3** (Suppl.), 141.

Holden, A. V. (1973). "Effects of Pesticides on Fish." In C. A. Edwards (Ed.), *Environmental Pollution by Pesticides.* Plenum Press, London, p. 213.

Hunt, E. G. and Bischoff, A. I. (1960). Inimical effects on wildlife of periodic DDD applications to Clear Lake. *Calif. Fish Game* **46**, 91.

Hurlbert, S. H. (1975). Secondary effects of pesticides on aquatic ecosystems. *Residue Rev.* **57**, 81.

Jefferies, D. J. (1975). "The Role of the Thyroid in the Production of Sublethal Effects by Organochlorine Insecticides and Polychlorinated Biphenyls." In F. Moriarty (Ed.), *Organochlorine Insecticides: Persistent Organic Pollutants.* Academic Press, London, p. 131.

Kaufman, D. D. and Kearney, P. C. (1976). "Microbial Transformation in the Soil." In L. J. Audus (Ed.), *Herbicides. Physiology, Biochemistry, Ecology*, Vol. 2, 2nd ed. Academic Press, London, p. 29.

Kaufman, D. D. and Plimmer, J. R. (1972). In R. Mitchell (Ed.), *Water Pollution Microbiology*, John Wiley & Sons, New York, Vol. 1, p. 173.

Kearney, P. C. and Plimmer, J. R. (1970). In *Pesticides in the Soil: Ecology, Degradation and Movement*. Michigan State University Press, East Lansing, Michigan, p. 65.

Kearney, P. and Plimmer, J. R. (1972). Metabolism of 3,4-dichloroaniline in soils. *J. Agric. Food Chem.* **20**, 584.

Kearney, P. C., Woolsen, E. A., Plimmer, J. R., and Isensee, A. R. (1969). Decontamination of pesticides in soils. *Residue Rev.* **29**, 137.

Kenaga, E. E. (1972). Guidelines for environmental study of pesticides: Determination of bioconcentration potential. *Residue Rev.* **44**, 73.

Kenaga, E. E. (1975). "Partitioning and Uptake of Pesticides in Biological Systems." In R. Haque and V. H. Freed (Eds.), *Environmental Dynamics of Pesticides*. Plenum Press, New York, p. 217.

Kerr, S. R. and Vass, W. P. (1973). "Pesticide Residues in Aquatic Invertebrates." In C. A. Edwards (Ed.), *Environmental Pollution by Pesticides*. Plenum Press, London, p. 134.

Kerswill, C. J. and Edwards, H. E. (1967). Fish losses after forest spraying with insecticides in New Brunswick, 1952–62. *J. Fish. Res. Board Can.* **24**, 709.

Khan, S. U. (1980). *Pesticides in the Soil Environment*. Elsevier, Amsterdam.

Khan, M. A. Q., Stanton, R. H., and Reddy, G. (1974). "Detoxication of Foreign Chemicals by Invertebrates." In M. A. Q. Khan and J. P. Bederka, Jr. (Eds.), *Survival in Toxic Environments*. Academic Press, New York, p. 177.

Khan, M. A. Q., Gassman, M. L., and Ashrafi, S. H. (1975). "Detoxication of Pesticides by Biota." In R. Haque and V. H. Freed (Eds.), *Environmental Dynamics of Pesticides*. Plenum Press, New York, p. 289.

Klein, W. and Scheunert, I. (1978). "Biotic Processes." In G. C. Butler (Ed.), *Principles of Ecotoxicology*, SCOPE 12. John Wiley & Sons, New York, p. 37.

Kohn, G. K. (1980). "Bioassay as a Monitoring Tool." In F. A. Gunther and J. D. Gunther (Eds.), *Residue Reviews. Residues of Pesticides and Other Contaminants in the Total Environment*, Vol. 76. Springer-Verlag, New York, p. 99.

Korte, F. (1978). "Abiotic Processes." In G. C. Butler (Ed.), *Principles of Ecotoxicology*, SCOPE 12. John Wiley & Sons, New York, p. 11.

Lichtenstein, E. P. (1980). "'Bound' Residues in Soils and Transfer of Soil Residues in Crops." In F. A. Gunther and J. D. Gunther (Eds.), *Residue Reviews. Residues of Pesticides and Other Contaminants in the Total Environment*, Vol. 76. Springer-Verlag, New York, p. 147.

Lichtenstein, E. P. and Schulz, K. R. (1964). The effects of moisture and microorganisms on the persistence and metabolism of some organophosphorus insecticides in soils, with special emphasis on parathion. *J. Econ. Entomol.* **57**, 618.

Livingston, R. J. (1977). Review of current literature concerning the acute and chronic effects of pesticides on aquatic organisms. *CRC Crit. Rev. Environ. Control* **7**, 4, 325.

Macek, K. J., Petrocelli, S. R., and Sleight III, B. H. (1979). "Considerations in Assessing the Potential for, and Significance of, Biomagnification of Chemical Residues in Aquatic Food Chains." In L. L. Marking and R. A. Kimerle

(Eds.), *Aquatic Toxicology*, ASTM STP 667. American Society for Testing and Materials, Philadelphia, p. 251.

Matsumura, F. (1973). "Degradation of Pesticide Residues in the Environment." In C. A. Edwards (Ed.), *Environmental Pollution by Pesticides*. Plenum Press, London, p. 494.

Matsumura, F., Boush, G. M., and Misato, T. (Eds.) (1972). *Environmental Toxicology of Pesticides*. Academic Press, New York.

McEwen, F. L. and Stephenson, G. R. (1979). *The Use and Significance of Pesticides in the Environment*. John Wiley & Sons, New York.

Metcalf, R. L. (1971). "The Chemistry and Biology of Pesticides." In R. White-Stevens (Ed.), *Pesticides in the Environment*. Vol. 1, Part I. Marcel Dekker, New York, p. 1.

Moilanen, K. W., Crosby, D. G., Soderquist, C. J., and Wong, A. S. (1975). "Dynamic Aspects of Pesticide Photodecomposition." In R. Haque and V. H. Freed (Eds.), *Environmental Dynamics of Pesticides*. Plenum Press, New York, p. 45.

Moreland, D. E. (1980). Mechanisms of action of herbicides. *Ann. Rev. Plant Physiol.* **31**, 597.

Moriarty, F. (Ed.) (1975). *Organochlorine Insecticides: Persistent Organic Pollutants*. Academic Press, London.

Moriarty, F. (1978). "Terrestrial Animals." In G. C. Butler (Ed.), *Principles of Ecotoxicology*, SCOPE 12. John Wiley & Sons, New York, p. 169.

Naqvi, S. M. and Ferguson, D. E. (1970). Levels of insecticide resistance in freshwater shrimp, *Palaemonetes kadakiensis*. *Trans. Am. Fish. Soc.* **99**, 696.

O'Brien, R. D. (1967). *Insecticides: Action and Metabolism*. Academic Press, New York.

OECD (1981). Guidelines for Testing of Chemicals. Organisation for Economic Co-operation and Development, Publications Office, Paris.

Parris, G. E. (1980). "Environmental and Metabolic Transformations of Primary Aromatic Amines and Related Compounds." In F. A. Gunther and J. D. Gunther (Eds.), *Residue Reviews. Residues of Pesticides and Other Contaminants in the Total Environment*, Vol. 76. Springer-Verlag, New York, pp. 1–30.

Pimentel, D. (1971). Ecological Effects of Pesticides on Non-target Species. Exec. Off. President, Off. Sci. Technol. Sup't. Doc. Washington (Sup't Doc'ts Stock No 4106-0029).

Pimentel, D. and Goodman, N. (1974). "Environmental Impact of Pesticides." In M. A. Q. Khan and J. P. Bederka, Jr. (Eds.), *Survival in Toxic Environments*. Academic Press, New York, p. 25.

Reinert, R. E. (1972). Accumulation of dieldrin in an algae (*Scenedesmus obliquus*), *Daphnia magna* and the Guppy (*Poccilia reticulata*). *J. Fish. Res. Board Can.* **29**, No. 10, 1413.

Risebrough, R. W., Huggett, J. R., Griffin, J. J., and Goldberg, E. D. (1968). Pesticides: Transatlantic movements in the Northeast Trades. *Science* **159**, 1233.

Robinson, J. (1973). "Dynamics of Pesticide Residues in the Environment." In C. A. Edwards (Ed.), *Environmental Pollution by Pesticides*. Plenum Press, London, p. 459.

Robinson, J. and Roberts, M. (1968). *Physico-chemical and Biophysical Factors Affecting the Activity of Pesticides.* Monograph No. 29. Society for Chemistry and Industry, London.

Robinson, J., Roberts, M., Baldwin, M., and Walker, A. I. T. (1969). The pharmacokinetics of HEOD (dieldrin) in the rat. *Food Cosmetic Toxicol.* **7**, 317.

Rudd, R. L. (1964). *Pesticides and the Living Landscape.* University of Wisconsin Press, Madison.

Sargent, J. A. (1976). "Relationship of Selectivity to Uptake and Movement." In L. J. Audus (Ed.), *Herbicides. Physiology, Biochemistry, Ecology*, Vol. 2, 2nd ed. Academic Press, London, p. 303.

Scott, T. G., Willis, Y. L., and Ellis, J. A. (1959). Some effects of a field application of dieldrin on wildlife. *J. Wildl. Manage.* **23**, 409.

Seiber, J. N., Woodrow, J. E., Shafik, T. A., and Enos, H. F. (1975). "Determination of Pesticides and Their Transformation Products in Air." In R. Haque and V. H. Freed (Eds.), *Environmental Dynamics of Pesticides.* Plenum Press, New York, p. 17.

Seume, F. W. and O'Brien, R. D. (1960). Potentiation of the toxicity to insects and mice of phosphorothionates containing carboxyester and carboxyamide groups. *Toxicol. Appl. Pharmacol.* **2**, 495.

Shaw, G. R. and Connell, D. W. (1983). Factors influencing concentrations of polychlorinated biphenyls in organisms from an estuarine ecosystem. *Aust. J. Mar. Fresh. Res.* **33**, 1057.

Shimabukuro, R. H., Frear, D. S., Swanson, H. R., and Walsh, W. C. (1971). Glutathione conjugation, an enzymatic basis for atrazine resistance in corn. *Plant Physiol.* **47**, 10.

Sjoden, P.-O. and Soderberg, U. (1978). Phenoxyacetic acids: Sublethal effects. *Ecol. Bull. (Stockholm)* **27**, 149.

Spencer, W. F. and Cliath, M. M. (1973). Pesticide volatilization as related to water loss from soil. *J. Environ. Qual.* **2**, 284.

Spencer, W. F. and Cliath, M. M. (1975). "Vaporization of Chemicals." In R. Haque and V. H. Freed (Eds.), *Environmental Dynamics of Pesticides.* Plenum Press, New York, pp. 61–78.

Stanley, C. W., Barney, J. E., Helton, M. R., and Yobs, A. R. (1971). Measurements of atmospheric levels of pesticides. *Environ. Sci. Technol.* **5**, 430.

Stickel, L. F. (1973). "Pesticide Residues in Birds and Mammals." In C. A. Edwards (Ed.), *Environmental Pollution by Pesticides.* Plenum Press, London, p. 254.

Thomson, J. M. (1970). The effects of pesticides on fish. *Queensland Littoral Soc. Newsletter* **39**, 1.

Tinsley, I. J. (1979). *Chemical Concepts in Pollutant Behaviour.* John Wiley & Sons, New York.

Tucker, R. K. and Crabtree, D. G. (1970). *Handbook of Toxicity of Pesticides to Wildlife.* U.S. Fish and Wildlife Service, Bureau of Sport Fisheries and Wildlife, Resource Publication No. 84.

Tucker, R. K. and Haegele, M. A. (1971). Comparative acute oral toxicity of pesticides to six species of birds. *Toxicol. Appl. Pharmacol.* **20**, 57.

van Emden, H. F. and Williams, G. F. (1974). Insect stability and diversity in agroecosystems. *Ann. Rev. Entomol.* **19**, 455.

Van Velzen, A. C., Stiles, W. B., and Stickel, L. F. (1972). Lethal mobilization of DDT by cowbirds. *J. Wildl. Manage.* **36,** 733.

Wain, R. L. and Smith, M. S. (1976). "Selectivity in Relation to Metabolism." In L. J. Audus (Ed.), *Herbicides. Physiology, Biochemistry, Ecology*, Vol. 2, 2nd ed. Academic Press, London, p. 279.

Walker, C. H. (1975). "Variations in the Intake and Elimination of Pollutants." In F. Moriarty (Ed.), *Organochlorine Insecticides: Persistent Organic Pollutants.* Academic Press, London, p. 73.

Walker, A. and Smith, A. E. (1979). Persistence of 2,4,5-T in a heavy clay soil. *Pestic. Sci.* **10,** 151.

Ware, G. W. (1980). "Effects of Pesticides on Non-target Organisms." In F. A. Gunther and J. D. Gunther (Eds.), *Residue Reviews. Residues of Pesticides and other Contaminants in the Total Environment*, Vol. 76. Springer-Verlag, New York, p. 173.

Way, J. M. and Chancellor, R. J. (1976). "Herbicides and Higher Plant Ecology." In L. J. Audus (Ed.), *Herbicides. Physiology, Biochemistry, Ecology*, Vol. 2, 2nd ed. Academic Press, London, p. 345.

Wheatley, G. A. (1973). "Pesticides in the Atmosphere." In C. A. Edwards (Ed.), *Environmental Pollution by Pesticides.* Plenum Press, London, p. 365.

Woodwell, G. M., Wurster, C. F., and Isaacson, P. D. (1967). DDT residues in an East Coast estuary: A case of biological concentration by a persistent pesticide. *Science* **156,** 821.

Woodwell, G. M., Craig, P. P., and Johnson, H. A. (1971). DDT in the biosphere: Where does it go? *Science* **174,** 1101.

8

PETROLEUM AND
RELATED HYDROCARBONS

Petroleum hydrocarbons are major pollutants in the oceans, and substantial quantities are released into the atmosphere, particularly adjacent to urban complexes. Many attempts have been made to quantify the inputs of petroleum into the oceans as shown in Table 8.1 (GESAMP 1977; Anon, 1975) and to a lesser extent evaluate the ecological significance of these discharges. (Korte and Boedefeld, 1978; Connell and Miller, 1981a, b).

Hydrocarbons are widely distributed throughout the oceanic, atmospheric, and terrestrial environment (Malins, 1977) but can originate from petroleum or be of recent origin. It is often difficult to determine the origin of hydrocarbons in an environmental sample based on composition. This difficulty reflects the many different sources and fluxes of hydrocarbons which determine the magnitude and rate of entry of these substances into any particular sector of the environment. Discharges and losses of crude and refined petroleum become part of this material flow through interactions between natural and human-induced processes. Consequently, any ecotoxicological assessment of the impact of petroleum hydrocarbons in global environmental systems depends on (1) knowledge of the material balance and flows of petroleum operations and (2) production rates and pathways of recent hydrocarbons between global compartments and reservoirs.

Korte and Boedefeld (1978) have reviewed the global impact of the petroleum industry and developed a comprehensive worldwide material balance for 1974, the selected reference year for which the most recent information was available on a worldwide basis.

In a little more than 100 yr of operations, global production of crude oil has increased by several orders of magnitude. For example, crude oil production in 1875 was 1.4×10^6 tonnes compared with 2861×10^6 tonnes in 1974 and current production of about 3100×10^6 tonnes. Refined products totalled 2754×10^6 tonnes in 1974 with fuels as the major products (2148×10^6 tonnes) and the remainder nonfuels such as napthas, bitumen, petroleum coke, lubricants, white spirits, and waxes (336×10^6 tonnes).

Petroleum operations involve a sequence of material losses and chemical conversions with consequent discharges to different compartments of the ecosphere. Specific estimates of material losses and discharges of petroleum into the three

Table 8.1.   Range of Estimates for Petroleum Inputs to the Marine Environment[a]

Source	Quantity (tonnes $\times 10^6$ yr^{-1})
Marine Operation Losses	
LOT tankers	0.03–0.31
Non-LOT tankers	0.41–1.00
Bilges, bunkering, and other normal ship operations (all ships)	0.05–0.61
Offshore Accidental Discharges	
Tanker accidents	0.10–0.22
Accidents, other ships	0.02–0.35
Pipeline accidents	<0.01
Offshore Oil Production	0.08–<0.38
Natural Marine Oil Seeps	0.2–7.00
Atmospheric Deposition	0.4–9.00
Land Based Discharges	
Refineries	0.20–0.30
Terminal transfer operations	0.01–0.25
Pipeline accidents	<0.01–0.03
Runoff (urban and river)	1.90
Industrial wastes	0.08–1.98
Automotive wastes	0.50–4.4
Aviation wastes	0.05
Municipal wastes	0.2–(11.8)[b]
Total	1.90–11.4

[a] Compiled from Connell (1981).
[b] Not included in total.

major environments – marine, terrestrial, and atmospheric – are summarized in Figure 8.1. Combustion accounts for about 88.6% of global production. Thus, petroleum is mainly converted to gaseous combustion products such as water and oxides of carbon, nitrogen, and sulfur which enter the natural global cycles of these compounds.

Petroleum transferred to the environment in the form of hydrocarbons comprises about 2.3% of global petroleum production (1974) if hydrocarbons released due to incomplete combustion, or formed during combustion, are not considered (Korte and Boedefeld, 1978). Anthropogenic emissions of nonmethane hydrocarbons to the atmosphere amount to 68×10^6 tonnes per year, according to the National Academy of Sciences (Anon, 1975), compared with the estimate by Korte and Boedefeld (1978) of petroleum evaporation of 43.6×10^6 tonnes per year. Miller

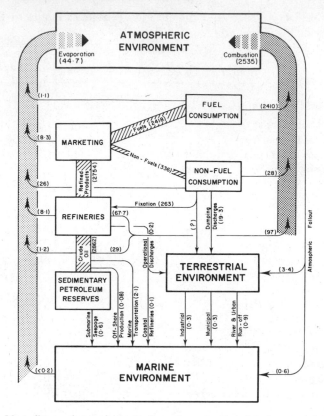

Figure 8.1.   Mass flows of petroleum hydrocarbons in the global system indicating usage patterns and discharges to the marine, atmospheric, and terrestrial environments (tonnes × 10^2). [From Miller and Connell (1982). Reprinted with permission from Gordon and Breach Science Publishers, Inc.]

and Connell (1982) have suggested that the difference of about 20×10^6 tonnes per year may be derived from combustion sources.

    Estimated annual losses of nonfuel petroleum to the terrestrial environment (19.3×10^6 tonnes) have been categorized as mainly lubricants (17.0×10^6 tonnes), white spirits (1.0×10^6 tonnes), and bitumen (1.0×10^6 tonnes). These calculations assume 263×10^6 tonnes of nonfuels are either transferred to other industries for processing (naptha and petroleum coke), recycled (lubricants), or fixed in the form of bitumen with losses to the environment assumed to be 1.0×10^6 tonnes per year.

    In the marine environment, direct discharges through marine operations are estimated to be about 1.4×10^6 tonnes from transportation and 0.2×10^6 from accidental spills (see Table 8.2). Total global anthropogenic discharges or losses of petroleum hydrocarbons are probably in the range $(64.5-89.5) \times 10^6$ tonnes which amounts to almost 2.3–3.2% of total production (1974) (see Table 8.2). Addition of natural submarine seepage (0.6×10^6 tonnes per year) means a final estimate for total environmental inputs of petroleum at $(65-90) \times 10^6$ tonnes per year (Miller

Table 8.2.   Estimates of Total Global Emissions and Discharges of Petroleum (tonnes $\times 10^6$ yr^{-1})

	Korte and Boedefeld (1978)	National Academy of Sciences (Anon., 1975)
Atmospheric emissions	43.6 (evaporation only)	68 (evaporation plus combustion sources)
Marine operations, losses and accidental spills	1.6	2.2
Terrestrial disposal	19.3	(19.3)[a]
Total	64.5	89.5

Source: Miller and Connell (1982). Reprinted with permission from Gordon and Breach Science Publishers, Inc.

[a] No estimate carried out by National Academy of Sciences. Estimated derived from Korte and Boedefeld assumed to be best available figure.

and Connell, 1982). By omitting atmospheric emission, or fallout, from these calculations, nongaseous discharges of petroleum would account for (21–22) $\times 10^6$ tonnes (see Table 8.2).

From a global perspective, a more complete representation of material transfers and losses of petroleum can be described in the form of a flow diagram. Figure 8.1 incorporates estimates, by Korte and Boedefeld (1978), for 1974 and petroleum inputs to the marine environment computed by the National Academy of Sciences (Anon, 1975).

COMPOSITION AND ENVIRONMENTAL OCCURRENCE OF PETROLEUM AND RECENT HYDROCARBONS

Petroleum

Both crude and refined petroleum vary widely in physical properties and chemical composition, depending on their origin and, in the case of refined products, the nature of the refining process. These factors also produce differences in important environmental properties such as solubilization, volatilization, photochemical and microbial oxidation, and biological toxicities (Connell and Miller, 1981a, b).

Crude petroleum contains complex mixtures of hydrocarbons as well as relatively small amounts of nitrogen-, sulfur-, and oxygen-containing organic compounds, asphaltenes, and various trace metals (uncomplexed and complexed forms) (Kallio, 1976; Posthuma, 1977). The hydrocarbons can be divided into two classes related to their chemical structure: the alkanes (n-normal, branched, and cyclo) and aromatic compounds (napthaeno; mono-, di-, and poly, i.e., PAH) (see Figure 8.2).

Crude petroleum can be converted by physical and chemical processes into a wide range of refined products including gasoline, kerosene, heating oils, diesel oils,

Table 8.3.   Hydrocarbons in Environmental Samples[a]

Sample	n-Paraffins	Saturates	Unsaturates	Aliphatics
Seawater (μg L^{-1})				
North America	0.024–7.32	1–280	–	–
Atlantic Ocean	0.22–0.145	1–239	–	–
Europe	0.2–11.8	–	–	–
Mediterranean Sea	0.3–9.5	–	–	–
Unexposed Marine Organisms (μg g^{-1} w/w)				
Molluscs	0.088–0.35	0.1–34	34–100	0.46–2
Fish	0.005–22.6	0–16.2	–	0.9–55
Exposed Marine Organisms (μg g^{-1} w/w)				
Molluscs	0.87–220	3.4–46	–	–
Fish	0.75–90	–	–	–

[a] Compiled from Connell and Miller (1981a, b).

lubricating oils, waxes, and asphalts. In refined products, the major hydrocarbons are alkanes, naphthenes, aromatic compounds, and alkenes contained in numerous distillates and blends. Alkenes are not common constituents of crude petroleum but are formed by the "cracking" process often used in petroleum production.

Extensive analytical information is available on the geographical distribution of hydrocarbons. However, comparison of results from the various studies is difficult because of the variation in sampling procedures and analytical techniques. As a general rule, aliphatic and aromatic hydrocarbons occur throughout estuarine, coastal, and oceanic environments with the highest levels in estuarine and intertidal habitats (Clark and MacLeod, 1977). A collation of some of the results available is shown in Table 8.3.

Recent Hydrocarbons

Hydrocarbons of recent origin (alkanes, isoprenoids, alkenes, and aromatics) occur widely in terrestrial and marine organisms. These compounds may be synthesized by organisms, translocated from food, or formed following ingestion of precursors from food and abiotic sources. Alkanes (straight and branched) and isoprenoids (e.g., pristane and squalene) tend to predominate in organisms although unsaturated hydrocarbons may be significant in bacteria and algae (see, e.g., Ackman, 1972; Kolattukudy, 1976; Koons et al., 1965).

Aromatics	PAH	Petroleum	Total Hydrocarbon
1–59	–	$(0–269) \times 10^3$	1–12,700
<1–1	–	0.37–0.56	1–22
–	–	–	<50–1,000
–	–	–	–
0.1–8	$\leq 0.2 \times 10^{-3} –2.3$	–	1–12
0–22.2	0.009–0.184	–	4–14
0.6–372	0–16	7.4–236	–
–	–	29–88	–

Polyaromatic Hydrocarbons (PAHs)

PAHs contain two or more benzene rings (see Figure 8.2) and, in contrast to the two groups discussed previously, have many diverse origins. There are three main routes of formation in the environment: (1) pyrolysis of organic matter, (2) generation in sedimentary organic matter and fossil fuels, and (3) biosynthesis by organisms. The first two routes are the most important while the last is a possibility not yet clearly demonstrated by the available evidence (Hites et al., 1977; La Flamme and Hites, 1978).

Pyrolysis and incomplete combustion of organic matter yields PAHs in the products. A number of detailed mechanisms have been proposed but generally the complex molecules in organic matter, for example, carbohydrates and proteins, yield comparatively low molecular weight free radicals which rearrange and combine to give PAHs. Soot consists of a large number of linked aromatic rings and is also often produced. This substance is not physiologically active but contains PAHs which often are. The conditions of pyrolysis or combustion strongly influence the type and quantity of PAHs produced (Neff, 1979).

Crude and refined petroleum products show a large variation in aromatic and PAH content but total aromatic hydrocarbons are usually in the range 0.2–7.4% (Gilchrist et al., 1972; Jewell et al., 1972). Synthetic crude petroleum produced by coal liquification and shale oil pyrolysis generally has a higher PAH content.

PAHs can occur in waste waters, such as domestic and industrial effluents,

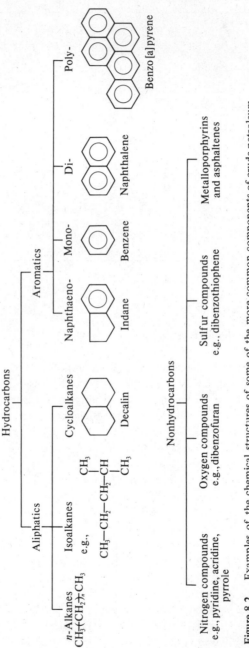

Figure 8.2 Examples of the chemical structures of some of the more common components of crude petroleum.

sewage, and urban runoff, and in atmospheric discharges from fossil fuel combustion, as well as combustion or pyrolysis of all kinds. Thus these substances are
ubiquitous in the environment where they often occur in a sequence of concentrations related to a known source. Some examples of their occurrence are shown in
Table 8.3. Many types of PAHs occur in the environment but usually those with
three to six benzene rings, some with alkyl side chains, are most common.

Maher et al. (1979) and Bagg et al. (1981) have results indicating highest concentrations of PAHs in marine sediments adjacent to urban areas. This may be a general
pattern whereby PAHs tend to be concentrated in aquatic sediments adjacent to
urban areas.

TRANSPORT AND TRANSFORMATIONS

The hydrocarbons exhibit low solubility in water and are strongly lipophilic. For
example, the simplest PAH (naphthalene) has a water solubility of about 30 mg L^{-1}
and this decreases with increasing molecular weight of PAH (McAuliffe, 1966). The
hydrocarbons are rapidly adsorbed onto particulate matter in aquatic areas. For
example, Connell (1982) found that in the Hudson Raritan Estuary the petroleum
hydrocarbons present were substantially adsorbed onto the bottom sediments. In
addition, a variety of other processes can occur which are illustrated in Figure 8.3.

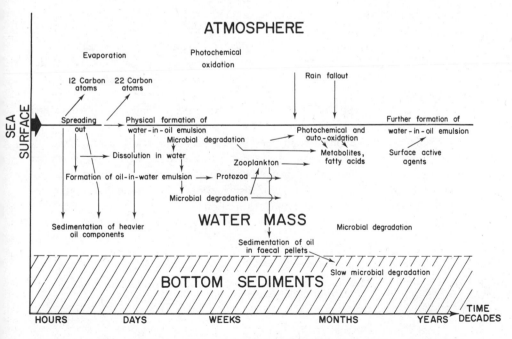

Figure 8.3.   Diagrammatic summary of the fate of petroleum discharged to an aquatic area.
[From GESAMP (1977). Reprinted with permission from the Food and Agriculture Organization
of the United Nations.]

Photochemical Processes

Photochemical transformations of crude and fuel oils involve (1) formation of reactive radicals, (2) formation of potentially toxic intermediate peroxides and hydroperoxides, and (3) formation of oxidation products such as carboxylic acids, esters, oxygenated aromatics, carbonyl compounds, and carbon dioxide (GESAMP, 1977; Larsen et al., 1976; Hansen, 1977). Preferential oxidation of aromatic hydrocarbons and sulfur compounds in surface films of crude oils on seawater irradiated at 250 nm has been observed (Hansen, 1977).

Microbiological Processes

Many marine microorganisms, such as bacteria and fungi, can metabolize either completely or partially the hydrocarbons from petroleum oils and fractions. Microbial oxidation is dominated by bacterial action which appears to be species dependent. These organisms are capable of oxidizing aromatic hydrocarbons from monocyclic to polycyclic compounds although biodegradation of highly condensed ring structures remains to be established (see Figure 8.4). The general mechanisms for bacterial oxidation of aromatics involve the incorporation of both atoms of molecular oxygen into aromatic hydrocarbons and the formation of cis-dihydrodiols, followed by further oxidation to catechol-type compounds (Gibson, 1977). Catechols can undergo enzymic cleavage of the benzene ring between or adjacent to the hydroxyl groups. Oxidation can proceed to completion or terminate at various oxygenated metabolites (e.g., salicylic acids, naphthenic acids and naphthyl alcohols) in the presence of an alternative growth substrate. In most cases, the higher molecular weight PAH cannot be used as the sole carbon source (Varanasi and Malins, 1977).

The distribution of hydrocarbon-utilizing bacteria and fungi, in marine waters, is worldwide, but populations appear to be extremely low except in oil-contaminated areas. Within aerobic sediments, higher levels of activity are observed, possibly facilitated by the presence of organic carbon substrates. Oxidation of oil in the natural environment is dependent on factors such as temperature, salinity, concentrations of inorganic nutrients, extent of dispersion in water, abundance and types of microorganisms, and chemical composition of oil.

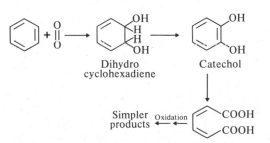

Dihydro
cyclohexadiene

Catechol

Simpler products ← ← Oxidation

Figure 8.4.   Microbial Oxidation of Benzene.

Many marine and estuarine microorganisms rapidly degrade certain fractions of petroleum hydrocarbon mixtures and oils (e.g., n-alkanes and isoalkanes). The alkanes are degraded by oxidation of one of the terminal methyl groups to a carboxylic acid and then stepwise degradation, eventually with complete conversion to carbon dioxide and water. The n-alkanes are oxidized most readily and the rate decreases with increasing branching. But it appears the rate of degradation of some fractions (e.g., cycloalkanes and PAHs) can be exceptionally slow, with the resultant formation of relatively persistent or refractory fractions (GESAMP, 1977). For example, Sanders et al. (1980) have found partly degraded fuel oil was present in estuarine sediments 5 yr after a spill.

UPTAKE AND METABOLISM BY LARGER BIOTA

Uptake of the hydrocarbons is governed by similar principles to other lipophilic substances. The water solubility is low and the n-octanol to water partition coefficient correspondingly high. The bioaccumulation factor (i.e., ratio of tissue concentration to water concentration) has been measured with the clam (*Rangia cuneata*) for naphthalene (6.1), phenanthrene (32.0), chrysene (8.2), and benzo[a]pyrene, B(a)P (8.7). Uptake rates are a function of the concentration in water and are influenced by chemical structure and environmental parameters such as temperature and salinity, although only a few studies have considered these factors (Connell and Miller, 1981a).

Absorption occurs through respiratory surfaces, the gastrointestinal tract, and external surfaces, with the hydrocarbons being generally deposited in lipid-rich tissues (GESAMP, 1977). However, the relative significance of petroleum hydrocarbon uptake through different routes of entry, and modification by specific physiological mechanisms, is not indicated by available data. Relatively few studies on the uptake of hydrocarbons by benthic plants have been undertaken.

Many marine organisms, particularly vertebrates, possess metabolic pathways capable of activation and/or deactivation of absorbed petroleum hydrocarbons. The general mechanism involves formation of metabolites, usually by cytochrome P-450 mediated mixed-function oxidase systems (MFO) and transformation by microsomal epoxide hydrase and glutathione-S-transferases, into more hydrophilic conjugates (see Figure 8.5). Consequently, enzymic studies of organisms have focused on the following aspects (Varanasi and Malins, 1977; Connell and Miller, 1981a).

1. The activities and distribution of MFO in marine species. With vertebrates (e.g., fish), this activity is concentrated in the microsomal hepatic system, but a different distribution pattern is observed among invertebrates (e.g., green gland and pyloric stomach in crabs) possessing MFO activities.

2. The formation of reactive intermediate arene or alkene oxides from aromatic hydrocarbons and alkenes, respectively. Certain of these electrophilic metabolites chemically interact with cellular macromolecules and are implicated as primary carcinogens, mutagens, and cytotoxins.

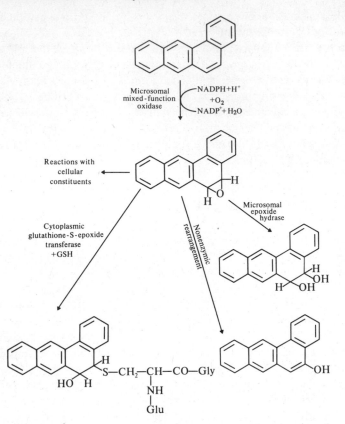

Figure 8.5.   Pathways involved in the metabolism of benzo(a)anthracene.

3.   Detoxification mechanisms which convert the epoxides to diols by the action of microsomal epoxide hydrases, and glutathione-S-transferases which catalyze glutathione conjugates (see Figure 8.5).

4.   Induction of mixed-function oxidase activity in marine species (e.g., fish) by petroleum hydrocarbons and other lipophilic substances.

Metabolic systems in marine organisms offer several potential explanations for the behavior of petroleum hydrocarbons. The metabolic level and activity of MFO in marine organisms may, in fact, determine the sensitivity or tolerance of marine organisms to sublethal levels of petroleum hydrocarbons, especially aromatics and alkenes. It has been postulated that a metabolic threshold level exists below which organisms function normally and accumulation of petroleum hydrocarbons does not occur (GESAMP, 1977; Miller, 1982). In some organisms (e.g., bivalves), limited MFO activity may explain rapid hydrocarbon saturation of lipid compartments.

Resistance to degradation, or persistance, is an important characteristic influencing the impact of a substance in an organism or the environment. Estimates of persistance can be made by assuming first-order kinetics for degradation allowing

Table 8.4.   Persistence of Some Hydrocarbons in Aquatic Organisms as Half-Lives[a]

Organism	Compound	Half-Life (days)
Oysters	Naphthalenes	2
(*Crassostrea virginica*)	Anthracene	3
	Fluoranthene	5
	Benzo[a]anthracene	9
	B(a)P	18
Mussels	B(a)P	16
(*Mytilus edulis*)	Aliphatics and aromatics	2
	Aliphatics	4
	Aromatics	48–60
Copepod	Naphthalene	1.5
(*Calanus helgolandicus*)		
Water flea	Anthracene	0.4
(*Daphnia pulex*)		
Rainbow trout	Naphthalene	Depending on
(*Salmo gairdneri*)	Methyl naphthalene	exposure time
		0.3–38
Sea mullet	*n*-Alkanes	18
(*Mugil caphalus*)		(max)

[a] Compiled from Connell (in press) and Miller (1982).

the calculation of "half-lives" [see Chapter 2, Equation (5)]. "Biological" half-lives for petroleum hydrocarbons in marine animals vary between field and laboratory estimates – the latter ranging from about 2 to 7 days (Connell and Miller, 1981a), while the field estimates are significantly longer (see Table 8.4). However, the depuration kinetics of residual cycloaliphatic and aromatic fractions in many marine animals could be better represented by second-order reaction kinetics (see Ellegehausen et al., 1980; Moriarty, 1975).

BIOLOGICAL TRANSFER AND BIOMAGNIFICATION

It has been suggested that petroleum hydrocarbons may undergo biomagnification since they exhibit a high partition coefficient (octanol to water) and are comparatively persistent in marine organisms (see Chapter 2). The studies by Blumer and his associates (see Blumer, 1967; Blumer et al., 1969) provide evidence of the stability and transfer of some biogenic hydrocarbons in marine food chains.

Teal (1977) has thoroughly reviewed the information on food chain transfer of hydrocarbons and concluded that some natural hydrocarbons in copepods (Blumer, 1967; Blumer et al., 1969) and fishes (Ackman, 1972) must have come from food web transfer. Straughan (1977) observed that although natural hydrocarbons may apparently be concentrated by the process of biomagnification, there is no convincing evidence of food chain magnification of petroleum hydrocarbons. A similar conclusion has been reached by Mackie et al. (1974), Zitko (1975), and GESAMP (1977).

However, petroleum contains many different hydrocarbons having different properties. For example, Connell and Bycroft (1978) have suggested that the normal alkanes are less likely to biomagnify than other hydrocarbons present. Stegeman and Teal (1973) have shown that, for long periods, oysters retained a portion of their body burden of hydrocarbon after being placed in clean water. Somewhat similar results, for example, by Lee (1975), for other organisms have led to the suggestion that a proportion of the petroleum hydrocarbons may not be lost during the life of the animal. The more resistant portion, which presumably would consist of compounds other than the normal alkanes, would potentially be available for biomagnification.

This is consistent with the findings of Miller and Connell (1980) who suggest that hydrocarbons, depleted in n-alkanes, occurred in seabirds as a result of biomagnification of petroleum hydrocarbons in low concentrations in the abiotic environment.

LETHAL TOXICITY

The alkanes in petroleum exhibit comparatively little toxicity or other adverse physiological impact. These effects are usually due to the aromatic substances present. These compounds are generally the most water soluble components and thus attention has been focused on the water soluble fractions (WSF) of petroleum products as well as petroleum itself.

Comparisons between estimated toxicities of soluble aromatics to classes of marine organisms is presented in Table 8.5. In general, No. 2 fuel oil was more toxic

Table 8.5.   Toxicities of Soluble Aromatics to Classes of Marine Organisms

Class of Organism	Estimated Ranges of Lethal Concentrations of Soluble Aromatics (ppm)		
	Moore and Dwyer (1974)	Crude Oils (96 hr LC_{50})	No. 2 Fuel Oil
Flora	10–100	ID[a]	ID
Finfish	5–50	1–>10	1–>10
Crustaceans	1–10	1–>10	0.1–10
Bivalves	5–50	1–>10	0.5[b]–>5
Gastropods	1–100	1–>10	1–>5
Other benthic invertebrates	1–10	>5–10 ID	>1 ID
Larvae (all species)	0.1–1	0.1–5	ID
Juveniles	NR[c]	5–>10	1–10

Source: Miller (1982). Reprinted with permission from John Wiley & Sons Ltd.

[a] ID = inadequate data.

[b] Tentative result.

[c] NR = not reported.

Table 8.6.   Adult Sensitivities to Crude Oils and No. 2 Fuel Oil

| Species Tested | Temperature Range (°C) | 96 hr LC$_{50}$ (ppm) | |
		Crude Oils	No. 2 Fuel Oil
Vertebrates (Teleosti)			
Silverside minnow (*Menidia beryllina*)	18–22	5.5	
Gulf killifish (*Fundulus simulus*)		16.8	
Sheepshead minnow (*Cyrinodon variegatus*)		>19.8	>6.9(3.1)
Pink salmon (*Oncorhyncus gorbuscha*)	3.6–10	2.92	0.81
Dolly Varden smolts (*Salvelinus malma*)		2.94	2.29
Saffron cod (*Eleginus gracilis*)		2.28	2.93
Tube-snouts (*Aulorhynchus flavidus*)		1.34	–
Invertebrates (Crustacea; Decapoda)			
Opossum shrimp (*Mysidopsis almyra*)	18–22	–	
Prawn (*Leander terrvicornis*)		6.0	
Grass shrimp (*Palaemonetes pugio*)		>16.8	3.5
Brown shrimp postlarvae (*Penaeus aztecus*)		>19.8	4.9
Dock shrimp (*Pandalus danae*)	3.5–5.4	0.81	1.11
Humpback shrimp (*Pandalus goniurus*)		1.98	1.69
Scooter shrimp (*Evalus fabrieii*)		1.46	0.53
Coonstripe shrimp (*Pandulus hypsinotus*)		2.72	–
Pink shrimp (*Pandulus borcalis*)		2.43	0.21

Source: Miller (1982). Reprinted with permission from John Wiley & Sons Ltd.

than the crude oils tested which is consistent with the results of many toxicity studies on oils. Considerable variation in sensitivity has been noted with warm-water species (tested at 18–22°C) of fish and crustaceans. Consistent with major oil field development there has been increased interest in subarctic communities. Tropical marine species have not been thoroughly investigated and it is inappropriate to extrapolate findings on temperate species because of the differences in their physiological and behavioral adaptations, as well as in the physical–chemical properties of oils.

Dissolved PAHs are toxic to aquatic organisms in concentrations usually ranging from 0.1 to 0.5 ppm, although results occasionally lie outside this range. Concentrations in the environment are generally comparatively low and sublethal effects are usually the most significant. Most LC$_{50}$ estimates for marine organisms exposed to oil–water mixtures and solutions are derived from nonstandardized static bioassays. For most adult or mature organisms, a lethal response can be expected in the range of 1–100 $\mu g\,mL^{-1}$ soluble aromatic derivatives (SAD) for exposure periods of several hours to 48 hr; and for the larval stages at concentrations of

Table 8.7.   Life Stage Sensitivities of Marine Organisms to Water Soluble Fractions of Crude Oil and No. 2 Fuel Oil

Species Tested	Stage Tested	96 hr LC$_{50}$ (ppm) Crude Oils	96 hr LC$_{50}$ (ppm) No. 2 Fuel Oil
Vertebrates (fish)			
Cod (*Gadus morhua*)	Eggs (0.5–10 days)	2.6–55	–
Invertebrates			
Brown shrimp (*Penaeus aztecus*)	Postlarvae	–	6.6
	Early juveniles	–	3.7
	Late juveniles	–	2.9
White shrimp (*Penaeus setifirms*)	Postlarvae	–	1.4
	Juveniles	–	1.0
Grass shrimp (*Palaemontes pugio*)	Larvae	–	1.2
	Postlarvae	–	2.4
	Adults	–	3.5
Coonstripe shrimp (*Pandalus hypsinotus*)	Stages I and II	4–7.9	–
	Stage I (molting)	0.9	–
	Stages I–VI	0.2–1.8	–
King crab (*Paralithodes camtschatica*)	Stage I	2.0	–
	Stage I (molting)	1.3	–
Polychaetes (*Neanthes arenaceodentata*)	Juveniles	15–19.9	4–8.4
	Adults	12.5–17.6	2–4.2

Source: Miller (1982). Reprinted with permission from John Wiley & Sons Ltd.

$0.1–1 \,\mu g \,mL^{-1}$ soluble aromatic derivatives (see Moore and Dwyer, 1974). Over longer exposure periods some toxicity levels may decrease to a tenth or less.

A summation of lethal toxicity of crude and fuel oils is shown in Table 8.6. A number of observations can be made:

1.   The refined (No. 2 fuel) oil was generally more toxic than the crude oils.

2.   Fish and crustaceans (decapods) appear to be among the most sensitive species to WSF of both crude oils and No. 2 fuel oil, although several species are relatively tolerant to crude oils (e.g., Sheepshead minnow).

3.   Cold-temperature fish and crustaceans (decapods) (3.5–10°C) were more sensitive than similar warm-temperature species (18–22°C).

4.   Intertidal species were generally more tolerant, probably because of their ability to insulate themselves from the static exposures used in the 96 hr tests (Rice et al., 1977).

Sensitivity to oil at different life stages is less clear because of very limited comparative data. From Table 8.7, the only apparent sensitivities to oil (WSF) occurred at the larval stage of the brown shrimp and molting stages of several other crustaceans (decapods). Neff et al. (1976) and Rice et al. (1977) concluded that generalizations cannot be made about larvae versus adult sensitivity. Egg sensitivities appear to be more perplexing because of limited and contradictory evidence of sensitivity, and tolerance toward several types of oil.

SUBLETHAL EFFECTS

Experimental investigations of sublethal responses have demonstrated both the sensitivity and tolerance of ecologically important species to low persistent concentrations of soluble petroleum hydrocarbons (GESAMP, 1977; Michael, 1977; Connell and Miller, 1981a). As a general rule, a decreased level of physiological activity is indicated by such findings as decreased photosynthetic, filter feeding, survival, and fecundity rates. Feeding and mating behavior in some species is affected. In addition, low concentrations of petroleum hydrocarbons in seawater and sediments can enter edible organisms and cause tainting. Tainting from a variety of sources and types of petroleum has been shown to occur in fish, crustacea, and molluscs (Connell and Miller, 1981b).

Sublethal concentrations range from less than 10 ppb to over 1000 ppb of SAD (see Table 8.8). Behavioral changes and uptake of aliphatic and aromatic hydrocarbons are detectable at the lowest concentrations, but the position with other sublethal responses is ambiguous at these concentrations. Larval and juvenile forms tend to be the most sensitive at concentrations under 1000 ppb of SAD.

Significant physiological and behavioral effects occur with many invertebrates, as well as larval and juvenile forms of fish. Also embryotoxicity and disruption of ionic regulation has been observed among seabirds.

PAHs in low concentrations have been shown to decrease growth, development, and feeding rates of aquatic organisms (Neff, 1979). Some PAHs are well known

Table 8.8.   Range of Effects (Responses) in Marine Organisms for Soluble Aromatic Hydrocarbons

Bioaccumulation

    Behavioral patterns

        Growth and reproduction

            Lethality (larval and juvenile stages)

                Lethality (adults)

0	10	100	1000	100,000
			Concentration (ppb)	

Source: Miller (1982). Reprinted with permission from John Wiley & Sons Ltd.

Table 8.9.   Relative Carcinogenicity of Some PAHs to Laboratory Mammals[a]

Compound	Carcinogenicity
Anthracene	—
Pyrene	—
Benzo[a]pyrene (B(a)P)	+++
Benzo[e]pyrene	—
Coronene	—
Dibenzo[ai]pyrene	+++
Benzo[a]anthracene	+

[a] Compiled from Neff (1979).

carcinogens with mammals and examples of carcinogenic and noncarcinogenic PAHs are shown in Table 8.9. Generally carcinogenic PAHs have four, five, or six rings and angular structures, but alkylation can modify these characteristics.

Many PAHs are also mutagenic or teratogenic. Abnormal growth and offspring have been observed in the laboratory and field situations with aquatic organisms (Neff, 1979). It is also important to note that many of the O, N, and S compounds identified in petroleum are toxic or have other detrimental effects. Very little is known concerning the biological effects of these substances when discharged into marine areas, as petroleum components.

ECOSYSTEMS EFFECTS

Environmental levels of SAD in different aquatic habitats are poorly described. In general, the data are either nonspecific or too inadequate to draw firm conclusions about existing sublethal concentrations in water and sediments. In marine areas, available measurements suggest a probable range of < 10–100 ppb of SAD in seawater and, probably, intertidal water. Levels of SAD approaching 100 ppb are generally detected in polluted estuarine and coastal environments (Clark and MacLeod, 1977).

Given this range, the following effects on marine organisms could be predicted for long-term exposure from the information available:

1. Bioaccumulation of hydrocarbons in body tissues and possible tainting of edible tissues. This may also include transfer of aromatic hydrocarbons through the blood–brain barrier in vertebrates.
2. Behavioral alterations, for example, perception in some species.
3. Metabolic stimulation and/or suppression; reduction in growth.
4. Reduced reproduction success in some species; evidence of deformities.
5. Mortalities among sensitive larval and juvenile forms.

Various factors influence the manifestations of sublethal (and lethal) toxicities. These include species differences in tolerances and responses, stages of development, genetic characteristics, pathological states, and simultaneous exposure to more than one toxicant.

Current assessment of the ecological impacts from petroleum hydrocarbon exposure rely on interpretations derived from two sources of information: (1) field observations and measurements, often in the absence of baseline studies, which generally describe effects caused by accidental oil spills, and (2) dose–response relationships for experimental organisms and conditions selected to represent, in effect, complex interrelationships between heterogeneous groups of marine organisms and the abiotic components of their habitats. Interpretation, of course, is limited by our level of understanding of fundamental ecological processes and interactions in marine environments.

The impact of oil, even in chronic amounts, affects a wide diversity of marine organisms, from plants through to invertebrates and vertebrates. However, considerable variation in tolerances and sensitivities are observed among species and between life stages. Generally, the effects increase according to the following habitat classification (see Sanders et al., 1980):

Pelagic < Subtidal < Intertidal

One notable exception is the high mortality of seabirds, such as auks and sea ducks, in pelagic regions as a result of chronic oil exposure.

The response of aquatic ecosystems to petroleum exposure can be placed into two general categories:

1. Acute exposure incidents where simplification of the ecosystem structure generally occurs and succession reverts back at least one stage. In some oil spills successional changes in marine communities may be more substantial, especially in areas of high mortalities (Michael, 1977; Bourderu and Treshow, 1978).

2. Chronic exposures where gradual modifications of community structure (diversity and influxes of opportunistic species) and basic processes (e.g., nutrient fluxes and production) occur. For example, Grassle et al. (1981) found that low concentrations of No. 2 fuel oil caused a reduction in total macrofaunal numbers after 5 months, particularly suspension feeders. Michael et al. (1975) and Michael (1977) and other workers have found a variety of effects including reduction in number of species, higher proportion of opportunistic species than control areas, and reduction in the numbers of individuals.

McGrath (1974) conducted an investigation of the benthic fauna in Raritan Bay, New Jersey. He reported depressed communities and the lack of amphipods in the genus *Ampelisca* which were dominant 10 yr earlier. These effects were attributed to PHCs. Genus *Ampelisca* has been shown to be a sensitive indicator of petroleum pollution (Blumer et al., 1970). Alterations also include a decrease in energy flow in aquatic ecosystems (Gilfillan et al., 1976).

Connell (1981) points out that enrichment of bottom sediments in aquatic areas with petroleum can lead to reduction in the dissolved oxygen content of the interstitial water. In this situation somewhat similar effects to those observed with organic discharges, as described in Chapter 5, can be expected, acting in conjunction with the other toxic effects described above. For example, Sanders et al. (1980) observed temporary increases in polychaete populations after an oil spill.

Recovery periods appear to be relatively short-term for some communities (e.g., planktonic), but substantial for some others. In particular, the few long-term studies of acute oil spills indicate recovery periods of up to 10 or more years for sedimentary environments in temperate estuarine regions (Anon, 1978). Predicted recovery rates for certain sensitive systems (e.g., polar regions or coral reefs) are expected to be much slower, although considerable uncertainties exist.

Current assessments of the impact of petroleum in the marine environment have concentrated on the responses or apparent tolerances of individual organisms or selected populations to petroleum hydrocarbons rather than the response of the ecosystem components as a whole.

REFERENCES

Ackman, R. G. (1972). Pristane and other hydrocarbons in some fresh-water and marine fish oils. *Lipids* 6, 520.

Anon (1975). *Petroleum in the Marine Environment*. National Academy of Sciences, Washington, D.C.

Anon (1978). Symposium on recovery potential of oiled marine northern environments. *J. Fish. Res. Board Can.* 35, 499–795.

Bagg, J., Smith, J. D., and Maher, W. A. (1981). Distribution of polycyclic aromatic hydrocarbons in sediments from southeastern Australia. *Aust. J. Mar. Freshwater Res.* 32, 65–73.

Blumer, M. (1967). Hydrocarbons in the digestive tract and liver of a basking shark. *Science* 156, 390.

Blumer, M., Robertson, J. C., Gordon, J. E., and Sass, J. (1969). Phytol-derived C19 di- and triolefinic hydrocarbons in marine zooplankton and fishes. *Biochemistry* 8, 4067.

Blumer, J., Sass, J., Souza, G., Sanders, H., Grassle, F., and Hampson, G. (1970). The West Falmouth oil spill: Persistence of the pollution eight months after the accident. Woods Hole Oceanographic Institution. Ref. No. 70-44.

Bourderu, P. and Treshow, M. (1978). "Ecosystem Response to Pollution." In G. C. Butler. (Ed.), *Principles of Ecotoxicology*. John Wiley & Sons, New York, p. 313.

Clark, Jr., R. C. and Macleod, Jr., W. D. (1977). "Inputs, Transport, Mechanisms and Observed Concentrations of Petroleum in the Marine Environment." In D. C. Malins (Ed.), *Effects of Petroleum on Arctic and Sub-arctic Marine Environments and Organisms, Nature and Fate of Petroleum*, Vol. 1. Academic Press, New York, p. 91.

Connell, D. W. (1981). Petroleum Hydrocarbons in the Hudson Raritan Estuary. Marine Sciences Research Center, State University of New York, Stony Brook, Working Paper 3.

Connell, D. W. (1982). An approximate petroleum hydrocarbon budget for the Hudson Raritan Estuary, New York. *Mar. Pollut. Bull.* **13**, 89.

Connell, D. W. (in press). Ecotoxicology on polychlorinated biphenyls, dioxins and polycyclic aromatic hydrocarbons. Proceedings of an Ecotoxicology Workshop, November 1982, Canberra, Australian Academy of Science.

Connell, D. W. and Bycroft, B. M. (1978). Maximum biological half-lives of *n*-alkanes (C_9–C_{13}) in the sea mullet (*Mugil cephalus*). *Chemosphere* **10**, 779.

Connell, D. W. and Miller, G. J. (1981a). Petroleum hydrocarbons in aquatic ecosystems – behaviour and effects of sublethal concentrations: Part 1. *CRC Crit. Rev. Environ. Control* **11**, 37.

Connell, D. W. and Miller, G. J. (1981b). Petroleum hydrocarbons in aquatic ecosystems – behaviour and effects of sublethal concentrations: Part 2. *CRC Crit. Rev. Environ. Control* **11**, 105.

Ellegehausen, H., Guth, J. A., and Esser, H. O. (1980). Factors determining the bioaccumulation potential of pesticides in the individual compartments of aquatic food chains. *Ecotoxicol. Environ. Safety* **4**, 134.

GESAMP (1977). (Joint Group of Experts on the Scientific Aspects of Marine Pollution). *Impact of Oil on the Marine Environment. Rep. Study No. 6.* Food and Agriculture Organization, Rome.

Gibson, D. T. (1977). "Biodegradation of Aromatic Petroleum Hydrocarbons." In D. A. Wolfe (Ed.), *Fate and Effects of Petroleum Hydrocarbons in Marine Organisms and Ecosystems*, Pergamon Press, New York, p. 36.

Gilchrist, C. A., Lynes, A., Steel, G., and Whitlam, B. T. (1972). The determination of polyaromatic hydrocarbons in the environment by glass gas chromatography. *Anal. Chem.* **50**, 243.

Gillfillan, E. S., Mayo, D., Hanson, S., Donovan, D., and Jiang, L. C. (1976). Reduction in carbon flux in *Myarenaria* caused by a spill of No. 6 fuel oil. *Mar. Biol.* **37**, 115.

Grassle, J. F., Elmgren, R., and Grassle, J. P. (1981). Response of benthic communities in MERL experimental ecosystems to low level chronic additions of No. 2 fuel oil. *Mar. Environ. Res.* **4**, 279.

Hansen, H. P. (1977). Photodegradation of hydrocarbon surface films. *Rapp. P-v. Reun Cons. Int. Explor. Mer.* **171**, 101.

Hites, R. A., LaFlamme, R. E., and Farrington, J. W. (1977). Sedimentary polycyclic aromatic hydrocarbons: The historical record. *Science* **198**, 829.

Jewell, D. M., Ruberto, R. G., and Davis, B. E. (1972). Systematic approach to the study of aromatic hydrocarbons in heavy distillates and residues by elution adsorption chromatography. *Anal. Chem.* **44**, 2318.

Kallio, R. E. (1976). "The Variety of Petroleums and Their Degradations." In *Sources, Effects and Sinks of Hydrocarbons in the Aquatic Environment.* The American Institute of Biological Sciences, Washington, D.C., p. 214.

Kolattukudy, P. E. (1976). "Biogenesis of Non-Isoprenoid Aliphatic Hydrocarbons." In *Sources, Effects and Sinks of Hydrocarbons in the Aquatic Environment.* The American Institute of Biological Sciences, Washington, D.C., p. 120.

Koons, C. B., Jamieson, G. W., and Cieroszko, L. S. (1965). Normal alkane distributions in marine organisms. Possible significance to petroleum origin. *Bull. Am. Assoc. Pet. Geol.* **49**, 301.

Korte, J. and Boedefeld, E. (1978). Ecotoxicological review of global impact on petroleum industry and its products. *Ecotoxicol. Environ. Safety* **2**, 55.

LaFlamme, R. E. and Hites, R. A. (1978). The global distribution of polycyclic aromatic hydrocarbons in recent sediment. *Geochim. Cosmochim. Acta* **42**, 289.

Larson, R. A., Blankenship, D. W., and Hunt, L. L. (1976). "Toxic Hydroperoxides: Photochemical Formation from Petroleum Constituents." In *Sources, Effects and Sinks of Hydrocarbons in the Aquatic Environment*. The American Institute of Biological Sciences, Washington, D.C., p. 298.

Lee, R. F. (1975). "Fate of Petroleum Hydrocarbons in Marine Zooplankton." In *Proceedings of the Joint Conference on Prevention and Control of Oil Pollution*. American Petroleum Institute, Washington, D.C., p. 549.

Mackie, P. R., Whittle, K. J., and Hardy, R. (1974). Hydrocarbons in the marine environment I. *n*-Alkanes in the Firth of Clyde. *Estuarine Coastal Mar. Sci.* **2**, 359.

Maher, W. A., Bagg, J., and Smith, D. J. (1979). Determination of poly-cyclic aromatic hydrocarbons in marine sediments using solvent extraction, thin layer chromatography and spectrofluorimetry. *Int. J. Environ. Anal. Chem.* **7**, 1.

Malins, D. C. (Ed.) (1977). *Effects of Petroleum on Arctic and Sub-arctic Marine Environments and Organisms. Nature and Fate of Petroleum*, Vol. 1. Academic Press, New York.

McAuliffe, C. (1966). Solubility in water of paraffin, cycloparaffin, olefin, acetylene, cyclolefin, and aromatic hydrocarbons. *J. Phys. Chem.* **70**, 1267.

McGrath, R. A. (1974). "Benthic Macrofaunal Census of Raritan Bay." In *Hudson River Ecology*, 3rd Symposium. Hudson River Environmental Society, New York.

Michael, A. D. (1977). "The Effects of Petroleum Hydrocarbons on Marine Populations and Communities." In D. A. Wolfe (Ed.), *Fate and Effects of Petroleum Hydrocarbons in Marine Organisms and Ecosystems*. Pergamon Press, New York, p. 129.

Michael, A. D., Van Raalte, C. R., and Brown, L. S. (1975). "Long Term Effects of an Oil Spill at West Falmouth, Mass." In *Proceedings of 1975 Conference on Prevention and Control of Oil Pollution*. American Petroleum Institute, U.S. EPA, U.S. Coast Guard, Washington, D.C.

Miller, G. J. (1982). Ecotoxicology of petroleum hydrocarbons in the marine environment. *J. Appl. Toxicol.* **2**, 88.

Miller, G. J. and Connell, D. W. (1980). Occurrence of petroleum hydrocarbons in some Australian seabirds. *Aust. Wildl. Res.* **7**, 281.

Miller, G. J. and Connell, D. W. (1982). Global production and fluxes of petroleum and recent hydrocarbons. *Int. J. Environ. Studies* **19**, 273.

Moore, S. F. and Dwyer, R. L. (1974). Effects of oil on marine organisms: A critical assessment of published data. *Water Res.* **8**, 819.

Moriarty, F. (1975). "Exposure and Residues." In F. Moriarty (Ed.), *Organochlorine Insecticides: Persistent Organic Pollutants*. Academic Press, London, p 29.

Neff, J. M. (1979). *Polycyclic Aromatic Hydrocarbons in the Aquatic Environment.* Applied Science Publishers, London.

Neff, J. M., Anderson, J. W., Cox, B. A., Laughlin, Jr., R. B., Rossi, S. S., and Tatum, H. E. (1976). "Effects of Petroleum on Survival, Respiration and Growth of Marine Animals." In *Sources, Effects, and Sinks of Hydrocarbons in the Aquatic Environment.* American Institute of Biological Sciences, Washington, D.C., p. 515.

Posthuma, J. (1977). The composition of petroleum. *Rapp. P-v. Reun Cons. Int. Explor. Mer.* **171**, 7.

Rice, S. D., Short, J. W., and Karinen, J. F. (1977). "Comparative Oil Toxicity and Comparative Animal Sensitivity." In D. A. Wolfe (Ed.), *Fate and Effects of Petroleum Hydrocarbons in Marine Organisms and Ecosystems.* Pergamon Press, New York, p. 78.

Sanders, H. L., Grassle, F., Hampson, G. R., Morse, L. S., Garner-Price, S., and Jones, C. C. (1980). Anatomy of an oil spill: Long-term effects from the grounding of the barge *Florida* off West Falmouth, Massachusetts. *J. Mar. Res.* **32**, 265.

Stegeman, J. J. and Teal, J. M. (1973). Accumulation, release and retention of petroleum hydrocarbons by the oyster, *Crassostrea virginica. Mar. Biol.* **22**, 37.

Straughan, D. (1977). "Biological Survey of Intertidal Areas in the Straights of Magellan in January, 1975, Five Months after the Metvia Oil Spill." In D. A. Wolfe (Ed.), *Fate and Effects of Petroleum Hydrocarbons in Marine Organisms and Ecosystems.* Pergamon Press, New York, p. 247.

Teal, J. M. (1977). "Food Chain Transfer of Hydrocarbons." In D. A. Wolfe (Ed.), *Fate and Effects of Petroleum Hydrocarbons in Marine Organisms and Ecosystems.* Pergamon Press, New York, p. 71.

Varanasi, V. and Malins, D. C. (1971). "Metabolism of Petroleum Hydrocarbons: Accumulation and Biotransformation in Marine Organisms." In D. C. Malins (Ed.), *Effects of Petroleum on Arctic and Sub-arctic Marine Environments and Organisms,* Vol. 2. Academic Press, New York, p. 175.

Zitko, V. (1975). Aromatic hydrocarbons in aquatic fauna. *Bull. Environ. Contam. Toxicol.* **14**, 5.

9

POLYCHLORINATED BIPHENYLS AND OTHER SYNTHETIC ORGANIC COMPOUNDS

POLYCHLORINATED BIPHENYLS (PCB)s

In the history of synthetic chemical compounds, PCBs are not new. A patent was taken out by Schmidt and Schultz in 1881 for the manufacture of these substances. However, commercial manufacture was not started until 1930 in the United States. In 1971, PCBs were produced in nine countries, mainly members of the OECD, as well as the USSR and Czeckoslovakia, and total production in that year amounted to 48,400 tonnes, excluding the USSR and Czeckoslovakia (Anon, 1973). A wide variety of trade names are used to describe the PCBs used in commerce and some of these are shown in Table 9.1. Peak production of PCBs in the United States, where Monsanto is the sole manufacturer, occurred in 1970 (Hutzinger et al., 1974). After that year, the Monsanto company instituted voluntary restrictions on the sale of PCBs for environmental reasons.

The action by Monsanto was to restrict sales to uses in electrical capacitors, electrical transformers, vacuum pumps, and gas transmission turbines. These uses, it was expected, would restrict emissions to the environment. Prior to 1970, uses in addition to those mentioned previously included hydraulic fluids, plasticizers, adhesives, heat transfer systems, lubricants, inks, and a variety of others (Hutzinger et al., 1974).

The major entry of PCBs into the environment results from vaporization during burning, leaks, disposal of industrial fluids, and disposal in dumps and landfills (Peakall, 1975). The cumulative production of PCBs since 1930 is calculated at about 1 million tonnes and approximately half of this quantity has been disposed of in landfills and dumps and believed to be released slowly from this situation (WHO, 1976).

The properties of the PCBs are somewhat similar to the organochlorine insecticides and therefore similar isolation techniques are used in the preparation of samples for analysis. Prior to the mid-1960s, gas chromatographic analyses of samples isolated in this way led to the detection of "spurious" chromatographic peaks. It is likely that there was much confusion between unrecognized PCB gas

Table 9.1.   The World's Major Producers of PCBs

Producer	Country	Tradename of PCB
Monsanto	United States and Great Britain	Aroclor®
Bayer	Germany	Clophen®
Prodelec	France	Phenoclor and Pyralene®
Kanegafuchi	Japan	Kanechlor®
Mitsubishi–Monsanto	Japan	Santotherm®
Caffaro	Italy	Fenclor®
Sovol	U.S.S.R.	
Chemko	Czechoslovakia	

Source: Hutzinger et al. (1974). Reprinted with permission from CRC Press, Inc.

chromatographic peaks and those due to organochlorine insecticides in the scientific literature prior to this period. In 1966 it was established that the "spurious" peaks were in fact due to PCBs (Jensen, 1972; Widmark, 1967). This has initiated extensive investigations throughout the world on the occurrence of PCBs in the environment. This work continues today and much of the data obtained have been summarized by Wasserman et al. (1979) and Richardson and Waid (1982) and is shown in Tables 9.2 and 9.3.

These data clearly indicate that PCBs are widespread in the global environment. However, it is interesting to note that Olafson (1978) after an extensive survey was

Table 9.2.   Examples of concentrations of PCBs in the Abiotic Environment[a]

Environmental Sector/Organism	Area	Concentration
Atmosphere	United States, Canada Sweden, Japan	$0.01-50 \text{ mg m}^{-3}$ Areas close to discharges up to $12 \mu g \text{ m}^{-3}$
Soils	United States, Canada Sweden, Japan	$< 1 \text{ mg kg}^{-1}$ Up to 510 mg kg^{-1} in areas close to discharges
Bottom sediments	United States, freshwater areas	Medium values $0.001-0.160 \text{ mg kg}^{-1}$
Bottom sediments	Great Lakes, North America	$0.004-0.1 \text{ mg kg}^{-1}$ (averages)
Bottom sediments	United States, estuarine and marine coastal areas	$0.002-13 \text{ mg kg}^{-1}$ Up to 486 mg kg^{-1} close to discharges
Freshwater	United States	Not detectable to $0.0005 \text{ mg kg}^{-1}$
Marine surface water	North Atlantic Ocean	$0.0008-0.041 \text{ mg L}^{-1}$

[a] Compiled from Richardson and Waid (1979).

Table 9.3.   Examples of PCB Concentrations in Biota (1970–1982)[a]

Organism Groups	Area	Concentration[b]
Plankton	Atlantic Ocean	0.03–1.5 (whole organism)
Invertebrates	Europe	0.01–1.70 (whole organism)
Invertebrates	Australia	N.D. to 0.29 (whole organism)
Fish	United States	Up to 360.0
Fish	Europe	Up to 3.9
Fish	Japan	1.0–10.0
Fish	Australia	Up to 6.03
Seals	Canada	3.0–22.0
Seals	England, Baltic, North, and Irish Seas	Mean 212 (blubber)
Whales	Atlantic Coast, Canada	0.012–0.185
Cetaceans and dolphins	Canada	0.070–14.7 (blubber)
Birds	North America	Up to 14 (overlap)
Seabirds	New Zealand	0.5–3.3 (adipose tissue)
Seabirds	Australia	Up to 15.7
Birds	Europe	Up to 740 (adipose tissue)
Terrestrial animals	North America, Europe, Japan	Up to 45 (adipose tissue)

[a] Compiled from Wasserman et al. (1979) and Richardson and Waid (1982).
[b] Wet weight; mg kg^{-1} in muscle tissue unless otherwise noted.

unable to detect PCBs in corals, fish, and molluscs from the Great Barrier Reef. Thus, some extensive areas are as yet unaffected by PCB contamination. Nevertheless, Peakall (1975) has produced data that indicate PCBs are widespread and occur in remote areas.

Chemical Nature and Properties

The commercial synthesis of PCBs involves the progressive chlorination of biphenyl to give a variety of products with differing chlorine content. Chlorine can be substituted in any of the biphenyl ring positions 2 to 6 or 2' or 6' to give compounds with 1 to 10 chlorine atoms (see Figure 9.1). A total of 210 isomers are possible with 102 probably formed. Examples of PCBs are shown in Figure 9.2. Polychlorinated terphenyls (PCTs) are also prepared in much more limited quantities (see Figure 9.3).

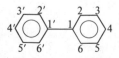

Figure 9.1.   Substitution positions for chlorine on biphenyl to give the PCBs.

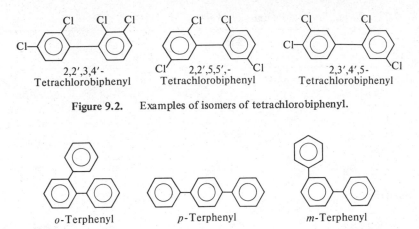

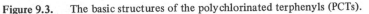

Figure 9.2.   Examples of isomers of tetrachlorobiphenyl.

o-Terphenyl   p-Terphenyl   m-Terphenyl

Figure 9.3.   The basic structures of the polychlorinated terphenyls (PCTs).

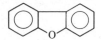

Figure 9.4.   Basic structure of the polychlorinated dibenzofurans.

The commercial products are graded according to chlorine content. For example, with the Monsanto Company the trade name Aroclor is used, characterized by a four digit code. The first two digits indicate the type of product involved with 12 indicating PCBs, 54 indicating PCTs, and 25 and 44 indicating mixtures of PCBs and PCTs (Hutzinger et al., 1974). The last two digits indicate the chlorine content as a percentage of the total (see Table 9.4). Figure 9.4 contains gas chromatographs of some common Aroclors and demonstrates the complexity of the mixtures of isomers which occur with the various commercial products.

Table 9.4.   Approximate Composition of Aroclors

No. of Cl Atoms in molecule	% of Chlorine Weight	Aroclor				
		1221	1242	1248	1254	1260
0	0	12.7				
1	18.8	47.1	3			
2	31.8	32.3	13	2		
3	41.3		28	18		
4	48.6		30	40	11	
5	54.4		22	36	49	12
6	59.0		4	4	34	38
7	62.8				6	41
8	66.0					8
9	68.8					1

Source: WHO (1976). Reprinted with permission from the World Health Organisation.

Table 9.5.   Some Physical Properties of PCBs

Property	Aroclor 1221	Aroclor 1248	Aroclor 1268
Appearance	Clear mobile oil	Clear mobile oil	White to off-white powder
Specific gravity	1.182–1.192	1.405–1.415	1.804–1.811
Distillation range ($^\circ$C corrected)	275–320	340–375	435–450
Solubility in nonpolar solvent	Very soluble	Very soluble	Very soluble
Solubility in water (mg L^{-1})	1.19–5.9	0.034–0.175	< 0.007
Vaporization rate (at 100°C with 12.28 cm^2 surface area, in g cm^{-2} hr^{-1} $\times 10^5$)	174	15.2	< 0.9

[a] Compiled from Hutzinger et al. (1974).

Some of the physicochemical properties are indicated in Table 9.5. There is a steady change in properties with increasing chlorine content and the other grades, not indicated in this table, have properties intermediate to those shown. As a general rule, the PCBs have low volatility, low water solubility, and high resistance to chemical and biological breakdown. In fact these substances are only destroyed at temperatures in excess of 800°C for 10 sec. This resistance to breakdown is a desirable property needed in many commercial products. For example, the PCBs have been added to plastics to improve flame resistance and increase life span. They also have high dielectric constants making them suitable for use in electrical equipment.

The polybrominated biphenyls (PBBs) have similar properties to the PCBs. These substances have been implicated in some environmental problems but are not as widely used as PCBs and do not occur in the environment as widely as the PCBs.

The PCBs contain a variety of impurities which are mostly polychlorinated dibenzofurans (see Figure 9.5) and chlorinated naphthalenes (WHO, 1976). The concentrations of the dibenzofurans in commercial products have been shown to range from the not detectable to 8.4 mg kg^{-1}. These impurities may have an influence on the toxicity exhibited by commercial PCB mixtures.

Sources of PCBs and Factors Influencing their Distribution in the Environment

Table 9.6 shows the fate of PCBs manufactured in North America in 1972. Current patterns may be different but these figures give an indication of the ultimate fate of the PCBs produced. The first three disposal routes provide the greatest entry to

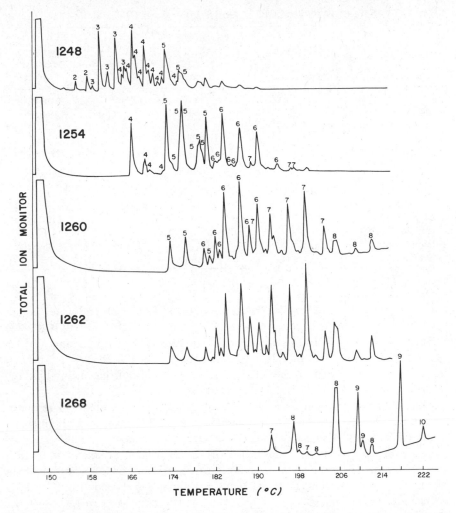

Figure 9.5. Capillary column chromatograms of some Aroclors. Numbers above the peaks designate the number of chlorine atoms per biphenyl molecule. [From Stalling and Huckins (1971). Reprinted with permission from the Association of Official Analytical Chemists.]

the environment. It has been suggested that disposal in dumps, and so on, in sealed containers will result in a low rate of release from this source.

Although the volume is low, incineration and evaporation of PCBs containing a low number of chlorine atoms is believed to be a significant release path. But the release path through leakage and dispersal of industrial fluids (WHO, 1976) is believed to be one of the most significant regarding the occurrence of PCBs in organisms.

PCBs released into the atmosphere are adsorbed onto atmospheric particles but the quantity in the vapor phase can be important. Washout or rainout by natural

Table 9.6.   Fate of PCBs Produced in North America in 1972

Route	Approximate Rate (tonnes yr^{-1})	Percentage of Annual Production	PCB Aroclor Grade	Environmental Compartment Affected
Vaporization from plasticizers	2.0×10^3	4.5	48–60	Atmosphere
Vaporization during incineration	4.0×10^2	1	42	Atmosphere
Leaks and disposal of industrial fluids	5.0×10^3	13	42–60	Hydrosphere
Destruction by incineration	4.0×10^3	9	Mainly 42	–
Disposal in dumps and landfills	1.8×10^4	52.5	42–60	Lithosphere
Accumulation in service	7.0×10^3	20	42–54	Lithosphere

Source: Nisbet and Sarofilm (1972).

precipitation results in a transfer of atmospheric PCBs to soils and water bodies. The average residence time in the atmosphere is estimated at 2–3 days (WHO, 1976).

The deposition of PCBs in soils occurs at the rate of 1000–2000 tonnes yr^{-1} in North America and most of this occurs in urban areas (WHO, 1976). The half-life in soil is estimated to be about 5 yr with loss occurring mainly by two processes — evaporation and biotransformation.

PCBs discharged to aquatic areas are rapidly adsorbed onto particulate matter and ultimately deposited in bottom sediments, usually close to PCB discharge points (Kalmaz and Kalmaz, 1979). Dispersal and movement after this stage are primarily dependent on the movement of the associated sediments with little resolution into the water mass. Discharges of sewage and urban runoff generally contain low concentrations of PCBs. Industrial discharges also contribute to PCBs in the aquatic environment.

The concentration of PCBs in North Atlantic ocean surface water has shown a substantial decrease from 1972 to 1974. This decrease is attributed to the limitations in the use of PCB products after 1971. The evidence also suggests that the PCBs are adsorbed onto particular matter and deposited in deep ocean sediments (Harvey et al., 1974).

PCBs can also be taken up by biota and dispersed by the movement of organisms but this mechanism is believed to be of limited importance as a distribution mechanism (WHO, 1976).

Uptake and Metabolism by Organisms

The physical properties of PCBs, shown in Table 9.5, indicate that these substances are generally soluble in nonpolar solvents and thus lipids. Correspondingly, they

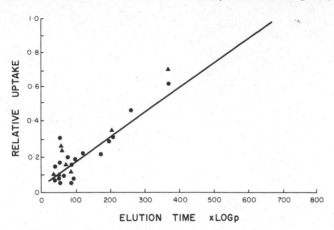

Figure 9.6. Plot of relative uptake of PCBs by aquatic orgnaisms vs. elution time (adsorption on carbon) multiplied by log p. [From Shaw and Connell (1983).]

also have very low solubility in water and can therefore be expected to rapidly transfer from water, or food, to the fatty tissues of animals and plants. Absorbed PCBs are associated with the fatty tissues in organisms (Ernst et al., 1976; WHO, 1976; Phillips, 1980; Shaw and Connell, 1980) and uptake is influenced principally by two factors; the n-octanol to water partition coefficient and the stereochemistry of the molecule. Figure 9.6 shows the relationship between uptake and a factor combining the partition coefficient and a quantitative stereochemical factor measured by chromatography on carbon.

The uptake process involves passage through membranes (gill or stomach) as well as partition between fatty tissues of the animal and water. Initially, this process requires adsorption of PCBs on membrane surfaces before passage through the membrane can occur. Compounds are adsorbed with different strengths depending on their stereochemistry (Shaw and Connell, 1983).

The major routes for uptake by marine organisms are from water via the gills and through food materials. WHO (1976) has summarized the information available on the influence of the food uptake route on PCB concentration in aquatic organisms. The food route can lead to biomagnification in food chains. However, the evidence is not clear generally as to the most important route with completely aquatic organisms. With birds the route through the gills is not applicable and food is the major source of PCBs, leading to biomagnification in many cases. For example, fish-eating birds, such as herring gulls and cormorants, have been found to have higher levels than birds feeding on invertebrates. Also, comparatively high concentrations have been found in eagle owls which are the top carnivores in Sweden (110 mg kg^{-1} in muscle) (WHO, 1976). The data reproduced in Table 9.3 show also a general tendency for top carnivores to have comparatively higher levels of PCBs. Similarly, predatory terrestrial birds tend to contain relatively high PCB concentrations. The levels in most other terrestrial organisms tend to be low and variable (WHO, 1976). The use of organisms as indicators of the presence of PCBs has been discussed by Phillips (1980).

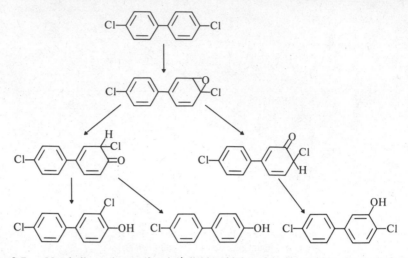

Figure 9.7. Metabolic pathways for 4,4'-dichlorobiphenyl in the rabbit. (From Safe et al., 1976). Reprinted with permission from the Royal Society of Chemistry.)

As a group the PCBs are very resistant to metabolic breakdown by organisms (e.g., WHO, 1976; Ballschmiter et al., 1978). These substances are also resistant to abiotic breakdown with the exception of photolysis (Hutzinger et al., 1972). Biological degradation occurs in organisms, nevertheless, and has been observed with PCBs containing less than four chlorine atoms. Field and laboratory studies have indicated that degradation in organisms occurs in the order, birds, mammals, and fish (Sundstrom et al., 1976). The conversion pathway for 4,4'-dichlorobiphenyl in the rabbit is shown in Figure 9.7. In the cases where metabolic breakdown of PCBs have been studied the formation of an arene oxide is usually involved.

It has been suggested that PCBs containing vicinal hydrogen atoms and with the 4,4' position vacant are the most susceptible to degradation (see Ballschmiter ct al., 1978). The results available indicate that substitution at the 3 and 4 positions occurs preferentially in the degradation process. Sundstrom et al. (1976) suggest that the stereochemistry of the individual PCB molecule also has an impact on the degradation process.

Physiological Effects

Lethal Toxicity

The lethal toxicity of PCBs to a range of organisms is indicated in Table 9.7. Generally toxicity to mammals is low (greater than $1\,g\,kg^{-1}$) compared with DDT and dieldrin (approximately $0.1\,g\,kg^{-1}$ for rats). The mink shows exceptional sensitivity within this group to PCBs (see Table 9.7). Rhesus monkeys are also sensitive, exhibiting symptoms of poisoning in 1 yr with 2.5–5.0 mg of PCB grade 1245 in food per day. At doses of 100–$300\,mg\,kg^{-1}$ in food per day, mortality occurs in 2–3 months (Richardson and Waid, 1979).

Table 9.7. Acute Toxicity of PCBs to Biota[a]

Organism	PCB Grade	Conditions	Test	Toxicity
Mammals				
Rats	1254	—	LD_{50}	$1.3\text{–}2.5 \text{ g kg}^{-1}$
Rats, mice, rabbits	1242	—	LD_{50}	$4.2\text{–}29 \text{ g kg}^{-1}$
Rats	1254	{ 100% mortality in 53 days	PCB in food	1000 ppm
		{ 50% mortality in 8 months	PCB in food	1000 ppm
Mink	1254	100% mortality in 105 days	PCB in food	3.6 ppm
Birds				
Birds (mallard, pheasant bob-white, and Japanese quail)	Six Aroclors (1232–1264)	50% in 5 days of toxic diet	LC_{50}	$0.745\text{–}5.0 \text{ g kg}^{-1}$
Chickens	PCBs	Several months in food	Some deaths	$30\text{–}250 \text{ mg kg}^{-1}$
Fish				
Gammarus oceanicus	1254	PCB colloidal and solubilized in an emulsion	LC_{50}	$0.001\text{–}0.1 \text{ mg L}^{-1}$
Goldfish	Clophen A50	5–21 days, respectively	LC_{50}	$4 \text{ and } 0.5 \text{ mg L}^{-1}$
Pinfish and spotfish	1254	12–18 days	LC_{50}	0.005 mg L^{-1}
Cutthroat trout (*Salmo clarki*)	1221–1260	96 hr	LC_{50}	$1.2\text{–}61.0 \text{ mg L}^{-1}$
Fat head minnows	1242–1254	Newley hatched, 96 hr	LC_{50}	$8 \text{ and } 15 \text{ } \mu\text{g L}^{-1}$
Crustaceans				
Pink shrimp	PCBs	48 hr	LC_{50}	0.1 mg L^{-1}

[a] Compiled from Peakall (1975), WHO (1976), and Richardson and Waid (1979).

259

As with mammals, birds indicate a wide degree of tolerance to PCBs (see Table 9.7) and, as a general rule, toxicity increases with increasing chlorine content. The PCBs exhibit less toxicity to birds than DDT and related compounds.

The toxicity of the PCBs to fish are shown in Table 9.7 but there can be difficulties in measuring the toxicity to aquatic organisms due to the very low solubility of the PCBs in water. As a result, some LC_{50} values were measured using suspensions or colloidal solutions. The PCBs are generally highly toxic to fish but in contrast to the birds the toxicity decreases with chlorine content, for example, Aroclor 1221 has a 96 hr LC_{50} of $1.2\,mg\,L^{-1}$ while Aroclor 1260 has a 96 hr LC_{50} of $61\,mg\,L^{-1}$. The various other Aroclors have a gradation in toxicity between these two figures as measured for the cutthroat trout (Mayer et al., 1977). There is a limited range of information on toxicity to aquatic invertebrates (see Table 9.7). It appears to be similar to fish but generally more toxic.

Sublethal Effects

With mammals, sublethal effects may be more important than lethal effects. Sublethal doses of PCBs lead to liver enlargement and damage (Peakall, 1975). In addition, exposure to these substances results in induction of all hepatic microsomal enzymes which is more marked with increasing chlorine content. The effects have been summarized by Wasserman et al. (1979). Stimulation of hepatic microsomal enzymes leads to potentiation of those activities by liver functions and a decrease of activities by those which detoxify in the liver (WHO, 1976).

Many mammals exhibit reduced weight gain and reduced growth with sublethal exposure to PCBs (Wasserman et al., 1979). Long-term experiments with mice and rats have indicated the tumorigenic potential of PCBs to cause liver tumors in test animals. It is believed that these tumors may be caused by the arene oxides which are produced by metabolism of PCBs (see Figure 9.7). However, there do not seem to be mutagenic or teratogenic effects in mammals (WHO, 1976; Richardson and Waid, 1979). On the other hand, Peakall (1975) has summarized a number of teratogenic effects in birds. The PCBs are also believed to have immunosuppressive effects which lead to a lower production of immunoglobulins and thus a greater susceptibility to some diseases (WHO, 1976).

Eggshell thinning in birds due to chlorinated hydrocarbons has been noted by a number of researchers (see Peakall, 1975). It has been suggested that this effect is due to altered steroid hormone levels leading to a lowered calcium metabolism and thus eggshell thinning. However, Peakall (1975) has suggested that PCBs do not affect eggshell thickness. Wasserman et al. (1979) have concluded that eggshell thinning in nature may be due to the presence of a number of xenobiotic substances acting together. However, PCBs in eggs can lead to death of the embryo before hatching (Wasserman et al., 1979).

A variety of sublethal effects have been noted with a wide array of aquatic organisms from phytoplankton to mammals (see Table 9.8). In addition, many sublethal effects have also been found with terrestrial mammals and are summarized in Table 9.9.

Table 9.8. Effects of PCBs on Reproduction in Aquatic Organisms

Species	Compound, Dosage	Effect
Chlamidomonas	Aroclor	Decreased cell density. Dose-related decrease in uptake of ^{14}C.
Phytoplankton	Aroclor 1242, 10 and 25 ppb	Net plankton production was inhibited > 50%, nannoplankton was not inhibited.
Phytoplankton	PCBs	Reduction of cell division rate.
Flagellated protozoan	Aroclor, 10.5 ppm	50% Depression of cell population growth.
	DDT, 425.0	50% Depression of cell population growth.
Daphnia magna Strauss	PCBs, 19.0 ppb	50% Reduction in the number of young.
Gammarus oceanicus	Aroclor 1254, 10 and 100 ppm in seawater for 10 hr	Survival and development were hindered in the embryos.
	Aroclor 1254, 21.0 ppb	Decrease in mating.
Eggs of coho salmon	Aroclor 1254, 4.4 ppb until 4 weeks after hatch. 15.0 ppb until 2 days before hatching	Decrease of egg hatchability, premature hatching, and decreased survival and growth in the young fishes.
Sea lions of southern California	Naturally exposed (1973)	A high rate of premature births in those with higher levels of PCBs.
Ringed seals from: Bosnian bay Choska inlet Ochotskan Sea	Naturally exposed (1976)	Pregnant seals had lower PCB and DDT residue levels. About half of the nonpregnant population showed evidence of implantations followed by resorption or abortion.
Seals, Gulf of Bothnia	Naturally exposed (1976)	Nonpregnant seals had higher PCB and DDT residues. They presented uterine changes.

Source: Wasserman et al. (1979). Reprinted with permission from the New York Academy of Sciences.

261

Table 9.9.   Effects of PCBs on Reproduction in Terrestrial Mammals

Species	Compound, Dosage	Effects
NMRI mice	PCBs, 20 mg kg^{-1} (1 injection)	Lengthening of the estrus cycle by 1 day per 3 cycles.
NMRI mice	Clophen A60, 25 μg day^{-1} for 62 days to the nursing mothers	Increase in the length of the estrus cycle. Decrease in the frequency of implanted ova/offspring.
NMRI mice	Clophen A60, 25 μg day^{-1} Chronic exposure	Lengthening of estrus cycle. Decrease in the frequency of implanted ova.
NMRI mice	PCBs 50 mg kg^{-1} wk^{-1} for 4 wk beginning the end of delivery	The frequency of implantation in the young decreased from 94% (controls) to 75%.
Rats	Aroclor 1242 and 1254, 100 ppm in diet	Low mating indices, decreased survival in pups.
	Aroclor 1260, 100 ppm in diet	No effect.
Rats	Aroclor 1254, 20 and 100 ppm in diet	Fewer offspring per litter.
	Aroclor 1260, 20 and 100 ppm in diet	No effect.
Rabbits (pregnant)	Aroclor 1254, 12.5 mg kg^{-1} day^{-1} for first 28 days	Abortions; fetotoxicity
Mink	Coho salmons and other Great Lake fishes	Embryonic mortality, death of newborn kits.
Mink	Aroclor 1254, 5–10 ppm	Complete failure to produce offspring.
Mink	Coho salmon of Lake Michigan	58% of whelped and 55% of the young were stillborn. Toxicity varied inversely with chlorine content.
Dogs	PCBs 22, 11 mg day^{-1} doses	Decrease in the number of whelps per bitch.
	PCBs 3.3 ppm + DDT 3.3 ppm	The number of delivering bitches, whelps per bitch, whelps living after 5 days, and weight of whelps was decreased. The number of stillborns was increased.
Monkeys	PCBs, 11 mg day^{-1} for 66 days (beginning 1 month before mating)	No whelp was born in this group.
Humans	Aroclor 1248, 2.5 ppm in diet Yusho patients	Lower fertility; diminished weight of young at birth. Stillbirths.

Source: Wasserman et al. (1979). Reprinted with permission from the New York Academy of Sciences.

Ecological Effects

There is a significant body of information available on the residues of PCBs which occur in the abiotic environment and biota. However, there is insufficient knowledge to give a clear interpretation of the ecological implications resulting from these residues. Thus, our knowledge of the ecological effects of residues is relatively undeveloped at this stage.

PCBs (Aroclor 1254) at concentrations of 1–10 ppb have been shown to lead to a reduction in size of individuals and biomass of phytoplankton in laboratory experiments. This effect also leads to a change in the community species composition with species having a small size favored. Consequent modifications to zooplankton communities may occur since zooplankton have been shown to be selective grazers, choosing food on the basis of size, shape, and species (O'Connors et al., 1978). Biggs et al. (1978) have suggested that the effect described above will lead to alterations in the flow of energy in ecosystems causing an alteration in the community structure at higher trophic levels.

In general, PCBs have not been found to exhibit a correlation with trophic level in aquatic ecosystems (Phillips, 1980). But, nevertheless, both terrestrial and aquatic birds and mammals at the top trophic levels tend to have higher concentrations of PCBs (see Table 9.3). This can lead to the exposure of organisms at the high trophic levels to comparatively high concentrations of these substances. These concentrations have been shown to lead to a lack of breeding success due to a variety of different factors (see Tables 9.8 and 9.9). In addition, Peakall (1975) has reported large kills of seabirds associated with PCB residues. It has been reported that restrictions on PCB use has led to reductions in the presence of residues in the brown pelican and an improvement in their breeding success in the northeastern United States (Blus et al., 1977).

DIBENZO-*p*-DIOXINS (DIOXINS)

The term "dioxins" is commonly used to describe a family of substituted dibenzo-*p*-dioxins. They are a class of substances never intentionally released to the environment which are formed as a result of contamination of commercial chemical products. In addition, these substances may possibly be formed from commercial chemical products in the environment. They occur in very low concentrations in the environment but due to their very high toxicity and physiological activity, particularly of one member of this family, 2,3,7,8-TCDD (2,3,7,8-tetrachlorodibenzo-*p*-dioxin), this group of substances has attracted considerable attention.

TCDD is formed in the manufacture of the herbicide 2,4,5-T and has been involved, or suspected to be involved, in some spectacular incidents. For example, at Seveso, Italy, a 2,4,5-T manufacturing plant accident resulted in the contamination of the surrounding area with TCDD and a variety of other substances. This caused substantial mortalities of wildlife and domestic animals and suspected effects on human health (Whiteside, 1977; Cattabeni et al., 1978; Kriebel, 1981). Also the

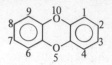

Figure 9.8.   Structure of dibenzo-*p*-dioxin with positions of substituents being 1 to 4 and 6 to 9.

use of 2,4,5-T and 2,4-D in Vietnam in the control of vegetation for military purposes resulted in wide areas being treated. Suspected human health problems have involved both civilian and military personnel. Finally, there have been widespread enquiries throughout the world into the suspected effects of dioxins, in the herbicides 2,4,5-T and 2,4-D, in relationship to birth abnormalities in women.

Nature and Physicochemical Properties of Dioxins

The basic structure of these substances is shown in Figure 9.8, which indicates the central six-membered ring containing two oxygens, 1,4- or para-dioxin). Each substituent position 1 to 4 or 6 to 9 can hold chlorine or hydrogen atoms or an organic group. Those substituents of most interest are the ones resulting in the formation of the chlorinated dibenzo-*p*-dioxins, particularly 2,3,7,8-TCDD. The first chlorinated dioxin was prepared in 1872 by Merz and Weith (Rappe, 1978). It was the octachlorinated compound, although later other chlorinated dioxins and brominated compounds were prepared in the 1930s and 1950s (Rappe, 1978). TCDD was prepared in the laboratory in 1957 (Rappe, 1978) and many accidents have been reported during laboratory preparations and also in commercial plants.

There are 75 possible chlorinated dioxins with different numbers of isomers for each number of chlorines (monochloro, 2; dichloro, 10; trichloro, 14; tetrachloro, 22; pentachloro, 14; hexachloro, 10; heptachloro, 2; octochloro, 1). Esposito et al. (1980) have reported that 40 of these compounds have been prepared synthetically and their physicochemical properties recorded. Examples of the substitution patterns of some members of this family are shown in Figure 9.9. Other substituents apart from chlorine are also possible (e.g., Figure 9.9).

Some physical properties of 2,3,7,8-TCDD and OCDD (see Figure 9.9) are shown in Table 9.10. TCDD and OCDD show high thermal stability with decomposition temperatures in excess of 700°C. TCDD is a colorless crystalline solid at room temperature. It has a lack of reactive groups which leads to its high thermal stability. TCDD also shows a general lack of solubility in both polar and nonpolar solvents but it is much more soluble in nonpolar than polar solvents. Esposito et al. (1980) have reported that it is extremely lipophilic.

Sources, Formation, and Occurrence

There is no evidence to indicate that any of this group of substances are formed in living organisms and they are not constituents of the normal environment (Esposito

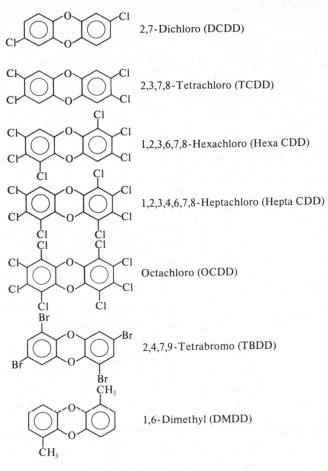

2,7-Dichloro (DCDD)

2,3,7,8-Tetrachloro (TCDD)

1,2,3,6,7,8-Hexachloro (Hexa CDD)

1,2,3,4,6,7,8-Heptachloro (Hepta CDD)

Octachloro (OCDD)

2,4,7,9-Tetrabromo (TBDD)

1,6-Dimethyl (DMDD)

Figure 9.9.   Structures of some members of the dioxin group.

Table 9.10.   Some Physical Properties of Two Chlorodibenzo-*p*-Dioxins

Property	2,3,7,8-TCDD	OCDD
Melting point (°C)	305	130
Decomposition temperature (°C)	>700	>700
Solubility (g L^{-1}):		
o-Dichlorobenzene	1.4	1.83
Chloroform	0.37	0.56
n-Octanol	0.048	–
Methanol	0.01	–
Water	2×10^{-7}	–

Source: Esposito et al. (1980).

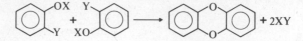

Figure 9.10. Generalized pathway for the formation of dioxins.

et al., 1980). The requirements for one route of formation from industrial chemicals has been defined by Esposito et al. (1980) as follows:

1. The precursor must contain an ortho-substituted benzene ring with the substituents containing an oxygen directly attached to the ring.
2. It must be possible for two substituents to react with each other to form an independent compound (see Figure 9.10).

In laboratory experiments, ortho-chlorinated phenols have been used as precursors. Figure 9.11 shows a variety of laboratory preparations of dioxins. The reaction involves two steps as illustrated in Figure 9.12 for 2,4,5-trichlorophenol. Predioxins have been isolated from commercial as well as laboratory experiments and Figure 9.12 also indicates other condensation reactions which can occur. There are many different condensation reactions possible giving a set of competitive reactions. As a result, the dioxins are usually formed in trace quantities together with a wide variety of other possible products. Generally, temperatures of 180–400°C are needed to form the dioxins. It has also been noted that pressure, UV light, and the use of catalysts can lead to the stimulation of formation in some cases.

Esposito et al. (1980) have identified a wide variety of chemicals known to or

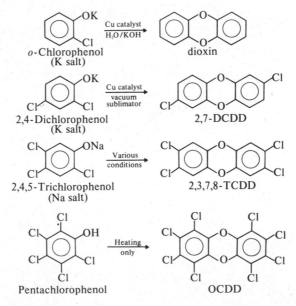

Figure 9.11. Formation of dioxins under a variety of conditions. (From Esposito et al., 1980.)

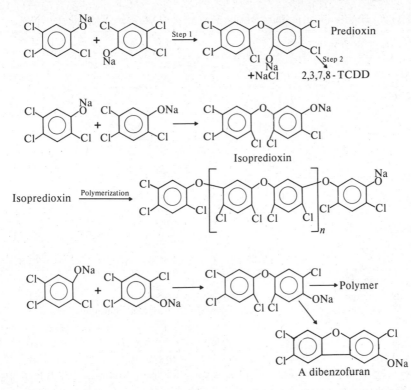

Figure 9.12.   Different condensation reaction paths for 2,4,5-trichlorophenol. (Adapted from Esposito et al., 1980.)

likely to result in the formation of dioxins. Pesticides have been identified as the most significant group, particularly the phenoxy acetic acid group (2,4-D, 2,4,5-T, silvex, and erbon). The concentrations and frequency of occurrence of a number of dioxins in a variety of pesticides are shown in Table 9.11. The 2,4-D and 2,4,5-T used in Vietnam contained 1–47 ppm TCDD. A reaction sequence indicating the formation of 2,3,7,8-TCDD in 2,4,5-T, silvex and ronnel is shown in Figure 9.13. During the commercial synthesis of these products, similar reactions can be expected with related pesticides.

In addition to environmental distribution in pesticides, dioxins can be formed as combustion products from burning vegetation treated with 2,4,5-T and 2,4-D. Also the dioxins are reported to occur in emissions from incinerators, flyash, and so on (e.g., Olie et al., 1977; Buser et al., 1978) but the source of dioxins in these emissions is not altogether clear. It has been suggested that they could be formed by a combination of naturally occurring phenols in the presence of chlorinated products (e.g., DDT, polyvinylchloride, etc.). Alternatively, thermal decomposition of chlorophenols used to treat paper and wood or pesticide-treated wastes may result in the formation of these products (e.g., Rappe et al., 1978).

Data collected by the Dow Chemical Company are reported in Table 9.12 and these show the distribution of dioxins in the vicinity of a Dow Plant in Michigan.

Table 9.11.   Higher Chlorinated Dioxins Found in Commercial Pesticides

Pesticide	Chlorodibenzo-*p*-dioxin Detected[a]				Sample	
	Tetra	Hexa	Hepta	Octa	Number Contaminated	Number Tested
Phenoxyalkanoates						
2,4,5-T	++	++	—	—	23	42
Silvex	+	—	—	—	1	7
2,4-D	—	+	—	—	1	24
Erbon	—	—	—	++	1	1
Sesone	—	+	—	—	1	1
Chlorophenols						
Tri-	—	+	+	+	4	6
Tetra-	—	++	++	++	3	3
Penta-(PCP)	—	++	++	++	10	11
Others[b]	—	++	++	+	5	22

Source: Esposito et al. (1980).

[a] Concentration range ++, >10 ppm; +, 0.5–10 ppm; —, <0.5 ppm.

[b] DMPA, ronnel, and tetradifon were found to contain chlorodioxin contamination.

The origin of these residues is a topic for debate but it has been suggested that they have originated as a natural consequence of combustion. Alternatively, the U.S. EPA has suggested that the Dow Plant is a possible source (Rawls, 1979; Esposito et al., 1980).

Accidents have resulted in comparatively large emissions of dioxins to the

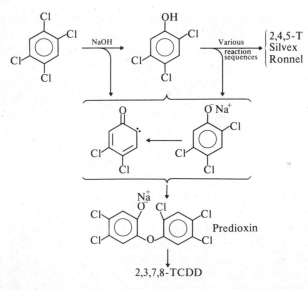

Figure 9.13.   Reaction sequence for the formation of 2,3,7,8-TCDD during the manufacture of some pesticides.

Table 9.12.   Dioxin Concentrations Near Dow Plant (Parts per Billion)[a]

	Tetra(TCDD)[b]		Hexa	Hepta	Octa
	2,3,7,8-Fraction[c]	Others			
Soil samples					
Dow plant	16	17	280	3200	20,500
	100	18	120	650	6,300
Rural sites	N.D.[d]		N.D.	N.D.	N.D.
	N.D.		N.D.	N.D.	0.1
Metropolitan sites	N.D.		1.2	1.6	2.0
	0.03		0.31	3.3	22.0
Dust samples					
Dow research building	1.0	0.5	18	240	960
	2.3	2.3	35	1200	7,500
Midland, Michigan	0.03[d]		0.2	2.3	19
	0.04		0.4	3.9	31
Metropolitan sites	N.D.		0.09	0.8	3.5
	N.D.		0.3	1.0	3.8
	0.04		0.34	3.2	8.2

Type of Chlorinated Dibenzo-p-Dioxin

Source: Rawls (1979). Reprinted with permission from the American Chemical Society.

[a] Table shows data from only a few of the samples taken, including samples with the highest total dioxin concentration taken from each type of site.

[b] N.D., no detectable peak at 2.5 times background.

[c] This fraction contains 2,3,7,8-TCDD and 11 of the 21 other TCDD isomers.

[d] TCDD isomers were separated only for samples taken at the Dow plant.

environment. As mentioned previously, the best known accident occurred in 1976 at Seveso, Italy, in a chemical manufacturing plant. This resulted in the formation of a cloud of material from a burst trichlorophenol reaction vessel. Consequently, an area of about 2000 hectares in the vicinity of the plant was affected and soil concentrations of TCDD of < 0.750 to $5000 \mu g \, m^{-2}$ were recorded. Extensive mortalities of animals and plants and human health problems were reported. Evacuation of the area was carried out and a decontamination program initiated (Pocchiari, 1978; Cattabeni et al., 1978; Mattiaschk, 1978).

Wastes from chemical manufacturing can contain dioxins which may be dispersed during disposal by landfill operations, by incineration, or in waste water. Incineration has been discussed previously but landfill has the potential to contaminate groundwater and enter surface waters. Waste water in some cases of chemical plant operation has led to the occurrence of low dioxin levels in biota in adjacent areas (Esposito et al., 1980).

Factors Affecting Distribution, Transport, and Degradation in the Environment

Most of the studies of dioxins have been on 2,3,7,8-TCDD. However, from the information available it would appear generally that the dioxins are tightly bound to soils and thus are not moved out of the upper layers after deposition. The strength of binding has been found to be related to the organic matter content of the soils. Kearney et al. (1973) have suggested as a result of this that dioxins generally present no threat to groundwater supplies. The movement of dioxins in soils thus occurs as a result of movement of soil particles. There is some evidence, however, of vertical movement in soils which are low in organic matter.

The dioxins, and in particularly TCDD, have very low water solubilities but various studies have indicated that leachate from soils can yield low concentrations in the associated water body. Dioxin levels can exceed water solubility concentrations due to their absorption onto organic and suspended particles (Matsumura and Benezet, 1973). The loss of TCDD by volatilization is low due to the low vapor pressure of this substance. However, there may be loss to the atmosphere by wind suspension of particles on which dioxin is adsorbed.

Most studies of degradation of dioxins in soils and water are on TCDD and Esposito et al. (1980) have concluded that biodegradation in soils is yet to be unequivocally demonstrated. In aquatic systems the TCDD is bound to organic matter and sediments with different strengths. Some of the more readily available TCDD can be degraded, usually within a month, but the remainder is much more resistant to degradation and persists over more extended periods of time. These degradation patterns are mediated by microorganisms (Esposito et al., 1980).

Photolysis investigations have indicated that the dioxins decompose under the action of solar ultraviolet light. This is possibly the most important mechanism for the removal of dioxins in the natural environment. The photodegradation process occurs on inert surfaces and the surfaces of plants and soils (Wong and Crosby, 1978). It requires an organic hydrogen donor and in fact has been used as a decontamination procedure. For example, hydrogen donors (e.g., vegetable oils) sprayed onto contaminated areas and exposed to sunlight result in the decomposition of the dioxins present (Wipf et al., 1978). Liberti et al. (1978) found that the rate of decomposition is dependent on the ultraviolet radiation intensity and complete decomposition can occur within 1 hr ($2\,mW\,cm^{-2}$) or 72 hr ($20\,\mu W\,cm^{-2}$).

The process occurs by progressive removal of chlorines from the dibenzo-p-dioxin nucleus and replacement with hydrogen. Finally, all the chlorines are removed and parent dibenzodioxin remains and is susceptible to further decomposition.

Uptake and Metabolism by Organisms

The major source of dioxins in the natural environment occurs as a result of the use of herbicides in agriculture. There have been many investigations into the uptake of dioxins by plants indicating that they are taken up by plants only in low concentrations, usually measured in $\mu g\,kg^{-1}$.

Because dioxins have very low solubility in water and are lipophilic, this leads to bioaccumulation in organisms. The data in Table 9.13 indicate some of the bioconcentration factors observed in laboratory experiments. It has been suggested that at normal application rates for the herbicides 2,4,5-T and 2,4-D, the concentrations

Table 9.13.   Bioconcentration Factors of TCDD for Some
Biota in Model Ecosystems[a]

Organism	BCF $\left(\dfrac{\text{tissue conc. TCDD}}{\text{water conc.}}\right)$
Daphnia	2,198
Ostracoda	107
Mosquito larvae	2,846
Northernbrook silverside fish	54
Algae	6–2083
Snails	735–3731
Daphnia	1,762–7,125
Mosquito fish	676–4,875
Algae	2,000
Duck weed	4,000
Snails	24,000
Daphnia	48,000
Mosquito fish	24,000
Catfish	2,000

[a] Compiled from Esposito et al. (1980).

of TCDD expected would not result in significant bioaccumulation in organisms of any kind (Matsumura and Benezet, 1973). Nevertheless, there is considerable bioaccumulation potential in areas of relatively high concentrations. As mentioned previously, biodegradation occurs at a slow rate in soils and the aquatic environment with calculated half-lives for TCDD in soil of 1.3 and 2.9 yr (Westing, 1978).

In experiments with guinea pigs, it has been found that TCDD is moderately well absorbed from the gastrointestinal tract. As a result, the substance has a half-life in plasma of about 1 month (Nolan et al., 1979). A large proportion of the absorbed dioxin persists in the liver and is slowly excreted as polar metabolites. It has been found that iron deficiency with rats gives a measure of protection against the toxic effect of TCDD. Accumulation in the liver follows first-order uptake kinetics (Rose et al., 1976). TCDD elimination follows first-order kinetics in the rat approximated by a one-compartment model and half-lives have been measured at 17 and 31 days (Esposito et al., 1980).

Physiological Effects

TCDD is highly toxic to a wide range of biota (see Table 9.14). In all laboratory tests delayed mortality is observed, usually of the order of 2–8 weeks. Increases in the dose do not give a significant decrease in the time to death which is preceded by weight loss. The toxicity of the substance is not altered by the route of exposure

Table 9.14.   Single Toxic Oral Doses of 2,3,7,8-TCDD for Some Biota[a]

Species	LD_{50} ($\mu g\,kg^{-1}$)	Mean Time to Death (days)
Rabbit	10 and 115	Unknown and 6–39
Rat	22 to <100	9–42
Guinea pig	0.6 and 2	5–34
Mouse	114–284	20–25
Monkey	<70	28–47
Dog	>30 to <300	9–15
Chicken	25–50	12–21

[a] Compiled from Moore (1978).

and uptake can occur through the skin or through the gastrointestinal tract (Moore, 1978). There have been many deaths of wildlife and domestic animals in cases of exposure.

Dibenzodioxin itself is not a particularly toxic compound. The 2,3,7,8-tetrachloro compound, the corresponding tetrabromo compound (PBrDD), and hexachloro dibenzodioxin (HCDD) are the only dioxins which have marked toxicity. The chlorine-substitution pattern has a powerful influence on exhibited toxicity since only those compounds with substituents in the 2,3,7,8 positions are highly toxic. For example, the 2,3,7,8-tetrabromo compound is as toxic as the corresponding tetrachloro compound. Compounds with four chlorines but a different substitution pattern, for example, 1,2,3,4 and 1,3,6,8 are much less toxic (see Table 9.15). The dioxins, and particularly 2,3,7,8-TCDD, have a powerful influence on enzyme systems in many organisms. Induction of mixed function oxidases (MFO) is noted with this compound and many of its physiological activities are due to this capacity.

TCDD is capable of inducing at least three enzyme systems (see Table 9.15). Induction of the ALAS system results in disorganization of the hepatic porphyrin metabolism resulting in porphyria cepanea tardia. The aryl hydroxylase (AHH) system activity increases considerably with TCDD exposure. Also there is powerful induction of zoxazolamine hydroxylase which indicates that the substance has a high probability of being a carcinogen (Saint-Ruf, 1978). In fact, tests on laboratory animals indicate that TCDD could be a potent carcinogen (Esposito et al., 1980).

Teratogenic effects have been noted with mice in concentrations greater than $1\,\mu g\,kg^{-1}$ TCDD and in a variety of combinations with 2,4,5-T. Fetal toxicity in primates has been found to occur at doses as low as $50\,ng\,kg^{-1}$ and the effects of these substances on humans is the subject of continuing controversy. Results up to the present time have been summarized by Esposito et al. (1980).

TCDD has also been found to reduce the peripheral lymphocytes as well as α, β, and γ globulins in some mammals. This effect may suppress the immune reaction and lead to a predisposition to disease (Vos, 1978).

Table 9.15.   Some Toxicological Properties of Dioxins[a]

Compounds	LD_{50} (mg kg^{-1} rat)	Teratogenic Effect	Enzyme Induction		
			ALAS (Chick Embryo)	AHH (Chick Embryo)	Zoxazolamine Hydroxylase (Rat)
TCDD	0.04	+++	+++	1	+++
DD	>1000	0	−	−	0
2,7-DCDD	~2000	±	0	0	−
2,3-DCDD	>1000	0	0	0	0
2,3,7-TCDD	>1000		++	0.02	−
2,3,7-TBrDD	>1000		++	0.6	−
1,2,3,4-TCDD	>1000	0	0	0	−
1,3,6,8-TCDD	>100	0	+	0.2	−
TBrDD	≤1		−	1.0	+++
HCDD (mixture)	~100	++	−	0.8	−
OCDD	~2000	±	−	0	0

[a] Compiled from Saint-Ruf (1978).

Ecological Effects

Esposito et al. (1980) have reviewed the available ecological information and some of this is summarized in Table 9.16. The data indicate that TCDD may accumulate in exposed organisms. For example, at Seveso the concentrations in terrestrial organisms were higher than in the soil (see Table 9.16). Similar results have been obtained elsewhere indicating that bioaccumulation from soil and water into organisms can occur (see Table 9.16). None of this information, however, provides convincing evidence for biomagnification.

Laboratory and other experiments have provided evidence of the bioaccumulation of dioxins in organisms from the abiotic environment with the bioconcentration factors shown in Table 9.13. These bioconcentration factors are high compared to some insecticides which suggests that the potential for bioaccumulation in natural systems is considerable. However, the levels in water and soils are usually very low, originating primarily from herbicides.

The effects on natural ecosystems are difficult to evaluate from the available information. There appears to be the potential for bioaccumulation of significant concentrations of dioxins in areas where there is a comparatively high usage of herbicides or environmental factors which result in excessive concentrations. Esposito et al. (1980) have suggested that the levels found in Vietnam may be ecologically significant (see Table 9.16). Overall, human health effects have been the major area of concern with these compounds.

PHTHALATE ESTERS (PHTHALIC ACID ESTERS OR PAEs)

Liquids with suitable properties (plasticizers) are added to synthetic polymers to give improved flexibility, extensibility, and workability. Without the addition of plasticizers, many plastics are hard, brittle solids. The addition of plasticizers results in a lowering of the glass transition temperature to below room temperature. Plasticizers must be compatible with the appropriate polymer and compatability can be calculated from the physical properties of the plasticizers and the polymer. For example, dibutyl and di-2-ethylhexyl phthalates have been shown, from appropriate calculations, to be highly compatible with polyvinyl chloride (PVC).

PAEs are plasticizers used (see Figure 9.14) in virtually all applications of plastics, for example, building and construction, home furnishing, transportation equipment, apparel, and food and medical products (Graham, 1973). Entry to the

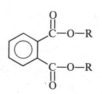

Figure 9.14.   General structural formula of PAEs where R can be isooctyl, 2-ethylhexyl, isodecyl, and a variety of other groups.

Table 9.16. Concentrations of TCDD Found in Some Biota[a]

Organism	Number Analyzed	Area	Exposure	Tissue	Concentration ($ng\,g^{-1}$)
Field mouse	14	Seveso, Italy	Chemical plant accident	Whole body	4.5 (0.07–49)
Hare	5			Liver	7.7 (2.70–13)
Toad	1			Whole body	0.2
Snake	1			Liver and adipose tissue	2.7
Earthworm	2			Whole body	16.0
Top 7 cm soil	—			Soil	3.5 (0.01–12)
Rats and mice	—	Eglin Air Force Base, Florida	Test site for Agent Orange disposal	Liver and fat	0.210–0.542
Control animals	—			Liver and fat	0.020
Beach mice	—	Eglin Air Force Base, Florida	Test site for herbicide Orange disposal	Liver	0.520 and 1.30
Top 6 in. soil	—			–	0.001–1.5
Control mice	—			Liver	0.020 and 0.085
Blue gill (*Lepomis puntatus*)	—	Pools and streams, Eglin Air Force Base, Florida	As above	Skin, gonads, muscle, and gut	0.004, 0.018, 0.004, and 0.085 (resp.)
Sailfish shiner and mosquito fish	—			Bodies minus heads, tails, and viscera	0.012
Catfish and carp	—	Dong Nai River, South Vietnam	Herbicide Orange treatment	–	0.320–1.02
Catfish and prawn	—	Saigon River, South Vietnam	Herbicide Orange treatment	–	0.034–0.089
Croaker and prawn	—	South Vietnam seacoast	Herbicide Orange treatment	–	0.014 and 0.110
Rainbow trout	6	Tittabawassee River, Michigan	In cages suspended in the receiving effluent from chemical plant and effluent	Whole fish	Not detectable to 0.05

(continued on next page)

275

Table 9.16. (continued)

Organism	Number Analyzed	Area	Exposure	Tissue	Concentration ($ng\,g^{-1}$)
Catfish			Environmental chemical plant discharge	Whole fish	0.07–0.23 and 0.04–0.15 (OCDD)
Fish	26	Tittabawassee, Grand, and Saginaw Rivers, Michigan	Environmental chemical plant discharge	—	Not detectable to 0.690

[a] Compiled from Esposito et al. (1980).

environment is believed to arise principally from volatilization during the incineration of plastics and water leachate from solid waste disposal.

Physicochemical Properties

The volatility of the PAEs is low, but significant as a form of entry into the environment. Most common PAEs have vapor pressure of 5×10^{-8} mm Hg at temperatures
from about 50 to 90°C (Gross and Colony, 1973). They are lipophilic and have a
low but significant solubility in water. They will migrate out of plastic into the
surrounding environment according to the following relationship:

$$\frac{Q}{S} = 2.26 \left(\frac{Dt}{L^2} \right)^{1/2}$$

where Q is the weight of plastic lost in t hours; S the total weight; D the diffusion
constant in cm hr^{-1}; and L the film thickness.

A key factor in this relationship is the diffusion constant which depends on the
affinity of the surrounding environment for the plasticizer and includes a number
of other factors such as temperature.

Uptake and Metabolism

Accumulation and metabolism occur principally in the liver. Metabolism of the
PAEs occurs at a rate which leads to substantial degradation within 24 hr in
mammals and fish (Stalling et al., 1973; Schulz and Rubin, 1973).

Residual concentrations in the natural environment have been detected in a
number of cases (see Table 9.17). Mayer and Sanders (1973) have reported bioconcentration factors of up to 720 after 14 days for various aquatic organisms. In the
scud *Gammarus pseudolimnaeus*, a bioconcentration factor of 3600 was measured.

Table 9.17.   Examples of Some PAE Concentrations in Environmental Samples

Sample	Location	PAE Concentration (ng g^{-1})
Water	Charles River (U.S.)	0.88–1.9
Water	Missouri River (U.S.)	4.99
Fish		800
Wastes	Great Lakes (U.S. and Canada)	0.04–300
Sediment		300
Freshwater fish	United States	N.d.–3200

Source: Stalling et al. (1973).

Physiological Effects

A low order of toxicity has been noted with a wide variety of PAEs. Mayer and Sanders (1973) have found that the toxicity to aquatic organisms is less than for many other organic environmental pollutants. For example, the oral LD_{50} values of the rat, mouse, and guinea pig range from 8 to 64 g kg^{-1}. Generally, the short-chain PAEs are more toxic than the long-chain compounds but no other structural relationships are apparent. For example, the 96 hr LC_{50} for dibutyl phthalate was found to range from 0.7 to 6.5 mg L^{-1} for fish. In comparison, di-2-ethylhexyl phthalate had LC_{50} values generally greater than 10 mg L^{-1}. Also, these substances adversely affect reproduction in some aquatic species at low concentrations, for example, 3–30 μg L^{-1}.

Ecological Effects

Aquatic organisms are more likely to be affected by the PAEs than terrestrial organisms. With terrestrial organisms food is the major uptake route and food contamination is required for uptake to occur. Humans are probably the terrestrial organisms most affected by PAEs. On the other hand, aquatic organisms are subject to the direct toxic influences and low concentrations affecting reproduction may have subsequent effects on ecosystems.

SURFACTANTS

Surfactants — surface active agents or tensides — are substances which cause lowering of the surface tension of liquids, particularly water. This causes the formation of bubbles and other surface effects which allow these substances to act as cleaning or dispersion agents in industry and for domestic purposes.

Soap is a surfactant which has been prepared and used over long periods of time. A major disadvantage with soap is its reaction with trace salts in water to give a loss of surfactant activity. In recent years, soaps have been replaced by synthetic surfactants prepared in commercial formulations as detergents. Detergents consist of surfactants plus a variety of other ingredients to assist the cleaning process, for example, builders, perfumes, bleaches, and so on.

Surfactants exhibit a wide variety of different chemical structures but can be conveniently classified into three major groups: anionic, cationic, and nonionic. This distinction is based on the nature of the polar grouping which gives the surfactant its characteristic properties. Of these groups, the cationic has relatively limited usage and will not be discussed further. Figure 9.15 shows the chemical structures of some commonly used surfactant types. The anionic alkyl aryl sulfonate type is the most commonly used at the present time. Of these, the alkyl benzene sulfonate (ABS) surfactants are prepared in the greatest quantities from petroleum. They enter waterways through treated and untreated sewage containing domestic or

Anionic surfactants

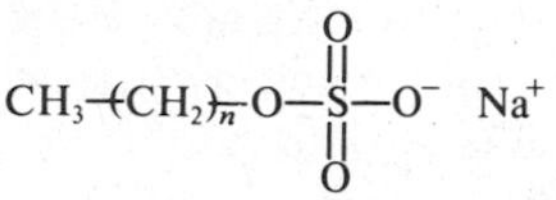

$$CH_3\text{-}(CH_2)_n\text{-}O\text{-}\overset{\displaystyle O}{\underset{\displaystyle O}{\overset{\|}{\underset{\|}{S}}}}\text{-}O^-\ Na^+$$

Primary alkylsulfate

$$\begin{matrix} CH_3\text{-}(CH_2)_n \\ \phantom{CH_3\text{-}(CH_2)_n}\text{CH} \\ CH_3\text{-}(CH_2)_n \end{matrix}\text{-}\!\!\bigcirc\!\!\text{-}\overset{\displaystyle O}{\underset{\displaystyle O}{\overset{\|}{\underset{\|}{S}}}}\text{-}O^-\ Na^+$$

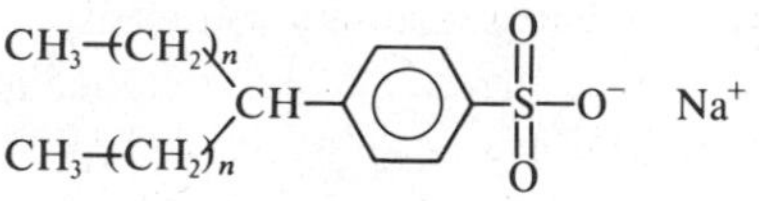

Alkylbenzene sulfonate
(ABS) (alkyl 9–15 C's)

Nonionic surfactants

$$CH_3\text{-}(CH_2)_n\text{-}O\text{-}(CH_2\text{-}CH_2\text{-}O)_n\text{-}H$$

Polyethoxylated alcohol
(alkyl 12–15 C's; approx,
9 ethoxy groups)

Figure 9.15.   Chemical structures of some commonly used surfactants.

industrial waste water and urban runoff. The nonionic surfactant type has been prepared in increasing quantities in recent years. The total usage of soaps and detergents in 1977 was 25.03 million tonnes (Taylor, 1980).

The most obvious environmental effect of surfactants is the appearance of foams on streams. This can also occur in sewage treatment plants and lead to a number of problems in the treatment process. Contamination of groundwater used for domestic supplies has occurred in many areas, for example, Long Island, New York.

Physicochemical Properties

Surfactant properties depend on a molecule having lipophilic and hydrophilic characteristics. At interfaces (e.g., fat and water or water and air), the surfactant molecules assemble leading to a lowering of surface tension. At these interfaces the occurrence of foaming leads to the production of large additional areas of interface and thus the accumulation of surfactants in the foam water and consequently a reduction of the concentration of surfactant in the water mass. This effect can lead to differences in concentration of orders of several thousandfold (Prat and Giraud, 1961). With ABS surfactants the threshold for permanent foam formation is about 0.3–$0.4\ mg\ L^{-1}$. Temperature, pH, the presence of other substances can all influence the concentration at which permanent foam occurs.

Although the ABS surfactants are used mainly in the form of the sodium salt, these substances occur in natural waterways as the calcium salt. This salt has a low water solubility and exists as an unstable suspension. It first assembles at interfaces such as air–water, fats–water, and bottom sediments–water, but ultimately it enters the bottom sediments as deposits. This leads to high concentrations of surfactants in sediments in areas receiving surfactant-containing waste water. Concentrations of $2.3\,g\,kg^{-1}$ of sediment (dry weight) have been observed in some areas. There is usually a pattern of decreasing concentrations in sediments at increasing distance from a surfactant-containing discharge (Prat and Giraud, 1961).

Surfactants in sediments behave as two fractions — a labile fraction and a more strongly adsorbed fraction. When sediments are resuspended the labile fraction is redistributed, leading to the occurrence of surfactants in the water mass and the lowering of surface tension. This can occur at weirs, rapids, riffles, and so on.

Some biodegradable lipophilic molecules may be temporarily protected from degradation by the presence of surfactants. Micelles containing the susceptible molecules become surrounded by surfactant molecules. The micelles consists of a spherically laminated structure in which the outer shell consists of the charged groups in the surfactants and the inner shell consists of the lipophilic portion of the molecule. The outer charged shell prevents contact with other micelles and forms a layer that can afford temporary protection to the internal lipophilic molecules. The outer layers must be degraded before attack of the internal molecules can occur (Prat and Giraud, 1961).

Surfactants can alter the hydraulic flow characteristics of porous media such as soils. The formation of micelles of calcium salts of ABS surfactants in natural systems allows the surfactant to be more readily precipitated than the sodium salts. Precipitation of the surfactant leads to the formation of a gel of calcium salts which can impede water flowing through porous systems (Emerson, 1974).

Prat and Giraud (1961) report that the surface layer of surfactant molecules at the air–water interface can inhibit oxygen transfer. This effect increases with increasing length of the alkyl chain in ABS surfactants. Oxygen transfer has been reduced by up to 70% but is difficult to predict under different environmental conditions.

Degradation

The most important process in natural systems is microbial degradation. Swisher (1971) has discussed the biodegradation of surfactants in some detail. Many soil and aquatic microorganisms have the capacity to attack surfactants through the degradation sequence shown in Figure 9.16.

The ABS surfactants retain the alkyl groups derived from the original petroleum. With the anionic surfactants the alkyl group usually contains 9–15 carbon atoms (see Figure 9.15). The alkyl groups contain many different structural types and there is a substantial body of information available on the effects of these different structural types on biodegradation. The presence of a quaternary carbon atom in the alkyl chain can slow down the degradation process since a hydrogen atom is not available for β oxidation. It has generally been found that branching of the alkyl chain results in increased resistance to biodegradation. On the other hand, straight-chain alkyl groups are relatively degradable. This property has been adapted in the

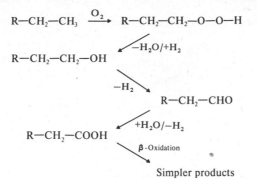

Figure 9.16. Reaction sequence for the degradation of ABS surfactants. R is the remainder of the surfactant molecule attached to an alkyl chain.

commercial preparation of surfactants. The older ABS surfactants with alkyl groups containing a mixture of branched chains are comparatively difficult to degrade and are described as "hard." More recently, the ABS surfactants have been prepared so as to contain the linear alkyl groups and so are described as linear alkyl benzene sulfonates (ALS). These are more readily degraded than the other type and are described as "soft."

Lundahl and Cabridenc (1978) have found that biodegradability changes with chain length. The biodegradability increases up to an alkyl chain length of approximately 15–16 carbon atoms and then decreases, showing an increase in biodegradability at longer chain lengths again.

Not a great deal is known about the mechanism of biodegradation of nonionic surfactants. It has been suggested that the general point of attack is the alkyl group (see Figure 9.15) and the degradation probably follows a similar sequence to the ABS surfactants (see Figure 9.16). Tobin et al. (1976) found that the alkyl group degrades rapidly and the original surfactant disappears but the polyethylate moiety remains for extended periods. They suggested that this remaining group was possibly toxic to aquatic life.

The U.S. manufacturing industry has set standards for biodegradability of synthetic surfactants. This requires a minimum of 90% degradation in 97 hr whereas the OECD has an 80% limit in the same time (Tobin et al., 1976).

Physiological Effects

In general, surfactants interact with membranes and enzymes. The effects can be moderated in plants by adsorption of surfactants and immobilization on cell walls (Fujita and Koga, 1976). The surfactants can cause an alteration to cellular ultrastructure (Healey et al., 1971). Lundahl and Cabridenc (1978), after investigation of a range of organisms, have suggested that toxicity arises from inhibition of enzymes or the selected transmission of ions through membranes.

Sublethal effects take a wide variety of different forms. Inhibition of growth in plants and fish (Mitrovic, 1972) and budding in hydra (Bode et al., 1978) have been

Table 9.18.   Examples of Surfactant Toxicity to Some Aquatic Organisms[a]

Organism	Surfactant	Conditions	Concentration
Hydra	Fatty alcohol ethoxylates	Lethal in 24 hr	2×10^{-1} mM
Algae	Dodecyl benzene sulfonates	IC_{50} in 4–24 hr	3–12.4 mg L^{-1}
Daphnia	As above	As above	12–13 mg L^{-1}
Phoxinus	As above	As above	4.5–12.8 mg L^{-1}
Fathead minnow	Alkyl benzene sulfonate	LC_{50} in 96 hr	4.5–23 mg L^{-1} (soft water) 3.5–12 mg L^{-1} (hard water)
Atlantic salmon, goldfish, and rainbow trout	Nonionic surfactants	LC_{50} in 96 hr	1–10 mg L^{-1}

[a] Data from Tobin et al. (1976), Lundahl and Cabridenc (1978), Bode et al. (1978), and Ehrichsen Jones (1964).

noted. Damage to respiratory epithelium of fish gills has been noted by many authors at sublethal concentrations (see Mitrovic, 1972). Cairns and Scheier (1962) observed damage to *Lepomis gibbosus* at 18 mg L^{-1} of ABS which persisted for at least 8 weeks. Damage to other sensitive external sensory organs has also been noted which may interfere with the search for food.

Synergistic effects with other substances have been widely reported. Generally, the presence of sublethal surfactants leads to more rapid uptake of lipophilic substances and also to enhanced toxicity of these substances (Mitrovic, 1972).

Some acute toxicities for aquatic organisms are shown in Table 9.18. Lundahl and Cabridenc (1978) found that the relationships between IC_{50} (immobilization concentration for 50% of organisms) and the number of carbon atoms in the homologue for different surfactant types were similar for algae, daphnia, and fish (see Figure 9.17). This suggests that the toxicity mechanism is similar in all cases. The toxicity of ABS surfactants increases with linearity of the alkyl group, due to deeper penetration of linear alkyl groups, and more area of the molecule available for interaction than with branched-chain molecules. The surfactant–protein interaction also increases as the hydrophobic tail grows and leads to increasing toxicity.

Reiff (1975) in experiments with rainbrow trout found that the toxicity showed a close correlation with surface tension. This correlation was much closer than the chemical analysis for surfactant concentration. It was suggested that this takes into account the effects of breakdown products from the original surfactant. Bode et al. (1978) reported that with hydra the lethal concentration coincided with a surface tension of 49 ± 4 dynes cm^{-1} since this level disrupts the cell membrane.

Ecological Effects

Most surfactants originate in the discharges from waste-water treatment plants giving concentrations in the receiving water usually ranging from zero to several

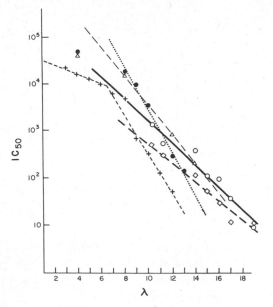

Figure 9.17. IC_{50} values of surfactants as related to chain length (carbon number $= \lambda$) for secondary alkane sulfonates with algae (□) and daphniae (○); primary alkane sulfonates with daphniae (△); primary alkyl sulfates with daphniae (●); carboxylates with daphniae (×). [From Lundahl and Cabridenc (1978). Reprinted with permission from Pergamon Press, Inc.]

parts per million (Arnoux and Caruelle, 1972). In accord with the comparatively short persistence time of surfactants in aquatic areas they are not accumulated to any extent nor does biomagnification in food chains occur. However, this may possibly occur to some extent with degradation products, although there is no direct evidence to show this at the present time.

It is difficult to delineate changes due to surfactants in aquatic areas from those due to other pollutants since surfactants usually occur in association with other polluting substances. Hynes and Roberts (1962), in studies of a stream in the United Kingdom receiving surfactant-containing waste water (2–4 ppm), were unable to detect any changes in the community structure due to the surfactants.

REFERENCES

Anon (1973). Polychlorinated biphenyls: Their use and control. Organisation for Economic Cooperation and Development, Paris.

Arnoux, A. and Caruelle, F. (1972). "Study of Marine Pollution by Anionic Detergents from Sewage of the City of Marseilles." in M. Ruivo (Ed.), *Marine Pollution and Sea Life*. Food and Agriculture Organization, Fishing News (Books Ltd.), London, p. 459.

Ballschmiter, K., Zell, M., and Neu, H. J. (1978). Persistence of PCB's in the Ecosphere: Will some PCB components "never" degrade? *Chemosphere* 2, 173.

Biggs, D. C., Rowland, R. G., O'Connors, H. B., Powers, C. D., and Wurster, C. F. (1978). A comparison of the effects of chlordane and PCB on the growth, photosynthesis and cell size of estuarine phytoplankton. *Environ. Pollut.* **15**, 253.

Blus, L. J., Neely, B. S., Lamont, T. G., and Mulhern, B. (1977). Residues of organochlorines and heavy metals in tissues and eggs of brown pelicans. *Pestic. Monit. J.* **11**, 40.

Bode, H., Ernst, R., and Arditti, J. (1978). Biological effects of surfactants, III. Hydra as a highly sensitive assay animal. *Environ. Pollut.* **17**, 175.

Buser, H. R., Bosshardt, H. B., and Rappe, C. (1978). Identification of poly-chlorinated dibenzo-*p*-dioxin isomers formed in fly ash. *Chemosphere* **7**, 165.

Cairns, J. and Scheier, A. (1962). The acute and chronic effects of standard sodium alkybenzene sulfonate upon the pumpkinseed sunfish, *Lepomis gibbosus* (Linn.) and the blue-gill sunfish, *L. macrochirus* (Raf.). *Proc. Ind. Waste Conf. Purdue Univ.* **17**, 14.

Cattabeni, S., Cavallaro, A., and Galli, G. (1978). *Dioxin–Toxicological and Chemical Aspects*. S.P. Medical and Scientific Books, New York, p. 222.

Emerson, W. W. (1974). The flow of solutions of sodium alkyl benzene sulfonate through porous media. *Environ. Pollut.* **7**, 39.

Erichsen Jones, J. R. (1964). *Fish and River Pollution*, Butterworth, London.

Ernst, W., Goerke, H., Eder, G., and Schaefer, R. G. (1976). Residues of chlorinated hydrocarbons in marine organisms in relation to size and ecological parameters. I. PCB, DDT, DDE and DDD in fishes and molluscs from the English Channel. *Bull. Environ. Contam. Toxicol.* **15**, 55.

Esposito, N. P., Teirnan, T. O., and Dryden, F. E. (1980). Dioxins, Industrial Environmental Research Laboratory, Office of Research and Development, U.S. Environmental Protection Agency, Cincinnati.

Fujita, T. and Koga, S. (1966). The binding of a cationic detergent by yeast cells to its germicidal action. *J. Gen. Appl. Microbiol.* **12**, 229.

Graham, P. R. (1973). Phthalate ester plasticizers: Why and how they are used. *Environ. Health Perspect.* EXP Issue **3**, 3.

Gross, F. C. and Colony, J. A. (1973). Ubiquitous nature and objectionable characteristics of phthalate esters in aerospace technology. *Environ. Health Perspect.* EXP Issue **3**, 37.

Harvey, G. R., Steinhaver, W. G., and Miklas, H. P. (1974). Decline of PCB concentrations in North Atlantic surface water. *Nature* **252**, 387.

Healey, P. L., Ernst, R., and Arditti, J. (1971). Biological effects of surfactants II. Influence on the ultrastructure of orchid seedlings. *New Phytol.* **70**, 477.

Hutzinger, O., Safe, S., and Zitko, V. (1972). Photochemical degradation of chloro-biphenyls (PCBs). *Environ. Health Perspect.* **1**, 15.

Hutzinger, O., Safe, S., and Zitko, V. (1974). *The Chemistry of PCBs*. CRC Press, Boca Raton, Florida, p. 269.

Hynes, H. B. N. and Roberts, F. W. (1962). The biological effects of synthetic detergents in the River Lee, Hertfordshire. *Ann. Appl. Biol.* **50**, 779.

Jensen, S. (1972). The PCB story. *Ambio* **1**, 123.

Kalmaz, E. V. and Kalmaz, G. D. (1979). Transport, distribution and toxic effects of polychlorinated biphenyls in ecosystems: Review. *Ecological Modelling* **6**, 223.

Kearney, P. C., Woolson, E. A., Isensee, A. R., and Helling, C. S. (1973). Tetrachlorodibenzodioxin in the environment: Sources, fate and decontamination. *Environ. Health Perspect.* **5**, 275.

Kriebel, D. (1981). The dioxins: Toxic and still troublesome. *Environment* **22**, 6.

Liberti, A., Brocco, D., Allegrini, I., and Bertoni, G. (1978). Field photodegradation of TCDD by ultra-violet radiations. In F. Cattabeni, A. Cavallaro, and G. Galli (Eds.), *Dioxin: Toxicological and Chemical Aspects*, S.P. Medical and Scientific Books, New York, p. 195.

Lundahl, P. and Cabridenc, R. (1978). Molecular structure — biological properties relationships in anionic surface-active agents. *Water Res.* **12**, 25.

Matsumura, F. and Benezet, H. J. (1973). Studies on the bioaccumulation and microbial degradation of 2,3,7,8-tetrachlorodibenzo-*p*-dioxin. *Environ. Health Perspect.* **5**, 253.

Matthiaschk, G. (1978). "Survey about Toxicological Data of 2,3,7,8-tetrachlorodibenzo-*p*-Dioxin (TCDD)." In F. Cattabeni, A. Cavallaro, and G. Galli (Eds.), *Dioxin: Toxicological and Chemical Aspects*. S.P. Medical and Scientific Books, New York, p. 123.

Mayer, F. L., Merhle, P. M., and Sanders, H. O. (1977). Residue dynamics and biological effects of polychlorinated biphenyls in aquatic organisms. *Arch. Environ. Contam. Toxicol.* **5**, 501.

Mayer, F. L. and Sanders, H. O. (1973). Toxicology of phthalic acid esters in aquatic organisms. *Environ. Health Perspect.* Exp Issue **3**, 153.

Mitrovic, V. V. (1972). "Sublethal Effects of Pollutants on Fish." In M. Ruivo (Ed.), *Marine Pollution and Sea Life*, Food and Agriculture Organization, Fishing News (Books Ltd.), London, p. 252.

Moore, J. A. (1978). "Toxicity of 2,3,7,8-Tetrachlorodibenzo-*p*-dioxin." In C. Ramel (Ed.), *Chlorinated Phenoxy Acids and Their Dioxins*. Swedish Natural Science Research Council, Stockholm, p. 134.

Nisbet, C. T. and Sarofilm, A. F. (1972). Rates and routes of PCBs in the environment. *Environ. Health Perspect.* EXP Issue **1**, 21.

Nolan, R. J., Smith, F. A., and Hefner, J. G. (1979). Elimination and tissue distribution of 2,3,7,8-tetrachlorodibenzo-*p*-dioxin (TCDD) in female guinea-pigs following a single oral dose. *Toxicol. Appl. Pharmacol.* **48**, A162.

O'Connors, H. B., Wurster, C. F., Powers, C. D., Biggs. D. C., and Rowland, R. G. (1978). Polychlorinated biphenyls may alter marine trophic pathways by reducing phytoplankton size and production. *Science* **201**, 737.

Olafson, R. W. (1978). Effect of agricultural activity on levels of organochlorine pesticides in hard corals, fish and molluscs from the Great Barrier Reef. *Marine Environ. Res.* **1**, 87.

Olie, L., Vermuelen, P. L., and Hutzinger, O. (1977). Chlorodibenzo-*p*-dioxins and chlorodibenzofurans are trace components of fly ash and flue gas of some municipal incinerators in the Netherlands. *Chemosphere* **8**, 455.

Peakall, D. B. (1975). PCBs and their environmental effects, *Crit. Rev. Environ. Control* **5**, 469.

Phillips, D. J. H. (1980). *Quantitative Aquatic Biological Indicators: Their Use to*

Monitor Trace Metal and Organochlorine Pollution. Applied Science Publishers, London, p. 488.

Pocchiari, F. (1978). "2,3,7,8-Tetrachlorodibenzo-*p*-Dioxin Decontamination." In C. Ramel (Ed.), *Chlorinated Phenoxy Acids and Their Dioxins.* Swedish Natural Science Research Council, Stockholm, p. 67.

Prat, H. and Giraud, A. (1961). *The Pollution of Water by Detergents.* Organisation for Economic Cooperation and Development, Paris, p. 85.

Rappe, C. (1978). "2,3,7,8-Tetrachlorodibenzo-*p*-Dioxin (TCDD) — Introduction." In F. Cattabeni, A. Cavallaro, and G. Galli (Eds.), *Dioxin: Toxicological and Chemical Aspects.* S.P. Medical and Scientific Books, New York, p. 9.

Rappe, C., Marklund, S., Buser, H. R., and Bosshard, H. P. (1978). Formation of polychlorinated dibenzo-*p*-dioxins (PCDDs) and dibenzofurans (PCDFs) by burning or heating chlorophenates. *Chemosphere* 7, 269.

Rawls, R. L. (1979). Dow finds support, doubt for dioxin ideas. *Chem. Eng. News* 57, 23.

Reiff, B. (1975). "Biodegradation and Aquatic Toxicity of Surfactants: A Laboratory Monitoring Method." In J. H. Koeman and J. J. T. W. A. Strik (Eds.), *Sublethal Effects of Toxic Chemicals on Aquatic Animals.* Elsevier Scientific Publishing, Amsterdam, p. 53.

Richardson, B. J. and Waid, J. S. (1979). PCBs in the Port Phillip region: The environmental significance of polychlorinated biphenyls (PCBs), Publication No. 248, Ministry for Conservation — Victoria, Environmental Studies Series, Melbourne.

Richardson, B. J. and Waid, J. S. (1982). Polychlorinated biphenyls (PCBs): An Australian viewpoint on a global problem. *Search* 13, 17.

Rose, J. Q., Wentzler, T. H., Hummel, R. A., and Gehring, P. J. (1976). The fate of 2,3,7,8-tetrachlorodibenzo-*p*-dioxin following single and repeated oral doses to the rat. *Toxicol. Appl. Pharmacol.* 36, 209.

Safe, S., Jones, D., and Hutzinger, O. (1976). Metabolism of 4,4'-dihalogeno-biphenyls. *J. Chem. Soc. Perkin Trans.* 1, 357.

Saint-Ruf, G. (1978). "The Structure and Biochemical Effects of TCDD." In F. Cattabeni, A. Cavallaro, and G. Galli, (Eds.), *Dioxin: Toxicological and Chemical Aspects.* S.P. Medical and Scientific Books, New York, p. 157.

Schmidt, H. and Schultz, G. (1881). Patent and manufacture of pentachloro-biphenyl. *Ann. Chem.* 7, 338.

Schulz, C. O. and Rubin, R. J. (1973). Distribution, metabolism and excretion of di-2-ethylhexyl phthalate in the rat. *Environ. Health Perspect.* EXP Issue 3, 123.

Shaw, G. R. and Connell, D. W. (1983). Physicochemical properties controlling polychlorinated biphenyl (PCB) concentrations in aquatic organisms. *Environ. Sci. Technol.* (in press).

Shaw, G. R. and Connell, D. W. (1980). Relationships between steric factors and bioconcentration of polychlorinated biphenyls (PCBs) by the sea mullet (*Mugil cephalus* Linneaus). *Chemosphere* 9, 731.

Stalling, D. L., Hogan, J. W., and Johnson, J. L. (1973). Phthalate ester residues — their metabolism and analysis in fish. *Environ. Health Perspect.* EXP Issue 3, 159.

Stalling, D. L. and Huckins, J. N. (1971). Gas–liquid chromatography–mass spectrometry characterization of PCBs (Aroclors) and chlorine-36 labelling of Aroclors 1248 and 1254. *J. Assoc. Offic. Anal. Chem.* **54**, 801.

Sundstrom, G., Hutzinger, O., and Safe, S. (1976). The metabolism of chlorobiphenyls – a review. *Chemosphere* **5**, 267.

Swisher, R. D. (1971). *Surfactant Biodegradation*. Marcel Dekker, New York.

Taylor, A. (1980). Soaps and detergents and the environment. *J. Am. Oil. Chem.* **57**, 859A.

Tobin, R. S., Onuska, F. I., Anthony, D. H. J., and Comba, M. E. (1976). Nonionic surfactants: Conventional biodegradation test methods do not detect persistent polyglycol products. *Ambio* **5**, 30.

Vos, J. G. (1978). "2,3,7,8-Tetrachlorodibenzo-*p*-Dioxin: Effects and Mechanisms." In C. Ramel (Ed.), *Chlorinated Phenoxy Acids and their Dioxins*. Swedish Natural Science Research Council, Stockholm, p. 165.

Wasserman, M., Wasserman, D., Cucos, S., and Miller, H. J. (1979). World PCBs Map: Storage and effects in man and his biologic environment in the 1970's. *Ann. N.Y. Acad. Sci.* **320**, 69.

Westing, A. H. (1978). "Ecological Considerations Regarding Massive Environmental Contamination with 2,3,7,8-Tetrachlorodibenzo-*p*-Dioxin." In C. Ramel (Ed.), *Chlorinated Phenoxy Acids and their Dioxins*. Swedish Natural Science Research Council, Stockholm, p. 285.

Whiteside, T. (1977). A Reporter at Large: The pendulum and the toxic cloud. *The New Yorker*, July 25, p. 30.

WHO (1976). Environmental Health Criteria II. Polychlorinated biphenyls and terphenyls. World Health Organization, Geneva, p. 85.

Widmark, G. (1967). *J. Assoc. Offic. Anal. Chem.* **50**, 1069.

Wipf, H. K., Homberger, E., Neuner, N., and Schenker, F. (1978). "Field Trials on Photodegradation of TCDD on Vegetation after Spraying with Vegetable Oil." In F. Cattabeni, A. Cavallaro, and G. Galli (Eds.), *Dioxin: Toxicological and Chemical Aspects*. S.P. Medical and Scientific Books, New York, p. 201.

Wong, A. S. and Crosby, D. G. (1978). "Decontamination of 2,3,7,8-Tetrachlorodibenzo-*p*-Dioxin (TCDD) by Photochemical Action." In F. Cattabeni, A. Cavallaro, and G. Galli (Eds.), *Dioxin: Toxicological and Chemical Aspects*. S.P. Medical and Scientific Books, New York, p. 185.

10

METALS AND SALTS

Natural processes, such as chemical weathering and geochemical activities, release the various elements in the earth's crust into the lithosphere, atmosphere, and hydrosphere. Transport and transformations of these elements, including metals and their salts, usually involve geochemical as well as biological recycling processes thereby forming the earth's biogeochemical cycles.

The influence of pollution on the global environment, through activities such as mining operations, burning of fossil fuels, agriculture, and urbanization, have accelerated the fluxes of trace metals and salts in the ecosphere. The present rate of global input of trace metals such as mercury, lead, zinc, and cadmium is in excess of the natural rate of biogeochemical cycling (Leckie and James, 1974). For some metals and salts, global contributions from anthropogenic sources may be small compared to natural fluxes, nevertheless, these may cause localized pollution such as with mining wastes and leachates.

This chapter identifies significant chemical, biological, and ecological consequences of excessive metal discharges entering the ecosphere, particularly aquatic systems, whereas Chapter 11 covers metallic pollutants in the atmosphere. Some aspects of the behavior of salts in the environment are also covered.

THE METALS

The term "metal" typically describes an element which is a good conductor of electricity and has high thermal conductivity, density, malleability, ductibility, and electropositivity. However, some elements (boron, silicon, germanium, arsenic, and tellurium), referred to as metalloids, possess one or more of these properties but are not sufficiently distinctive in their characteristics to allow a precise delineation as a metal or nonmetal. Furthermore, allotrophic forms of some borderline elements may also exhibit different properties (Wittmann, 1979).

Metals react as electron-pair acceptors (Lewis acids) with electron-pair donors (Lewis bases) to form various chemical groups such as an ion pair, a metal complex, a coordination compound, or a donor–acceptor complex. This type of equilibrium reaction can be generalized as follows:

$$M + L \rightleftharpoons ML$$

$$K_{ML} = \frac{[ML]}{[M][L]}$$

(*Note*: Charges on chemical species have been omitted for convenience.)

In the above equation, M represents the metal ion, L the ligand, ML the metal–ligand complex, and K_{ML} the equilibrium (or stability) constant.

Accordingly, the broad preference of a metal cation either for large, easily polarizable, low electronegativity ions (e.g., sulfide) or smaller, more electronegative anions (e.g., oxides) has been utilized as a classification system for metals. Ahrland et al. (1958) proposed that metals could be separated into three categories – class A, class B, and borderline – on the basis of the equilibrium constants for the formation of metal ion/ligand complexes in aqueous solution. The larger the magnitude of the equilibrium constant, the more stable is the metal complex in solution. For example, transition metals of the $3d$ electron series show an increase in complex stability in the following order: $Mn^{2+} < Fe^{2+} < Co^{2+} < Ni^{2+} < Cu^{2+} > Zn^{2+}$ (Irving–Williams Series). This approach has been further developed (see Pearson, 1968a, b) to classify electron-pair acceptors and donors into hard and soft acids and bases. For example, hard acids (e.g., Mg^{2+}, Ca^{2+}, Al^{3+}) bind strongly with hard bases (e.g., O^{2-} or CO_3^{2-}). Conversely, softer acids (e.g., Hg_2^{2+} or Hg^{2+}, Pb^{2+}) show a greater preference for soft bases (e.g., S^{2-}).

Such considerations are usually limited to inorganic equilibria rather than biological systems (Nieboer and Richardson, 1980). There is considerable difference of opinion on the classification of metals from a biological point of view. In particular, the term "heavy metals" is used widely in the scientific literature to describe toxic metals and has been used in this sense in this book. The definition of "heavy metals" has been based primarily on (1) the specific gravity of the metals (greater than 4 or 5), (2) location within the Periodic Table, for example, elements with atomic numbers 22–34 and 40–52, and lanthanides and actinides, and (3) specific biochemical responses in animals and plants (Murphy, 1981). Nieboer and Richardson (1980) proposed that the term "heavy metals" be replaced by a classification of metal ions into three biologically and chemically significant classes – oxygen-seeking, nitrogen/sulfur-seeking, and intermediate. The section on "Metal Toxicity" later in this chapter considers the application of this classification to metal ion toxicity. Throughout this book the term "metals" has been used broadly to refer to metallic elements and may include metalloids of environmental significance.

SOURCES OF METALS IN THE HYDROSPHERE

Metals enter the hydrosphere from a variety of sources, either natural or human induced. On a geological time scale natural sources such as chemical weathering and volcanic activities have been the major release mechanisms responsible for the chemical composition of freshwater and marine ecosystems.

Natural Sources

In freshwater systems, chemical weathering of igneous and metamorphic rocks and soils in drainage basins is the most important source of background levels of trace metals entering surface waters (Leckie and James, 1974). Considerable variation in background levels in surface water and bottom sediment are observed due to the presence of mineralized zones in the drainage basins. Decomposing plant and animal detritus also contribute small yet significant amounts of metals to surface waters and bottom sediments.

Precipitation and atmospheric fallout, within a drainage basin or directly on the water surface, are the second most important source for trace metals entering water bodies on a global basis. Dry fallout of metals is generally directly proportional to their concentrations in the particles but metals in precipitation vary with the amount and type of precipitation. Metals in the atmosphere attributed to natural sources are derived from (1) dusts from volcanic activities, (2) erosion and weathering of rocks and soil, (3) smoke from forest fires, and (4) aerosols and particulates from the surface of the oceans.

Natural inputs into the marine environment can be categorized according to Bryan (1976a) as: (1) coastal supply, which includes input from rivers and from erosion produced by wave action and glaciers; (2) deep sea supply which includes metals released by deep sea volcanism and those removed from particles or sediments by chemical processes; (3) supply which bypasses the nearshore environment and includes metals transported in the atmosphere as dust particles or as aerosols and also material produced by glacial erosion in polar regions and transported by floating ice.

Pollution Sources

Human activities are also a major source of metal introduction into the aquatic environment. Metal inputs arise from direct discharges of various kinds of contaminated wastes, disturbance of drainage basins, atmospheric precipitation, and fallout. Major inputs are summarized below (Wittmann, 1979):

1. *Mining Operations.* The exploitation of ore deposits invariably exposes fresh rock surfaces and large quantities of waste rock or soil to accelerated weathering conditions. For example, exposure of pyrite and other sulfide minerals to atmospheric oxygen and moisture results in oxidation of this mineral and the formation of "acid mine drainage water."

This is a complex process substantially mediated by *Thiobacillus* and *Ferrobacillus* bacteria (see Figure 10.1). Low pH conditions are produced together with various metallic compounds including ferric hydroxide which forms yellow-orange deposits in the affected area. Consequently, the release of acid mine drainage from active and abandoned mines, particularly coal mines, has been widely associated with serious water quality problems, involving relatively high levels of metals such as Fe, Mn, Zn, Cu, Ni, and Co.

(1) $FeS_2 + 2H_2O + 7O_2 \longrightarrow 4H^+ + 4SO_4^{2-} + 2Fe^{2+}$

(2) $4Fe^{2+} + O_2 + 4H^+ \longrightarrow 4Fe^{3+} + 2H_2O$

(3) $FeS_2 + 14Fe^{3+} + 8H_2O \longrightarrow 15Fe^{2+} + 2SO_4^{2-} + 16H^+$

(4) $Fe^{3+} + 3H_2O \longrightarrow Fe(OH)_3 + 3H^+$

Figure 10.1. Sequence of reactions involved in the oxidation of pyrites to form "acid mine drainage water."

In addition, ore processing, smelting, and refining operations can cause the dispersion and deposition of large quantities of trace metals such as Pb, Zn, Cu, As, and Ag into surrounding drainage basins (see Chapter 11) or direct discharge into aquatic environments.

2.   *Domestic Effluents and Urban Stormwater Runoff.*   Appreciable amounts of trace metals are contributed to domestic effluents by metabolic wastes, corrosion of water pipes (Cu, Pb, Zn, and Cd), and consumer products (e.g., detergent formulations containing Fe, Mn, Cr, Ni, Co, Zn, Cr, B, and As). Concentrations are often in the milligram per liter range but vary according to such factors as water usage patterns, time of year, and economic status of consumers. Waste-water treatment by the activated sludge process generally removes less than 50% of influent metals yielding effluents with significant trace metal loadings. The disposal of sewage sludge can also contribute to metal enrichment (Cu, Pb, Zn, Cd, and Ag) in receiving waters (Williams et al., 1974). For example, municipal waste water and the dumping of domestic and industrial sludges are the major artificial sources for Cd, Cr, Cu, Fe, Pb, and Hg loadings in the New York Bight (Mueller et al., 1976).

Stormwater runoff from urbanized areas has been increasingly recognized as a substantial source of trace metal contamination in receiving waters. For example, urban runoff into the New York Bight is estimated to contribute a substantial proportion of the total loads of Cu, Cr, Pb, and Zn, and to a lesser extent Cd, Hg, and Fe, as shown in Table 10.1. Metal concentrations in urban stormwater exhibit extreme variability which is mainly related to locality and time of sampling (Malmquist, 1975). Whipple and Hunter (1977) observed that the highest concentrations of Pb, Zn, and Cu occurred in urban runoff within 30 min of stormwater events.

The metal composition of urban runoff is dependent on many factors such as city planning, traffic, road construction, land use, and the physical characteristics and climatology of the watersheds. The use of common salt for deicing streets and highways during winter months in some temperate regions may increase the mobilization of metal ions (e.g., Hg^{2+}, Cd^{2+}, Zn^{2+}, and Pb^{2+}) through formation of chloride and hydroxy complexes (Hahne and Kroontji, 1973).

3.   *Industrial Wastes and Discharges.*   Many trace metals are discharged into the aquatic environment through industrial effluents as well as the dumping and leaching of industrial sludges. Metal concentrations in industrial waste waters are frequently in the milligram per liter range (see Table 10.2). In general, there is multipurpose usage of several metals within most industries although there are many examples of specific metal pollution related to certain industries.

Table 10.1. Percentages of New York Bight Metal Loads by Source

| | Direct Bight | | Waste Water | | Coastal Zone | | |
	Barge	Atmospheric	Municipal	Industrial	Gauged	Urban	Groundwater
						Runoff	
Cadmium	82	2	5	0.6	5	5	<1
Chromium	50	1	22	0.8	10	16	<1
Copper	51	3	11	9	10	16	<1
Iron	79	3	5	0.5	6	6	<1
Mercury	9	–	71	2	13	5	
Lead	44	9	19	3	6	19	<1
Zinc	29	18	8	2	21	22	<1

Source: Mueller et al. (1976).

Table 10.2.   Metals in Industrial Waste Waters

Industry	Average Concentrations in $\mu g\, L^{-1}$				
	Cu	Cr	Ni	Zn	Cd
Meat processing	150	150	70	460	11
Fat rendering	220	210	280	3,890	6
Fish processing	240	230	140	1,590	14
Bakery	150	330	430	280	2
Miscellaneous foods	350	150	110	1,100	6
Brewery	410	60	40	470	5
Soft drinks and flavorings	2,040	180	220	2,990	3
Ice cream	2,700	50	110	780	31
Textile dyeing	37	820	250	500	30
Fur dressing and dyeing	7,040	20,140	740	1,730	115
Miscellaneous chemicals	160	280	100	800	27
Laundry	1,700	1,220	100	1,750	134
Car wash	180	140	190	920	18

Source: Klein et al. (1974). Reprinted with permission from the Water Pollution Control Federation.

Leachate from sanitary landfills can have substantial levels of Cu (5 ppm), Zn (50 ppm), Pb (0.3 ppm), and Hg (60 ppb) but flows are generally low (Williams et al., 1974). Metal emissions from the combustion of fossil fuels are also a major source of airborne metal contaminants in natural waters and catchments. In addition, the burning of leaded fuels contributes significantly to urban lead deposition (Wittmann, 1979).

4.   *Agricultural Runoff.*   The diverse nature of agricultural activities and practices throughout the world makes it difficult to provide an overall evaluation of the significance of this metal source. Nevertheless, enormous quantities of sediment containing trace metals are lost from agricultural regions as a result of soil erosion (McElroy et al., 1975). Agricultural soils may become enriched with trace metals from animal and plant residues, phosphatic fertilizers, specific herbicides and fungicides, and through the use of sewage effluent or sludge as a plant nutrient source. On the other hand, trace metals in soils tend to be stabilized through oxidation, formation of insoluble salts, and absorption reactions, dependent on soil characteristics.

BEHAVIOR AND FATE IN ABIOTIC ENVIRONMENTS

Metals in Aquatic Systems

Natural waters and associated particulate matter are complex heterogeneous electrolyte systems containing numerous inorganic and organic species distributed between

Table 10.3.   Forms of Occurrence of Metal Species in Natural Waters

Form	Examples
Free metal ions	$Cu^{2+}(aq)$, $Fe^{3+}(aq)$, $Pb^{2+}(aq)$
Inorganic ion pairs	$Cu_2(OH)_2{}^{2+}$, $Pb(CO_3)_2{}^{2-}$
Inorganic complexes	$CdCl^+$
Organic complexes	Me–SR, Me–OOCR

$$
\begin{array}{c}
CH_2-C=O \\
\diagup \qquad \diagdown \\
NH_2 \qquad \qquad O \\
\diagdown \qquad \diagup \\
Cu \\
\diagup \qquad \diagdown \\
O \qquad \qquad NH_2 \\
\diagdown \qquad \diagup \\
O=C-CH_2
\end{array}
$$

Form	Examples
Metal species complex bound to high molecular weight organic material	Me–lipids Me–humic acid Me–polysaccharides "Lakes" "Gelbstoffe"
Metal species in the form of highly dispersed colloids	$Fe(OH)_3$, $Mn(IV)$ oxides, $Mn_7O_{13} \cdot 5H_2O$
Metal species sorbed on colloids	$Me_x(OH)_y$, $MeCO_3$, MeS, etc. on clays
Precipitates, organic particles, remains of living organisms	

Source: Stumm and Belinski (1972). Reprinted with permission from Pergamon Press, Inc.

aqueous and solid phases. Trace metals entering natural waters become part of this system and their distribution processes are controlled by a dynamic set of physico-chemical interactions and equilibria (Stumm and Morgan, 1970) (see Chapter 2).

The solubility of trace metals in natural waters is principally controlled by (1) pH, (2) type and concentration of ligands and chelating agents, and (3) oxidation state of the mineral components and the redox environment of the system (Leckie and James, 1974). In addition, dynamic interactions at solution–solid interfaces determine the transfer of metals between aqueous and solid phases. Thus, trace metals may be in a suspended, colloidal, or soluble form. In general, suspended particles are considered to be those greater than 100 μm in size, soluble particles are those less than 1 μm in size and colloidal particles are those in the intermediate range. The suspended and colloidal particles may consist of (1) compounds or heterogeneous mixtures of metals in forms such as hydroxides, oxides, silicates, or sulfides or (2) clay, silica, or organic matter to which metals are bound by absorption, ion exchange, or complexation (see Tinsley, 1979). The soluble forms are usually ions, simple or complex, or unionized organometallic chelates or complexes (Stumm and Morgan, 1970; Leckie and James, 1974).

Stumm and Bilinski (1972) have developed a general scheme illustrating metal speciation in natural waters, as given in Table 10.3. Metals in natural waters may exist simply in the form of free metal ions surrounded by coordinated water molecules, although the concentrations of anionic species (e.g., OH^-, Cl^-, SO_4^{2-}, HCO_3^-, organic acids, and amino acids) are usually sufficient to form inorganic or organic complexes with the hydrated metal ions by replacing the coordinated water. Other associations occur with colloidal and particulate material such as clays and hydrous iron and manganese oxides and organic material (Hart and Davies, 1978). Several types of interaction occur between metal ions and other species in aqueous solutions (Leckie and James, 1974; Stumm and Morgan, 1970) as outlined below:

1. *Hydrolysis Reactions of Metal Ions.*   Most highly charged metal ions (e.g., Th^{4+}, Fe^{3+}, and Cr^{3+}) are strongly hydrolyzed in aqueous solution and have low pK_1 values.

$$Fe(H_2O)_6^{3+} + H_2O \rightleftharpoons Fe(H_2O)_5OH^{2+} + H_3O^+$$

Hydrolysis may also proceed further by the loss of one or more protons from the coordinated water.

$$Fe(H_2O)_5OH^{2+} + H_2O \rightleftharpoons Fe(H_2O)_4(OH)_2^+ + H_3O^+$$

Many divalent metals (e.g., Cu^{2+}, Pb^{2+}, Ni^{2+}, Co^{2+}, and Zn^{2+}) hydrolyze also within the pH range of natural waters.

The hydrolysis of aqueous metal ions can also produce polynuclear complexes containing more than one metal ion, for example,

$$2FeOH^{2+} \rightleftharpoons Fe_2(OH)_2^{4+}$$

Polymeric hydroxo forms of metal ions (e.g., Cr^{3+}) may condense slowly with time to yield insoluble metal oxides or hydroxides. Polymeric species are important in moderate to high concentrations of metal salt solutions (e.g., Sillén, 1961).

2. *Complexation of Metal Ions.*   Metal ions also react with inorganic and organic complexing agents present in water from both natural and pollution sources. Dominant inorganic complexing ligands include Cl^-, SO_4^{2-}, HCO_3^-, F^-, sulfide, and phosphate species. These reactions are somewhat similar to the hydrolysis reactions of metal ions in that sequences of soluble complex ions and insoluble phases may result depending on the metal and ligand concentrations and pH.

Inorganic ligands are usually present in natural waters at much higher concentrations than the trace metals they tend to complex. Each metal ion has a speciation pattern in simple aqueous solutions that is dependent upon (1) the stability of the hydrolysis products and (2) the tendency of the metal ion to form complexes with other inorganic ligands. For example, Pb(II), Zn(II), Cd(II), and Hg(II) each form a complex series when in the presence of Cl^- and/or SO_4^{2-} at concentrations similar to those of seawater. The pH at which a significant proportion of hydrolysis products are formed is dependent upon the concentration of the ligand, for example, Cl^- competing with OH^- for the metal ion.

Metals can also bond to natural and synthetic organic substances by way of (1) carbon atoms yielding organometallic compounds, (2) carboxylic groups producing salts of organic acids, (3) electron-donating atoms O, N, S, P, and so on forming coordination complexes, or (4) π-electron donating groups (e.g., olefinic bonds, aromatic ring, etc.).

Sillén (1961), Morel and Morgan (1972), and Stumm and Brauner (1975) have investigated metal speciation in natural waters by computation of thermodynamic equilibria for model systems of metals and ligands. In the pH range of natural water, between 5 and 9.5, Morel et al. (1973) found that under aerobic conditions free metal ions occur mainly at low pH, and with increasing pH the carbonate and then the oxide, hydroxide, or even silicate solids precipitate.

Metal speciation is also controlled by oxidation–reduction conditions. Redox environments within natural waters are usually complex, in a nonequilibrium state, and may show marked variations and gradients between air–water and water–solid or water–sediment interfaces. Metal speciation is affected in two ways: (1) by direct changes in the oxidation state of the metal ions, for example, Fe(II) to Fe(III) and Mn(II) to Mn(IV), and (2) by redox changes in available and competing ligands or chelates. Typical redox environments found in aquatic systems can be characterized by the use of $p\epsilon$–pH stability field diagrams (see Chapter 2). Stability fields for metal phases as a function of $p\epsilon$ and pH at $25°C$ and 1 atm total pressure are widely available, for example, for lead and mercury (Leckie and James, 1974) and iron (Hem, 1970).

Metal Speciation in Freshwater and Seawater

Metal speciation in freshwater and seawater differ considerably due mainly to (1) different ionic strengths, (2) lower content of adsorbing surfaces in seawater, (3) different concentrations of trace metals, (4) different concentrations of major cations and anions, and (5) usually higher concentrations of organic ligands in freshwater systems (Sibley and Morgan, 1977; Förstner, 1979a). Figure 10.2 illustrates some of the differences between the two systems. In freshwater adsorbed metals on particulates dominate and soluble metal–ligand complexes are considerably more varied than in seawater. In mixtures of freshwater and seawater chlorocomplexes become the dominant species for Cu, Zn, Hg, and Co, while Ni tends to remain as the free ion and Cr forms hydroxide complexes. Adsorption is negligible for all these metals, because an increase in ionic strength decreases the density of metal ions on particle surfaces due to competitive exchange in the electric double layer (see Förstner, 1979a).

Few metals, for example, Cu(II) and Fe(II) (Stumm and Brauner, 1975), form significant amounts of organic complexes in seawater. In most cases abundant cations such as Mg^{2+} and Ca^{2+} compete for the organic functional groups (Förstner, 1979a).

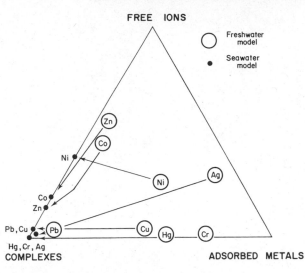

Figure 10.2.   Evaluation of the changes in chemical species of selected trace elements from a seawater/freshwater model. [From Sibley and Morgan (1977). Reprinted with permission from the Institute for Environmental Studies, University of Toronto.]

Interactions Between Aqueous and Solid Phases

The behavior of metals in natural waters is strongly influenced by the interactions between aqueous and solid phases, particularly water and sediments. Dissolved metal ions and complexes are rapidly removed from solution upon contact with the surfaces of particulate matter through several different types of surface bonding phenomena (see Table 10.4). The formation of metal substrates described in Table 10.4 leads to the deposition and enrichment of trace metals in sedimentary environments. Enrichment and remobilization of metals in sediments are dependent on factors such as chemical composition (e.g., the amount of dissolved iron and carbonate), salinity, pH, redox values, and the hydrodynamic conditions (see Stumm and Morgan, 1970; Hart and Davies, 1977; Förstner, 1979b).

Environmental factors affecting trace metal enrichment of aquatic sediments and their function as metal sinks are set out below (Förstner, 1979b):

1.   *Detrital Minerals.*   The presence of heavy minerals in silt and fine-sand fractions of sediment results in enrichment with trace elements, through surface adsorption processes. On the other hand, quartz, feldspar, and detrital carbonates tend to have the opposite effect.

2.   *Sorption.*   The following generalized sequence of the capacity of solids to sorb heavy metals was established by Guy and Chakrabarti (1975): $MnO_2 >$ humic acid $>$ iron oxide $>$ clay minerals. The sorption capacity of iron oxides (crystalline phase goethite) for trace metals is at least 10 times less than that of the manganese oxides (Suarez and Langmuir, 1976). Also Rashid (1971) estimates that with the total bonding capacity of humic substances, approximately two-

Table 10.4. Carrier Substances and Mechanisms of Trace Metal Bonding[a]

Minerals of natural rock debris, e.g., trace metals.		Metal bonding predominantly in inert positions.
Trace metals: hydroxides, sulfides, and carbonates	Precipitation as a result of exceeding the solubility product K_{sp} in water bodies (e.g., surface and interstitial waters; sewage and water treatment systems).	For most metals, three solid phases are possible: hydroxide, carbonate, and sulfide. In the presence of free O_2, M^{2+} is stable at pH values less than approximately 7–8. With increasing pH, the stable phase progresses from carbonate to hydroxide. For reduced conditions (– Eh), the sulfide remains the stable phase over a wide pH range.
Hydroxides and oxides of Fe/Mn	pH dependent ⎡ Physicosorption ⎣ Chemical sorption (exchange of H^+ in fixed positions)	The hydrous oxides of iron, manganese, and aluminum, particularly the Fe- and Mn-hydroxides and oxides under oxidizing conditions, readily sorb or coprecipitate cations and anions from solution. In this manner they act as a major sink for trace metals in aqueous systems. Under reducing conditions the sorbed metals are readily remobilized into solution and thus act as a major source for dissolved trace metals. In heavily polluted waters both mechanisms may be observed.
Sulfides of Fe	Coprecipitation as a result of exceeding the solubility product	The significance of iron sulfides on the coprecipitation of trace metals is relatively uncertain. It appears to be less significant than trace metal removal by hydrous iron oxides.

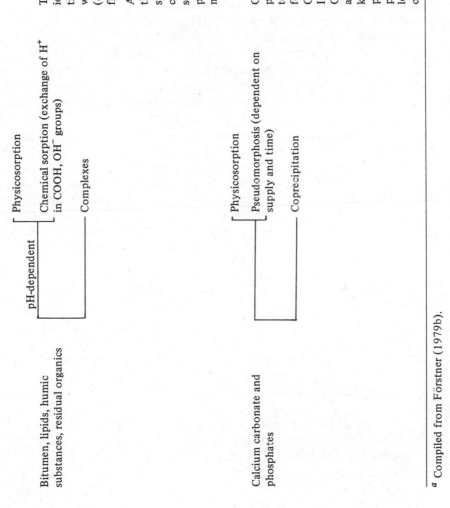

Bitumen, lipids, humic substances, residual organics

pH-dependent
— Physicosorption
— Chemical sorption (exchange of H^+ in COOH, OH^- groups)
— Complexes

The attractive forces between metal ions with soluble, colloidal, or particulate organic material range from weak (physical adsorption) to strong (complexation), e.g., humic and fluvic acids.

A number of processes will lead to the incorporation of metal–organic species onto or into sediments: coagulation and flocculation of soluble and colloidal material, direct precipitation, or adsorption onto sedimentary material, e.g., clay surfaces.

Calcium carbonate and phosphates

— Physicosorption
— Pseudomorphosis (dependent on supply and time)
— Coprecipitation

Carbonate adsorption and co-precipitation of trace metals appear to be important removal mechanisms for a number of metals, e.g., Zn, Co, Cd, and Pb, in alkaline environments. Ions adsorbed on the surface of $CaCO_3$ tend to enter the adsorbent and to form pseudomorphs. Little is known about the action of insoluble phosphates. However, marine phosphorites tend to contain enriched levels of uranium, arsenic, and cadmium.

[a] Compiled from Förstner (1979b).

thirds can be attributed to chemical sorption and organic complexations and one-third to cation exchange.

3. *Coprecipitation with Hydrous Fe/Mn Oxides and Carbonates*.   Under oxidizing conditions, hydrous iron and manganese oxides constitute a highly effective sink for metals (Lee, 1975). Groth (1971) demonstrated that Co, Zn, and Cu coprecipitated from natural lake water with Fe/Mn hydroxides at a rate of 67%, 86%, and 98%, respectively. Coprecipitation with carbonate may be an important depositional mechanism for Zn and Cd when carbonate is a major component, that is, when other substrates, particularly hydrous iron oxides or organic substances are less abundant.

4. *Complexation and Flocculation with Organic Matter*.   In systems rich in organic matter, the role of Fe/Mn oxides is much less important because of competition from the more reactive humic acids, organo-clays, and oxides coated with organic matter (Jonasson, 1977). Organic coatings greatly affect the adsorption capacities of sediment and suspended matter (Pillai et al., 1971; Sholkovitz, 1976). Metals complexed by humic acids become unavailable to form sulfides, hydroxides, and carbonates and thus prevent the formation of insoluble salts (Rashid and Leonard, 1973). Chemical and electrostatic processes result in flocculation of Fe, Al, and humates, particularly in marine estuaries (Eckert and Sholkovitz, 1976). Thus enrichment with metals has been found by Nissenbaum and Swaine (1976) in the humic substances from marine reducing environments.

5. *Trace Metal Precipitates*.   The amount of sediment formed in natural waters and from the direct precipitation of trace metal hydroxides, carbonates, and sulfides is difficult to distinguish from that coprecipitated with other substances. An example of the significance of the various bonding mechanisms and their respective substrates for metal to sediment associations has been quantitatively estimated for the estuarine zone in Table 10.5. The processes of flocculation of metal–organic complexes and coprecipitation with hydrous Fe/Mn oxides are of particular importance especially for metals introduced by human activities.

Mobilization of Trace Metals from Sediments

Under suitable conditions some metals associated with sediments and suspended particles are returned to the overlying water following remobilization and upward diffusion (Bryan, 1976a). This process may act as a significant source of trace metal contamination. At least five major processes, as outlined below, control the release of metals in this way (Förstner, 1979b).

1. *Elevated Salt Concentrations*.   At elevated concentrations the alkali and alkaline earth cations can compete for adsorption sites on the solid particles, thereby displacing the sorbed trace metal ions.

2. *Changes in Redox Conditions*.   A decrease in the oxygen potential in sediments can occur due to such conditions as advanced eutrophication. This results in a change in the chemical form of the metals and thus a

Table 10.5. Factors Affecting the Accumulation of Heavy Metals in Estuarine Sediments

	Detrital Minerals, Organics	Reactive Organics (Humic Acids, Bitumen)	Trace Metal Hydroxides, Carbonates, Sulfides	Hydrous Iron and Manganese Oxides	Calcium Carbonate
Incorporation in inert positions	xx	(x)			
Adsorption = physicosorption	x	x	(x)	(x)	(x)
Cation exchange = chemisorption	x	x	(x)	x	(x)
Precipitation			xx		
Coprecipitation				xxx	x
Complexation + flocculation		xxx			

Source: Förstner (1979b). Reprinted with permission from Springer-Verlag.

change in water solubility. Under reducing conditions trace metals in the interstitial waters occur as (a) sulfide complexes for Cd, Hg, and Pb; (b) organic complexes for Fe and Ni; (c) chloride complexes for Mn; and (d) hydroxide complexes for Cr. With the development of oxidizing conditions the solubility of metal ions is influenced by a gradual change from metallic sulfides to carbonate hydroxides, oxyhydroxides, oxides, or silicates (Lu and Chen, 1977).

3. *pH Changes.*   Reduction of pH leads to dissolution of carbonates and hydroxides, as well as to increased desorption of metal cations due to competition with hydrogen ions.

4. *Presence of Complexing Agents.*   Increased use of natural and synthetic complexing agents can form soluble stable metal complexes with trace metals that are otherwise adsorbed to solid particles.

5. *Biochemical Transformation.*   This can lead to either transfer of metals from sediments into the aqueous phase or their uptake by aquatic organisms and subsequent release via decomposition products.

After sedimentary deposition the organic matter, including metals, undergoes diagenesis, involving an increase in molecular weight and loss of some functional groups. A relatively stable and less reactive reservoir for heavy metals in aquatic sediments is formed. However, remobilization may occur through microbial processes.

TRANSPORT AND TRANSFORMATIONS IN BIOTA

Microbial–Metal Interactions

There are three major microbial processes affecting the environmental transport of metals: (1) degradation of organic matter to lower molecular weight compounds, which are more capable of complexing metal ions; (2) alterations to the physico-chemical properties of the environment and chemical form of metals by metabolic activities, for example, the oxidation–reduction potential and pH conditions; and (3) conversion of inorganic compounds into organometallic forms by means of oxidative and reductive processes (Förstner, 1979b). This third mechanism involves bacterial methylation of a number of elements, for example, Hg, As, Pb, Se, and Sn, in which methyl cobalamin appears to be the primary biological methylating agent. The biological cycle for mercury is given in Figure 10.3. Craig (1980) has extensively reviewed biological transformations of metals and their environmental cycles.

Uptake Processes

The initial uptake of metals by aquatic organisms can be considered in terms of three main processes: (1) from water through respiratory surfaces (e.g., gills); (2) adsorption from water onto body surfaces; and (3) from ingested food, particles,

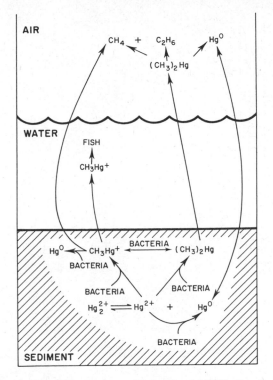

Figure 10.3. Biological cycle for mercury. [From Wood (1974). Reprinted with permission from the American Association for the Advancement of Science.]

or water through the digestive system. Mechanisms involved in uptake processes for metals are summarized in Table 10.6 for several major taxonomic groups of organisms.

In the case of photo- and chemoautotrophic organisms, metal uptake occurs directly from solution, or for higher plants additionally via the roots. Phytoplankton, for example, appear to rapidly absorb metals at the cell surface, from where they diffuse into the cell membrane and are adsorbed or bound to proteins (ion exchange sites) within the cell (see Davies, 1973). In general, Bryan (1976b) considers the uptake of heavy metals by aquatic plants to be a passive process that can be influenced indirectly by metabolism. However, several species of chemotrophs, for example, those active in the formation of acid mine drainage, can metabolize metals directly from inorganic compounds such as metal sulfides (Prosi, 1979).

In heterotrophic organisms the modes of metal entry are much greater than in autotrophic organisms and vary widely according to the species. Absorption from solution by most animals occurs by passive diffusion, probably as a soluble metal complex, through gradients created by adsorption at the body surfaces and binding by body constituents (Bryan, 1976b). Rates of absorption are influenced by changes in physicochemical factors (e.g., temperature, pH, salinity) and physiological

Table 10.6.   Uptake Processes for Metals in Aquatic Organisms

Class of Organisms	Source	Process/Mechanism	References
		Autotrophic	
Phytoplankton	Water	Ion exchange processes involving organic molecules such as proteins: Rapid adsorption at the cell surface, diffusion through cell membrane, and adsorption at ion exchange sites in cell, e.g., Zn in the diatom, *Phaeodactylum tricornutum*.	Bryan (1976b) Davies (1973)
Macroalgae	Water	Adsorption or an ion exchange process involving both cell proteins and polysaccharides, e.g., alginates in cell walls of seaweeds; Zn in brown seaweed, *Laminaria digitata*.	Gutknecht (1965) Bryan (1969)
	Sediments	Absorption of metals from interstitial or pore water via root system: Initial entry into root free space is probably passive via bulk flow of soil/sediment water. If absorption equals rate of bulk flow, then initial free space uptake is a function of transpiration rate. However, if absorption is greater than rate of bulk flow, then a depletion zone is created around the root and the concentration gradient thus established promotes metal diffusion from the soil to root. Low molecular weight chelates are important in this process.	Hughes et al. (1980)

Heterotrophic

Filter Feeders (e.g., bivalve molluscs and tunicates)	Water	Adsorption of metals from mucus sheets of ciliary feeding mechanisms: Adsorption of metals by attachment to mucus may promote diffusion through the body surface or absorption may occur when mucus is passed through digestive system, e.g., V in ascidians.	Pentreath (1973c)
	Food	In molluscs, e.g., oysters, metals are mainly obtained from ingested particles rather than from solution.	Kalk (1963); Kustin et al. (1975); Bryan (1976a)
Polychaetes	Water	Adsorption process involving diffusion through external body surface: In the burrowing polychaete, *Neris diversicolor*, the rate of Zn absorption is proportional to the degree of adsorption at the surface of the body rather than to the external concentration.	Bryan and Hummerstone (1973)
Crustaceans	Water	Adsorption on body surfaces, e.g., cuticle, followed by diffusion through surface, e.g., gill epithelium, probably attached to organic ligands and binding to internal proteins.	Bryan (1976b)
	Food	In larger crustacea, e.g., lobster, absorption from the diet via the stomach or digestive system appears to be more important. Adsorption from solution seems to be most important for shrimp and marine isopods.	
Fish	Water	Adsorption processes similar to larger crustacea, e.g., lobster. Pentreath (1973a, b) has suggested trace metals are absorbed by the gills through a passive process, since a favorable concentration gradient is probably achieved by adsorption of metals on mucus covering the gills and the accumulation of brief levels in the gill tissues.	Pentreath (1973a, b) Bryan (1976b)
	Food	Absorption from ingested food more important.	Pentreath (1973a, b) Renfro et al. (1975)

and behavioral characteristics of organisms. For many metals, rates of absorption are directly proportional to the levels of availability in the environment (Bryan, 1979).

Metal uptake from dietary sources in comparison to direct adsorption from solution is of fundamental importance with heterotrophic aquatic organisms. Available evidence is limited but indicates that food and particulates are more important sources of metals than water for large animals such as fish and lobster (Bryan, 1976b; 1979). Within polluted aquatic environments dietary preferences or feeding habits are significant because of metal enrichment in sediments, particulates, and detritus. Prosi (1979) has suggested that the following feeding habits should be considered in relation to metal concentrations: (1) phytophagous (e.g., gastropods, crustaceans); (2) filter feeding (e.g., zooplankton, barnacles, bivalves); (3) sediment feeding (e.g., poly- and oligochaetes); (4) detritus feeding (e.g., gastropods, isopods and amphiphods, chironomid larvae); and (5) carnivorous (e.g., zooplankton, polychaetes, gastropods, cephalopods, crustaceans, freshwater insect larvae, fish).

Excretion and Regulation

Although aquatic organisms readily absorb metals, it is their ability to regulate abnormal concentrations that determines tolerance and is a critical factor in survival. Some animals, such as fish and crustacea, are able to excrete high proportions of abnormal metal intake and consequently regulate the concentration in the body at fairly normal levels (Bryan, 1976b). This occurs more commonly with the essential, and relatively abundant, metals such as Cu, Zn, and Fe rather than nonessential metals such as Hg and Cd. Regulation or excretion occurs through the gills, gut, feces, and urine.

Nevertheless, there is an upper limit to the amount of metal which can be excreted by animals above which there is accumulation in body tissues (see Bryan, 1976a). In this way, differences in metal concentrations can be apparent within single species taken from the same contaminated waters. For example, Johnels et al. (1967) found that mercury concentrations in muscle tissues from Swedish pike increased with the size or age of the fish but this did not occur with fish from relatively uncontaminated waters. This and other similar studies indicate that an exposure threshold exists above which metal accumulation occurs.

Overall, aquatic plants and bivalves are relatively poor regulators of metals, particularly the nonessential ones. Losses may occur through diffusion in aquatic plants, or diffusion and several other mechanisms for molluscs. These include excretion as granules from kidneys in scallops, spheres pinched off from digestive cells in *Cardium edule*, and particulate forms from the mantle edge of oysters (Bryan, 1976b).

The rate of loss of metals, as radionuclides, in marine and estuarine animals is presented in Table 10.7. Generally, the loss curve consists of two components, fast and slow, with different half-times (see Chapter 2 – Models for Biological

Table 10.7. Some Biological Half-Times for Loss of Radionuclides in Marine and Estuarine Animals

| Species | Form of Radionuclide | Method of Introduction | Slow Component | | Conditions |
			% of Total Loss	Half-Time $Tb_{1/2}$ (days)	
Mammals					
Seal	*Pusa hispida*	$CH_3\,^{203}Hg$ proteinate	By mouth	45	500	1–15°C
Fish					
Flounder	*Platichthys flesus*	$CH_3\,^{203}HgNO_3$	By mouth	–	700	$S^0/_{00} = 6$
		$CH_3\,^{203}HgNO_3$	Injected	–	1200	
		$CH_3\,^{203}Hg$ proteinate	By mouth	–	780	
–	*Serranus seriba*	$CH_3\,^{203}HgNO_3$	By mouth	96	267	16–23°C
Croaker	*Micropogon undulatus*	$^{65}ZnCl_2$	Injected	90	138	4–15°C
		$^{51}CrCl_2$	Injected	6	70	7–10°C
Plaice	*Pleuronectes platessa*	$^{65}ZnCl_2$	From water	95	313	10°C
		$^{54}MnCl_2$	From water	81	329	10°C
Killifish	*Fundulus heteroclitus*	$^{65}ZnCl_2$	In food	–	75	10°C
		$^{65}ZnCl_2$	In food	–	35	30°C
Crustaceans					
Crab	*Carcinus maenas*	$CH_3\,^{203}HgNO_3$	Injected	94	400	16–23°C
Lobster	*Homarus gammarus*	$^{54}MnCl_2$	By mouth	100	11	13°C
Krill	*Euphausia pacifica*	^{65}Zn	By feeding	10	140	10°C

(continued on next page)

Table 10.7. (continued)

				Slow Component		
	Species	Form of Radionuclide	Method of Introduction	% of Total Loss	Half-Time $T_{b_{1/2}}$ (days)	Conditions
Molluscs						
Oyster	*Crassostrea gigas*	^{65}Zn	Absorbed in field	100	255	In field
Oyster	*Ostrea edulis*	^{203}HgCl$_2$	From water	–	44	In field
		^{65}Zn	Absorbed in field	100	890	In field
Mussel	*Mytilus edulis*	^{65}Zn	Absorbed in field	100	76	14–22°C
	Mytilus galloprovincialis	CH$_3$ ^{203}HgNO$_3$	Injection	80	1000	16–23°C
Polychaetes						
Ragworm	*Nereis diversicolor*	^{65}ZnCl$_2$	From water	70	14–17	20°C
	Hermione hystrix	^{65}ZnCl$_2$	From water	–	165–197	20°C

Source: Bryan (1976a). Reprinted with permission from Academic Press, Inc. (London) Ltd.

Accumulation). The slow component is considered to be a more representative value and less influenced by the mode of introduction (Bryan, 1976a). The limitation of using metal isotopes to predict loss rates of stable metals from tissues has been discussed by Bryan (1976a). In some contaminated animals loss rates have been slower or faster than predicted from isotopic rates.

Metal Tolerance and Biotransformations

Many contaminated organisms are able to tolerate metal concentrations which are in excess of known physiological needs, and, in some situations, levels occur at which enzyme inhibition would be expected. Metal-tolerant organisms may contain concentrations of metals two or three orders of magnitude higher than normal.

Detoxification mechanisms can involve storage of metals at inactive sites within organisms on a temporary or more permanent basis (see Table 10.8). Temporary storage is generally by binding of metals to proteins, polysaccharides, and amino acids in soft tissues or body fluids. Metallothionein, however, effectively stores cadmium in liver and kidney tissues. Storage sites such as bone, feathers, fur, or exoskeleton provide useful means for the elimination of some metals (e.g., Pb, Cd, and Hg).

Chemical transformations and associations are also important but are not well understood. Methylmercury uptake appears to be detoxified to some degree by demethylation and storage in tissues as less toxic inorganic forms. Many organisms may be able to convert inorganic arsenic and selenium to less toxic organic forms (Wood, 1974; Förstner, 1979b). The strong correlations between concentrations of mercury and selenium in large vertebrates have led to the suggestion that the selenium may inhibit the toxic action of mercury (Förstner, 1979b).

Bioaccumulation

The role and dynamics (see Chapter 2) of metals in aquatic organisms under both field and laboratory conditions is now under active investigation. The ability of a wide range of aquatic organisms to accumulate both biologically essential and non-essential metals is well established. For example, Jenkins (1980a, b) has presented data on the bioconcentration and bioaccumulation of many metals in plants and animals. Concentration factors (ratio of the metal concentration in the animal, $\mu g\,kg^{-1}$, to that in the ambient water, $\mu g\,L^{-1}$) for various aquatic species frequently range between 10^2 and 10^6 (Wright, 1978; Phillips, 1980). Callahan et al. (1979) concluded, in a review on priority pollutants, that bioaccumulation was a significant process in determining the fate of certain metals in biota. These were: As, Cd, Cr, Cu, Pb, Hg, and Zn but not Sb and Ni, and in the case of Be, Se, Ag, and Tl the position was uncertain. Considerable interest in the selection of biological indicator organisms for the management of metal pollution has provided a great deal of stimulus for bioaccumulation studies (see Phillips, 1980).

Table 10.8.   Different Sites for Metal Storage[a]

Species	Metals	Tissue	Comment
Mammals			
Zalophus californianus (sea lion)	Pb	Bone	Probably similar to humans
Phoca vitulina (seal)	Cd	Kidney	Probably Cd-metallothionein
Halichoerus grypus (seal)	Cd	Liver	Cd-metallothionein
Halichoerus grypus (seal)	Hg	Liver, fur, claws	Low proportion methylmercury–demethylation?
Delphinus delphis (dolphin)	Hg, Se	Liver	Two metals linearly correlated; not methylmercury
Fish			
Makaira ampla (Pacific blue marlin)	Hg	Liver, muscle	Low proportion methylmercury–demethylation?
Sebastodes caurinus (rock fish)	Cd	Liver	Induced Cd-metallothionein
Crustaceans			
Procambarus clarkii (freshwater crayfish)	Cu, Fe	Hepatopancreas	Large granules in Fe and Cu cells
Homarus vulgaris (lobster)	Mn	Exoskeleton	
Crangon vulgaris (shrimp)	Cu	Hepatopancreas	Excess Cu stored as granules
Lysmata seticaudata (shrimp)	Cd	Exoskeleton	50% total body Cd lost at moult
Molluscs			
Biophalaria glabrata (freshwater pulmonate)	Cu	Leucocytes	Phagocytosis of excess Cu
Oncomelania formosama (freshwater prosobranch)	Cu	Connective tissue	Crystals of carbonate deposited by wandering cells
Ostrea edulis (oyster)	Cu, Zn	Leucocytes	Phago- or pinocytosis of excess metals
Pecten maximus (scallop)	Zn, Mn, Pb	Kidney	Large (5 µm) granules inside cells
	Ag, Cd, Cu	Digestive gland	May occur in granules
Octopus vulgaris (octopus)	Hg	Digestive gland	Especially in contaminated conditions
Polychaete			
Nereis diversicolor (ragworm)	Cu, Pb	Epidermis	Fine granules in high-metal worms

[a] Adapted from Bryan (1976b). Reprinted with permission from Cambridge University Press.

Considerable inter- and intraspecies differences exist in the bioaccumulation capacity of individual metals, and different metals exhibit marked variation in kinetics in any one species. In addition, different chemical forms of any one metal may be absorbed and excreted at widely different rates (Phillips, 1980). The complexities of factors such as age (size, weight), seasonal variation, and lipid storage acting on trace metal accumulation have been extensively reviewed by Phillips (1980).

Generally, the relative abundance of essential metals in an organism reflects levels necessary to maintain biochemical functions, for example, enzyme systems (Bryan and Hummerstone, 1973). Where uptake of essential metals exceeds these levels, homeostatic mechanisms control body levels and tissue distributions. For example, in various decapod crustacea, total body levels of Zn and Cu appear to be regulated within definite limits, exhibiting concentration factors of about 10^4 for both metals (Phillips, 1977). However, with essential and nonessential metals, if uptake is excessive the homeostatic mechanisms are inhibited and bioaccumulation proceeds as uptake exceeds the loss rate.

Food Chain Transfer and Biomagnification

Many studies have suggested that metals biomagnify (see Chapter 2) substantially by analogy to the behavior of DDT in food chains. However, Prosi (1979) has concluded that aquatic food chain enrichment of metals does not occur and the biomagnification mechanism has been confused by oversimplification. In general, there has been a tendency to overlook several key factors relating to the uptake and accumulation of metals by aquatic organisms (Prosi, 1979). These are as follows:

1. The bioavailability of metals to animals at higher trophic levels is generally determined by transfer from water rather than food organisms.
2. Filter-feeding organisms are known to accumulate high levels of metals in their tissues but transfer only a small proportion to predatory organisms.
3. Sediments and detritus usually contain the highest metal concentrations in polluted systems and sediment- and detritus-feeding animals tend to accumulate higher metal concentrations than animals at higher trophic levels.
4. The life span of animals at higher trophic levels is usually greater than that of organisms at lower levels. Thus, age-related enrichment may be a significant factor influencing the level of metal enrichment at the higher trophic levels.
5. Preferential uptake and elimination of different metals and forms occurs.

One exception to these observations is mercury. Mercury concentrations in aquatic species are strongly related to position in the food chain (Ratkowsky et al., 1975), particularly in progressing from herbivores to large predators (Bryan, 1979). Young et al. (1980) found no evidence for biomagnification of Ag, Cd, Cr, Cu, Fe, Mn, Ni, Pb, and Zn in several marine ecosystems in southern California, but mercury, notably the organic forms, generally increased with trophic level. Cesium to potassium (Cs/K) ratios in organisms were used to indicate the trophic status of organisms within these ecosystems.

The high mercury concentrations observed in some individual species of finfish may be as much a function of time as trophic level. A strong positive power relationship has been established between mercury concentrations and body size (weight) in several large predatory fish (Walker, 1976; Mackay et al., 1975; Phillips, 1980). Bryan (1979) considered that both trophic position and size (age) are important factors in determining high mercury levels in some species of fish and seals. Also, metal concentrations in birds from various estuaries in New Zealand did not reflect pollution levels and differences were attributed to dietary intake (Turner et al., 1978).

TOXIC EFFECTS ON INDIVIDUALS

Mechanisms of Metal Toxicity

Ochiai (1977) has divided general toxicity mechanisms for metal ions into the following three categories: (1) blocking of the essential biological functional groups of biomolecules (e.g., proteins and enzymes); (2) displacing the essential metal ion in biomolecules; and (3) modifying the active conformation of biomolecules.

On the basis of the toxicity sequences presented in Table 10.9, Nieboer and Richardson (1980) concluded that there are similar patterns even for dissimilar

Table 10.9.   Toxicity Sequences for Metal Ions in a Range of Organisms

Organisms	Sequence[a]
Algae,[c] *Chlorella vulgaris*	Hg > Cu > Cd > Fe > Cr > Zn > Ni > Co > Mn
Fungi[d]	Ag > Hg > Cu > Cd > Cr > Ni > Pd > Co > Zn > Fe > Ca
Flowering plants,[c] barley	Hg > Pb > Cu > Cd > Cr > Ni > Zn
Protozoa,[c] *Paramecium*	Hg, Pb > Ag > Cu, Cd > Ni, Co > Mn > Zn
Platyhelminths,[c] *Polycelis*, a planarian	Hg > Ag > Au > Cu > Cd > Zn > H > Ni > Co > Cr > Pb Al > K > Mn > Mg > Ca > Sr > Na
Annelida,[c] *Neanthes*, a polychaete	Hg > Cu > Zn > Pb > Cd
Vertebrata,[c] stickleback	Ag > Hg > Cu > Pb > Cd > Au > Al > Zn > H > Ni > Cr Co > Mn > K > Ba > Mg > Sr > Ca > Na
Mammalia,[b,c] rat, mouse, rabbit	Ag, Hg, Tl, Cd > Cu, Pb, Co, Sn, Be, In, Ba, Mn, Zn, Ni, Fe, Cr > Y, La > Sr, Sc > Cs, Li, Al

Source: Nieboer and Richardson (1980). Reprinted with permission from Applied Science Publishers Ltd.

[a] In this table the atomic symbols represent tripositive ions for In, Al, Cr, La, Y, Sc, and Au; dipositive ions for Ni, Hg, Cu, Pb, Cd, Zn, Fe, Sn, Co, Mn, Mg, Ba, Be, Sr, and Ca; and monopositive ions for Ag, Tl, Cs, Li, H, Na, and K.

[b] Based upon a sequence synthesized from data on lethal doses (LD or LD_{50}) to small mammals administered i.v., i.p., sc, or orally. The data were abstracted from Venugopal and Luckey (1975).

[c] In these sequences the metal concentrations resulting in toxicity were expressed on a molar basis and this accounts for any discrepancies with the originally reported sequences. Metals administered as oxo anions were omitted.

[d] Concentration units were unavailable.

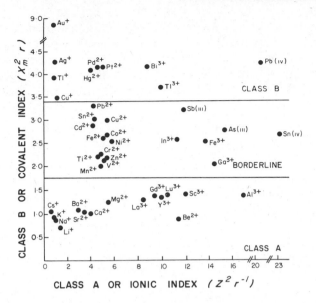

Figure 10.4. A separation of metal ions and metalloid ions – As(III) and Sb(III) – into three categories: class A, borderline, and class B. The class B index $X_m^2 r$ is plotted for each ion against the class A index Z^2/r. In these expressions, X_m is the metal-ion electronegativity, r its ionic radius, and Z its formal charge. (Oxidation states given by Roman numerals imply that simple cations do not exist even in acidic aqueous solutions.) [From Nieboer and Richardson (1980). Reprinted with permission from Applied Science Publishers, Ltd.]

organisms. Such similarities in metal toxicity sequences can be explained by their classification of metal ions according to binding preferences, that is, oxygen seeking – class A; nitrogen or sulfur seeking – class B and intermediate (see Figure 10.4). In this scheme, class B ions are more toxic than the borderline ions which are more toxic than the class A group. Accordingly, the relative toxicities can be explained as follows:

1.   The most toxic class B ions exhibit a broad spectrum of toxicity mechanisms: (a) They are most effective at binding to SH groups (e.g., cysteine) and nitrogen-containing groups (e.g., of lysine and histidine imidazole) at catalytically active centers in enzymes. (b) They can displace endogenous borderline ions (e.g., Zn^{2+}) from metallo-enzymes, causing inactivation of the enzymes through conformational changes. (c) Together with some of the borderline ions, they can form lipid-soluble organometallic ions, involving Hg, As, Sn, Tl, and Pb, capable of penetrating biological membranes and accumulating within cells and organelles. (d) Some metals in metallo-proteins exhibit oxidation–reduction activity, for example, Cu^{2+} to Cu^+, which can alter structural or functional integrity.

2.   Borderline ions frequently act to displace other endogenous borderline ions or class A ions from biomolecules. For example, the replacement of Zn^{2+} by Ni^{2+} in enzymes, such as carbonic anhydrase, results in the loss of enzyme activity. Also, Ca^{2+} displacement in membrane proteins results in functional disorder.

3.   Toxicity among class A ions is related to the substitution of an endogenous class A ion by another unstable class A ion. For example, the preferential substitution of Be^{2+} for Mg^{2+} in certain enzymes can result in their deactivation. Nieboer and Richardson (1980) have concluded that the toxicity of Be^{2+} is predictable from an inspection of Figure 10.4 where Be^{2+} is seen to have a high ionic index (Z^2/r) compared with Mg^{2+}.

Apparent anomalies in the above classification scheme are related to several factors, particularly the formation of insoluble complexes (e.g., by Al^{3+}) which can prevent or greatly reduce uptake or redistribution of the metal ion within an organism. Furthermore, the development of tolerance mechanisms for specific ions in various organisms can result in exclusion or innocuous accumulation of a potentially toxic ion.

Lethal Toxicity of Metals

There is a large volume of bioassay data on the acute toxicities of metals to various organisms. Lethal toxicity values for some common metals to selected marine, estuarine, and freshwater species are shown in Tables 10.10, 10.11, and 10.12. Bryan (1976a) has stressed the following limitations of lethal toxicity data in metal studies:

1.   In short-term experiments, for example 24, 48, and 96 hr, the mechanism of lethal toxicity may differ from that which results from long-term exposure. For example, at high concentrations and short exposure times, a metal may lethally disrupt the respiratory surfaces, whereas at lower concentrations and longer exposure, lethality may occur by accumulation in internal organs.

2.   In many of the larger species, a relatively slow rate of metal uptake occurs such that longer exposure periods are required before a value such as the asymptotic LC_{50} can be estimated (see Chapter 2). Long-term experiments may also involve absorption of metals from food as a primary source of metal uptake.

3.   Comparisons between long-term and short-term lethal concentrations confirm the limitations of using short-term exposures as guidelines for estimating metal concentrations which may be relatively harmless (see Bryan, 1976a). For example, a low but significant mortality was observed among rainbow trout exposed continuously for 4 months to constant zinc concentrations of 0.2 of the 5-day LC_{50} and among rudd exposed for $8\frac{1}{2}$ months to 0.3 of the 7-day LC_{50} (Alabaster and Lloyd, 1980).

4.   In addition, metal uptake and toxicity with aquatic organisms are strongly influenced by various physicochemical and biological factors (see Table 10.13). For example, length (or weight) in the freshwater shrimp, *Austrochiltonia australis*, has been quantitatively related to the observed 5-day LC_{50} for cadmium by the following equation (Lake et al., 1979);

$$5\text{day-}LC_{50} = 0.19\,e^{0.2L}$$

Table 10.10. Lethal Toxicity of Metals to Marine Organisms as 96 hr LC$_{50}$ (mg L^{-1})[a]

Classes of Organisms	Metals							
	Cd	Cr	Co	Cu	Hg	Ni	Pb	Zn
Fish	22–55	91	–	2.5, 3.2	0.8, 0.23	350	188	60, 60
Crustaceans	0.015–47	10	4.5	0.17–100	0.05–0.5	6, 47	–	0.4–50
Molluscs	2.2–35[b]	14–105[b]	–	0.14, 2.3	0.058–32[b]	72,[b] 320	–	10–50[b]
Polychaetes	2.5–12.1	2.0 to > 5.0	–	0.16–0.5	0.02–0.09	25, 72	7.7–20	1.8–55
Echindoderm[c] (1 specie only)	0.82	–	–	–	0.06	150	–	39

[a] Compiled from Jackim et al. (1970), Bryan (1976a), and Reisch et al. (1979).
[b] Gastropod, *Nassarius obsoletus*.
[c] *Asterias forbesi*.

Table 10.11. Lethal Toxicity of Metals to Freshwater Fish[a]

Pollutant	Species	LC$_{50}$ (96 hr mg L^{-1})	Exposure Type[b]	Temperature (°C)	Test Conditions
Cadmium	Goldfish	12.6	F	25	HD 99
Cadmium	Northern squawfish (*Ptychocheilus oregonensis*)				
	juvenile	1.092–1.104	F	12–17	pH 7.1–7.5
Cadmium sulfate	*Colisa fasciatus*	126	S	25	pH 7.3 HD 120
Chromium(VI)	Flathead minnow				
	juvenile	36.2	S	25	pH 7.7 HD 209
	juvenile	36.9	F	25	
Chromium trioxide	*Colisa fasciatus*	40	S	25	pH 7.3 HD 120
Copper	Pumpkinseed (*Lepomis gibbosus*)				
	juvenile	1.24–1.67	F	20	pH 7.2
	adult	1.74–1.94	F	20	HD 125
	Rainbow trout				
	juvenile	0.19–0.20	F	10	pH 7.8
	adult	0.21	F	10	HD 125
Copper	Bluntnose minnow (*Pimephales notatus*)	0.23	F	25	pH 7.9–8.3
Copper	Northern squawfish				
	juvenile	0.023	F	8	pH 7.1–7.5 HD 20–30
Copper	Northern squawfish	0.018	F	12	pH 7.1–7.5
Copper oxychloride	Mosquito fish	1.750	S	27	pH 7.5 HD 32
Copper sulfate	Golfish	0.30	F	21	pH 7.3 HD 52
Copper sulfate (hydrated)	*Poecilia reticulata*	0.16–0.48	R	17–22	pH 6.4–6.6
	Labeo rohita	0.087–0.295	R	17–22	HD 144–188
	Carp	0.117–0.53	R	17–22	

316

Compound	Species		S/F/R	Temp.	Conditions[b]
Copper sulfate	Puntius conchonius	0.57	S	13–22	pH 8.1 HD 310
Copper sulfate	American eel black eel stage	3.20	S	22	pH 7.2–7.6 HD 40–48
	glass eel stage	2.54	S	22	pH 7.2–7.6 HD 40–48
Copper sulfate	Mosquito fish	0.20	S	27	pH 7.5 HD 32
Copper and zinc	Chinook salmon fry		F	9–11	pH 6.6–7.4 HD 21–27
Mercuric chloride	Mosquito fish	0.18	S	27	pH 7.5 HD 32
Mercuric chloride	Poecilia reticulata	0.03–0.054	R	17–22	pH 7.8–8.1 HD 144–188
Mercuric chloride	Labeo rohita	0.015–0.184	R	17–22	pH 7.3 HD 160
Mercuric chloride	Channa punctatus	1.8	S	18	pH 7.5
Mercuric chloride	Chingatta (Channa gachua)	1.4	S	24–25	pH 7.5
Mercury	Heteropneustes fossiles	0.35	S	28	–
	Sarotherodon mossambica	0.075	S	28	–
Methoxyethyl mercuric chloride	Mosquito fish	0.91	S	27	pH 7.5 HD 32
Zinc	Bluegill	3.0–3.6	F	19–22	pH 6.8–7.5 HD 21–59
Zinc	Northern squawfish juvenile	3.5–3.7	F	10–12	pH 7.1–7.5 HD 20–30
Zinc sulfate	Puntius conchonius	33.3	S	13–19	pH 7.5 HD 310

[a] Compiled from Spehar et al. (1981).
[b] F indicates flow through aquaria conditions, S static, R renewal bioassay. HD is hardness as $CaCO_3$ mg L^{-1}.

Table 10.12. Lethal Toxicity (LC_{50} in mg L^{-1}) of Metals to Freshwater Invertebrates

Taxon	Cadmium	Chromium	Copper	Lead	Nickel	Zinc
Crustacea	0.005–0.55 (24–504)[a]	0.008–10.1 (24–96)	0.007–1.2 (24–504)	0.28–5.5 (48–672)	0.13–15.2 (24–504)	0.04–5.5 (48–504)
Insecta	<0.003–18.0 (24–672)	2.0–64.0 (24–96)	0.03–13.9 (24–336)	3.5–64.0 (168–336)	4.0–48.4 (24–96)	8.7–62.6 (24–336)
Gastropoda	—	—	0.013–3.0 (24–96)	—	11.4–26.0 (24–96)	0.62–20.2 (24–96)
Oligochaeta	0.0028–4.6 (24–96)	0.063–48.0 (24–96)	0.0064–2.6 (24–96)	2.75–49 (24)	0.08–16.2 (24–96)	0.11–60.2 (24–96)
Mollusca		0.8–17.3 (24–96)				
Rotifera		9.1–65.0 (24–96)				

Source: Compiled from Murphy (1979).

[a] Duration in hours.

Table 10.13.   Factors Influencing the Toxicity of Heavy Metals in Solution

Form of metal in water	inorganic organic	soluble	ion complex ion chelate ion molecule
		particulate	colloidal precipitated adsorbed
Presence of other metals or poisons	joint action no interaction antagonism	more-than-additive additive less-than-additive	
Factors influencing physiology of organisms and possibly form of metal in water	temperature pH dissolved oxygen light salinity		
Condition of organism	stage in life history (egg, larva, etc.) changes in life cycle (e.g., moulting, reproduction) age and size sex starvation activity additional protection (e.g., shell) adaptation to metals		
Behavioral response	altered behavior		

Source: Bryan (1976a). Reprinted with permission from Academic Press Inc. (London) Ltd.

where LC_{50} = milligrams of cadmium per liter and L = body length in millimeters. Prosi (1979) has reviewed the importance of various abiotic and biotic factors which influence metal uptake and toxicity in organisms.

Sublethal Effects

Sublethal effects of metals on aquatic organisms have been studied mainly under laboratory conditions. In some cases, field observations have been made on freshwater species or experiments carried out with caged animals (see Alabaster and Lloyd, 1980). Most sublethal toxicity appears to be biochemical in origin and often related to metabolic processes. It is usually short-term, variable, nonlinear in dose–response, and difficult to correlate with significant changes identified at the whole organism level (e.g., morphology, behavior, and reproduction). Much of the data on sublethal metal toxicity results from multiple and inconsistent experimental methods and lacks a mechanistic explanation for observed effects at the organism level.

Sublethal effects have been observed as changes in: (1) morphology/histology; (2) physiology (growth, development, swimming performance, respiration, circulation; (3) biochemistry (blood chemistry, enzyme activity, endocrinology); (4) behavior/neurophysiology; and (5) reproduction (Bryan, 1976a; Alabaster and Lloyd, 1980). In general, the following points can be made:

1. Histological or morphological changes in the tissues of various fish and crustacean species following sublethal exposure to metals (e.g., Cu, Cd, and Hg) are secondary effects of interference with enzyme processes involved in food utilization (Bryan, 1976a). Many diverse morphological changes are recorded including replacement of mucous cells of the gill epithelium by chloride cells, vertebral damage by Zn in freshwater minnow, as well as developmental and structural abnormalities, for example, with sea urchin eggs and larvae (Bryan, 1976a).

2. Suppression of growth and reproduction occurs widely among aquatic vertebrates and invertebrates exposed to relatively low metal concentrations. But freshwater invertebrates and plants tend to be more resistant than fish although considerable variation in sensitivity exists between and within taxonomic groups (Alabaster and Lloyd, 1980).

3. Reproduction in many aquatic organisms is affected in the parts per billion range for most aquatic toxic metal ions.

4. Effects of metal exposure on physiological processes are complex, variable, and tend to be unpredictable. Essential and nonessential metals may stimulate organisms at low concentrations and be inhibitory at higher concentrations.

5. Behavioral changes observed in laboratory experiments are diverse and difficult to relate to field conditions. However, metals can impair processes such as feeding, learning, swimming activity, and response to external stimuli (Bryan, 1976a).

The relationship between long-term and short-term effects is particularly important when dealing with behavioral changes. For example, with freshwater fish, Alabaster and Lloyd (1980) concluded that depression of chemoreceptor responses, respiration, and feeding activity, together with an increase in cough frequency at sublethal concentrations of copper, were only transitory.

ECOLOGICAL EFFECTS

Assessments of the ecological impact of metals are derived principally from field observations in aquatic systems receiving mining wastes, sewage, industrial discharges, or sludges containing sufficient metal content to which toxicity can be assigned. The most significant evidence of responses by aquatic organisms is derived from a relatively limited number of faunal studies in temperate freshwater systems (see Tables 10.14 and 10.15). Bryan (1976a) points out that no definite ecological effects have been reported in several marine areas where metal concentrations in seawater have increased by about a factor of 10, for example, Cu and Zn in Restrongurt Creek, Cd and Zn in the Bristol Channel, and Hg on the Italian coast. An increase of this order would be expected to produce some observable effects on the basis of evidence from sublethal toxicity studies (Bryan, 1976a). On the other hand, the detection of subtle ecological effects can only be carried out satisfactorily if adequate baseline studies or comparable reference areas are available to compensate for natural variations. Also, field studies of marine pollution are generally complicated by the presence of pollutant mixtures (e.g., industrial wastes and sewage) which confuse attempts to evaluate the impact of individual pollutants.

One of the few ecological studies of trace metal effects in a marine ecosystem has been recently reported by Ward and Young (1982). The community structure (species richness, species composition, and abundance of species) of epibenthic seagrass fauna near a large lead smelter was investigated. Decreased frequencies of 20 common species, mostly fish, were correlated with the concentration of contaminant metals (Cd, Cu, Pb, Mn, and Zn) in the sediments. It was also found that the frequencies of certain species, mostly crustaceans, correlated with particle size distributions. Ward and Young (1982) concluded that both contaminant metals and sediment particle size have substantial controlling effects on the community structure. Furthermore, contaminant metals showed a greater effect on fish than crustaceans, and opportunistic epibenthic species failed to exploit the contaminated area.

The direct ecological impacts of metal pollution are most readily characterized by studies on biota in rivers and lakes. In these situations consistent patterns of change occur and dose–response relationships can be estimated from field measurements and comparisons made with laboratory toxicity data (see Alabaster and Lloyd, 1980). For example, Weatherley et al. (1967) investigated mine drainage pollution (Cu, Fe, and Zn) of the Molongo River, Australia, and found that uncontaminated areas usually contain 30–45 species of benthic invertebrate fauna. Immediately below the discharge the number of species was reduced to approximately 4, and a slow recovery to normal species numbers was apparent downstream (see Figure 10.5). The number of individual animals present followed a similar pattern to the species numbers as would be expected with toxic substances (see Chapter 4).

Several general characteristics of metal toxicity within aquatic populations and communities can be suggested.

1. Metal ions and complexes exhibit a wide range of toxicity to marine and freshwater organisms. In general, mollusca, crustaceans, oligochaetes, and leeches appear to be the most sensitive taxa to zinc (Alabaster and Lloyd, 1980). Algae and insects show a wide divergence in effects but generally aquatic plants and invertebrates are more tolerant than fish (see Ward and Young, 1982).

2. Significant to severe modifications in community structure involving a

Table 10.14. Effects of Metals on Freshwater Fish Populations and Communities

Metal(s)	Location	Concentrations (mg L^{-1})	Effects	References
Zn (low Cu and some other metals)	Polluted Swedish river	0.15–0.28 (mean) 0.25–0.95 (mean)	High mortalities among caged yearling salmon and adult minnow.	Hasselrot (1965)
Zn (Cu also present)	Molongo River, New South Wales, Australia	0.1–0.5	Complete mortalities of caged rainbow trout (30–59 mm). Partial mortalities of brown trout (45–94 mm). 16 days exposure. (Hardness: 10–13 mg L^{-1} as CaCO$_3$.)	Weatherly et al. (1967)
Zn		2–7	Crucian carp were more resistant than trout.	
Cu		0.04–0.08	40% mortalities in 16 days exposure (Hardness: 57 mg L^{-1} as CaCO$_3$.)	
Zn	L. Burley Griffin, Molongo River system, Australia	0.1–0.4	Rainbow and brown trout survived but showed gill damage (similar to acute toxicity experiments) and proportion of fish in older age groups appeared abnormally low. (Hardness: 10–13 mg L^{-1} as CaCO$_3$.)	Weatherley et al. (1975)
Zn (trace Pb)	R. Ystwyth, U.K.	Up to 0.8 (normal) Up to 1.4	Mortalities observed among brown trout washed into main river by floods and also sea trout ascending lower reaches.	Jones (1940)
			Small numbers of brown trout present and occasional sea lamprey, minnow, stickleback, and bullhead absent.	Jones (1958)
Zn	Lakes and rivers, Norway	Up to 0.15 at higher concentrations of Zn (other metals also present)	Self-sustaining populations of salmonids, brown trout, Atlantic salmon, and Arctic Char absent. (Hardness: 2–20 mg L^{-1} as CaCO$_3$.)	Alabaster and Lloyd
Zn	R. Orkla, Norway	0.1 (median) 0.4 (95 percentile) 0.035 (median) 0.1 (95 percentile)	Salmonids able to migrate 5 km of river. (Hardness ∼ 39 mg L^{-1} as CaCO$_3$.)	Arnesen et al. (1977)

Substance	Location	Concentration (mg L⁻¹)	Comments	Reference
Zn	Willowbrook, U.K. (waste water from steelworks)	0.9 (annual 50 percentile) 3.7 (annual 95 percentile) 0.4 (annual 50 percentile) 2.2 (annual 95 percentile)	Roach, tench, and stickleback (upper reaches). Roach, chub, minnow, stickleback, eel, and stoneback (lower reaches).	Alabaster and Lloyd (1980)
Cu	L. Orta, Italy	0.07	2-day LC_{50} for caged rainbow trout. (5–12°C; pH 5.5–6.4; Hardness: 8–10 mg L^{-1} as $CaCO_3$.)	Calamari and Marchetti (1974)
Cu (and Zn)	Norwegian lakes	Up to 0.06	Salmonoids present.	Grande (1967)
Cu	Water taken from Exploits River, Newfoundland	0.125	4-day LC_{50} for Atlantic salmon parr. (Hardness: 8–10 mg L^{-1} as $CaCO_3$.) (Cf. ~ 30 µg L^{-1} as predicted from laboratory tests by Sprague and Ramsay, 1965.)	Wilson (1972)
Cu (and other metals e.g., Zn)	R. Churnet, U.K.	0.034 (annual 50 percentile) 0.083 (annual 75 percentile)	Brown trout, bulkhead, three-spine stickleback. Minnow present.	Alabaster and Lloyd (1980)
	R. Ouse, U.K.	0.050 (50 percentile)	Minnow, roach, stickleback, and chula present.	
	(sewage and industrial wastes)	0.130 (95 percentile)		
Cd (soluble)	R. Arrow, U.K. (down stream of sewage effluent)	0.008–0.025	Mean period of survival for caged rainbow trout was 8 days (other poisons were present at 0.1 of the combined 2-day LC_{50} values for rainbow trout).	Ministry of Technology (1970)
Cd (soluble)	R. Teon, U.K.	0.0026 (median) 0.0064 (95 percentile) 0.007 (median) 0.019 (95 percentile)	Brown trout present upstream of effluent outfall. Brown trout absent downstream of outfall (44-month period of study); other species present. Cu/Zn concentrations only 0.06/0.04 of the respective predicted 2-day LC_{50} for rainbow trout. (Hardness: 210 mg L^{-1} as $CaCO_3$.)	Alabaster and Lloyd (1980)
Cd	R. Thames and tributaries, U.K.	~ 0.0035 (mean value total Cd in 1975)	Good to very good populations of trout. (Hardness: ~ 300 mg L^{-1} as $CaCO_3$.)	Alabaster and Lloyd (1980)

Table 10.15. Effects of Metals on Freshwater Invertebrates

Metal(s)	Location	Concentration (mg L^{-1})	Effects	References
Zn	L. Burley-Griffin, Molongo River system, Australia	$\sim$ 0.1–0.4	Benthic fauna depauperate: only one molluscan specie, poor in species of Crustacea, Odonata, and Ephemeroptera.	Weatherley et al. (1975)
Zn (Cu and Fe)	Molongo River, Australia (Cu, Pb, Zn mine)	0.1–0.5	Severe reduction in fauna downstream of mining wastes; slow recovery to normal species composition over next 70 km (see Figure 10.5).	Weatherley et al. (1967)
Zn	R. Ystwyth, U.K.	$\sim$ 0.7–1.2	Mollusca, crustaceans, oligochaetes, and leeches absent.	Jones (1940)
Pb		$\sim$ 0.5		
Zn		0.2–0.7 (some years later)	Fauna still consisted mainly of lithophilic insects.	Jones (1958)
Zn	R. Sulz and its tributaries	> 0.25	*Gammarus pulix* and *Lymnen ovata* absent.	Herbst (1967)
		0.8–6.5	High density of insect larvae.	
		25	Larvae of *Baetis* present.	
		29	Larvae of the trichoptera, *Rhyacophila* sp.	
Cu	L. Orta, Italy	0.5	Fish and zooplankton absent but algal production on stones was high.	Alabaster and Lloyd (1980)
Cu	L. Orsvjøen, Norway	0.13	Chironomid larvae and a few planktonic crustaceans were present while *Gammarus lacustris*, snails, insect life, and fish were absent.	Alabaster and Lloyd (1980)
Zn		0.4		
Zn	South Esl: River, Tasmania, Australia (mining wastes)	$\sim$ 0.150	Crustacea, molluscs, and larvae of Odonata and Plecoptera absent downstream of mining areas.	Thorp and Lake (1974)
Cd		$\sim$ 0.040	Mayflies, beetles, and caddisflies were severely affected. Virtual recovery when metal concentration was 20–30% lower.	
		$\sim$ 0.030		

324

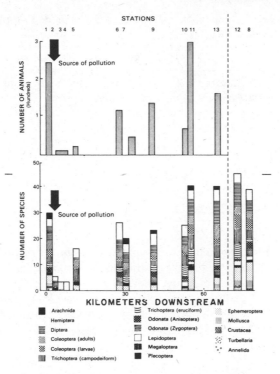

Figure 10 5. Composition of the fauna and number of animals in the Molongo River, Australia. (Sample station 1 is immediately upstream of the zone of major pollution and the water is almost unpolluted. Stream water is polluted at station 2, but the major source of pollution occurs at station 3. Sample stations 8 and 12 are in unpolluted tributaries.) [From Weatherley et al. (1967). Reprinted with permission from Australian National University Press.]

reduction in numbers of species, including complete absence of sensitive species, are relatively common, particularly in affected streams.

3. A reduction in the number of individuals of surviving species occurs with the amount of reduction related to the metal present.

This represents the classical pattern of ecological effects of toxicants on ecosystems (see Chapter 4). With toxic substances there is an overall reduction in number of species and individuals present in relation to the severity of the toxicant stress. The development of tolerance to metals in some species has increased their capacity to accumulate relatively high concentrations of metals and may result in some modification to community structure changes over time.

SALTS IN INDUSTRIAL WASTES

The metals discussed above can occur in association with a wide variety of anions. The toxicity and environmental effects of a salt are often due to the metal ion but

Table 10.16.   Some Salts and Related Substances Found in Waste from Chemical Industries

Industry	Substances in Waste Water
Electroplating, and iron and steel processes	Cyanides, chromate, copper, nickel, cadmium, zinc, tin, and silver
Leather tanning	Lime, chromium, sulfide, and organic matter
Coal gas manufacture	Phenols, ammonia, cyanides, thiocyanates, and sulfides
Battery manufacture	Sulfuric acid
Chemical manufacture	Acids, alkalis, phenols, amines, and other chemicals

can be derived from the anion, for example, potassium cyanide. The anions are chemically diverse and frequently contained in complex industrial wastes (see Table 10.16). The toxicity can vary from low, as occurs with chloride, to high, as with cyanide, which usually has LC_{50} values below $1 \, mg \, L^{-1}$ with aquatic organisms. There is little information on the ecotoxicology of these substances but it would be expected that each would be quite different corresponding to the diverse properties exhibited.

STREAM AND GROUNDWATER POLLUTION BY NATURAL SALTS

Comparatively nontoxic salts, for example, sodium, calcium, chloride, sulfate, and carbonate ions, occur naturally in all stream waters and in some streams comparatively high concentrations are present. Salt as a pollutant differs from most other pollutants in that it is generally directly derived from sources within the natural environment. Salts are present in rock and soil and leached out by water to enter streams either by above-ground runoff or by subsurface drainage.

Agricultural and other developments in a river system can increase the salt content of river water. Water used for irrigation of crops leaches salt from the soil and subsoil and the resultant drainage water, returned to the river, has a higher salinity. In addition, irrigation and the construction of dams can raise the level of groundwater in adjacent areas. This has two major effects. First, the increase in groundwater levels results in an increased flow of groundwater into rivers. If the groundwater is saline an increase in river water salinity occurs. Second, groundwater level increases may result in increased salinity at the land surface, killing natural vegetation and crops. This has caused particular problems in the more arid regions of North America and Australia (Halvorson and Black, 1974; Pels, 1978; Talsma and Philp, 1971; Holmes and Talsma, 1981).

In some cases, salinity problems result from clearing natural vegetation and introduction of crops (Malcolm and Stoneman, 1976; Mulcahy, 1978). If the natural vegetation consists of trees with comparatively deep root systems these can keep the water table low by evapotranspiration. Crops have a relatively shallow root system and allow the water table to rise. In many cases the groundwater is saline and intersects with the surface in low areas, rendering these areas unsuitable for vegetation. The ecological effects of salinity increases of the type mentioned above on aquatic systems have not been subject to intensive investigation and are not known at present.

REFERENCES

Ahrland, S., Chatt, J., and Davies, N. R. (1958). The relative affinities of ligand atoms for acceptor molecules and ions. *Q. Rev. Chem. Soc.* **12**, 265.

Alabaster, J. S. and Lloyd, R. (1980). *Water Quality Criteria for Freshwater Fish*. Butterworths, London.

Arnesen, R. T., Grande, M., Eversen, E. R., and Baalsrud, K. (1977). *Overvakings-undersokelser i nedre del av Orklavassdraget*. Blindern, Oslo, Norsk Institutt for Vannforskning.

Arthington, A. H., Conrick, D. L., Connell, D. W., and Outridge, P. M. (1982). The ecology of a polluted urban creek. Australian Water Resources Council Technical Paper, No. 68. Australian Government Publishing Service, Canberra.

Bryan, G. W. (1969). The absorption of zinc and other metals by the brown seaweed *Laminaria digitata. J. Mar. Biol. Assoc. U.K.* **49**, 225.

Bryan, G. W. (1976a). "Heavy Metal Contamination in the Sea." In R. Johnston (Ed.), *Marine Pollution*. Academic Press, London, pp. 185–302.

Bryan, G. W. (1976b). "Some Aspects of Heavy Metal Tolerance in Aquatic Organisms." In A. P. M. Lockwood (Ed.), *Effects of Pollutants on Aquatic Organisms*. Cambridge University Press, Cambridge.

Bryan, G. W. (1979). Bioaccumulation of marine pollutants. *Phil. Trans. R. Soc. Lond. Ser. B* **286**, 483.

Bryan, G. W. and Hummerstone, L. G. (1973). Adaption of the polychaete *Nerei diversicolor* to estuarine sediments containing high concentrations of zinc and cadmium. *J. Mar. Biol. Assoc. U.K.* **53**, 839.

Calamari, D. and Marchetti, R. (1974). Predicted and observed acute toxicity of copper and ammonia to rainbow trout (*Salmo gairdneri* Rich). *Prog. Water Technol.* **7**, 569.

Callahan, M. A., Slimak, M. W., Gabel, N. W., May, I. P., Fowler, C. F., Freed, J. R., Jennings, P., Durfee, R. L., Whitmore, F. C., Maestri, B., Mabey, W. R., Holt, B. R., and Gould, C. (1979). Water-Related Environmental Fate of 129 Priority Pollutants. Vol. 1: Introduction and Technical Background, Metals and Inorganics, Pesticides and PCB's. EPA-440/4-79-029A.

Craig, P. J. (1980). "Metal Cycles and Biological Methylation." In O. Hutzinger (Ed.) *The Natural Environment and the Biogeochemical Cycles*. Springer, Berlin, p. 169.

Davies, A. G. (1973). In *Radioactive Contamination of the Marine Environment*. International Atomic Energy Agency, Vienna.

Eckert, J. M. and Sholkovitz, E. R. (1976). The flocculation of iron, aluminium and humates from river waters by electrolytes. *Geochim. Cosmochim. Acta* **40**, 847.

Förstner, U. (1979a). "Metal Concentrations in River, Lake and Ocean Waters." In U. Förstner and G. T. W. Wittmann (Eds.), *Metal Pollution in the Aquatic Environment*. Springer-Verlag, Berlin, p. 71.

Förstner, U. (1979b). "Metal Transfer between Solid and Aqueous Phases." In U. Förstner and G. T. W. Wittmann (Eds.), *Metal Pollution in the Aquatic Environment*. Springer-Verlag, Berlin, pp. 197–270.

Grande, M. (1967). Effect of copper and zinc on salmonid fishes. *Adv. Water Pollut. Res.* **3**, 1, 97.

Groth, P. (1971). Untersuchungen über einige Spurenelemente in Seen. *Arch. Hydrobiol.* **68**, 305.

Gutknecht, J. (1965). Uptake and retention of cesium-137 and zinc-65 by seaweeds. *Limnol. Oceanogr.* **10**, 58.

Guy, R. D. and Chakrabarti, C. L. (1975). Distribution of metal ions between soluble and particulate forms. *Abstr. Int. Conf. Heavy Met. Environ.*, Toronto, Canada, p. D-29.

Hahne, H. C. H. and Kroontji, W. (1973). Significance of pH and chloride concentration on behaviour of heavy metal pollutants. Mercury(II), cadmium(II), zinc(II), and lead(II). *J. Environ. Qual.* **2**, 444.

Halvorson, A. D. and Black, A. L. (1974). Saline-seep development in dryland soils of northeastern Montana. *J. Soil Water Conserv.* **29**, 77.

Hart, B. T. and Davies, S. H. R. (1977). Physico-chemical forms of trace metals and the sediment-water interface. In H. L. Golterman (Ed.), *Interactions Between Sediments and Freshwaters*. Dr. W. Junk Publishers, Wagenigen, p. 398.

Hart, B. T. and Davies, S. H. R. (1978). A Study of the Physico-Chemical Forms of Trace Metals in Natural Waters and Wastewaters. Australian Water Resources Council Technical Paper No. 35. Australian Government Publishing Service, Canberra.

Hasselrot, T. B. (1965). A study of remaining water pollution from a metal mine with caged fish as indicators. *Vattenhygien* **21**, 11.

Hem, J. D. (1970). Study and interpretation of the chemical characteristics of natural water. *U.S. Geol. Surv. Water Supply Pap.* 2nd ed. **1473**, 363.

Herbst, H. V. (1967). Experimentelle Untersuchungen zur Toxizitat des Zinks. *Gerwässer und Abwässer* **44/45**, 37.

Holmes, J. W. and Talsma, T. (1981). *Land and Stream Salinity*. Elsevier Scientific Publishing, Amsterdam.

Hughes, M. K., Lepp, N. W., and Phipps, D. A. (1980). "Aerial Heavy Metal Pollution and Terrestrial Ecosystems." In A. MacFadyen (Ed.), *Advances in Ecological Research*, Vol. II. Academic Press, London, p. 217.

Jackim, E., Hamlin, J. M., and Sonis, S. (1970). Effects of metal poisoning on five liver enzymes in the killifish (*Fundulus heteroclitus*). *J. Fish. Res. Board Can.* **27**, 383.

Jenkins, D. (1980a). Biological Monitoring of Toxic Trace Metals. Vol. I. Biological Monitoring and Surveillance. National Technical Information Service PB81-103475.

Jenkins, D. (1980b). Biological Monitoring of Toxic Trace Metals. Vol. 2. Toxic Trace Metals in Plants and Animals of the World. National Technical Informational Service PB81-103483.

Johnels, A. G., Westermark, T., Berg, W., Persson, P. I., and Sjostrand, B. (1967). Pike (*Esox lucius* L.) and some other aquatic organisms in Sweden as indicators of mercury and contamination in the environment. *Oikos* 18, 323.

Jonasson, I. R. (1977). "Geochemistry of Sediment/Water Interactions of Metals, Including Observations on Availability." In H. Shear and A. E. P. Watson (Eds.), *The Fluvial Transport of Sediment-Associated Nutrients and Contaminants*. IJC/PLUARG, Windsor, Ontario, pp. 255–271.

Jones, J. R. E. (1940). A study of the zinc-polluted river Ystwyth in North Cardiganshire, Wales. *Ann. Appl. Biol.* 27, 368.

Jones, J. R. E. (1958). A further study of the zinc-polluted river Ystwyth. *J. Anim. Ecol.* 27, 1.

Kalk, M. (1963). Absorption of vanadium by tunicates. *Nature (London)* 198, 1010.

Klein, L. A., Lang, M., Nash, N., and Kirschner, S. L. (1974). Sources of metals in New York City waste-water. *J. Water Pollut. Control Fed.* 46, 2653.

Kustin, K., Ladd, K. V., and McLeod, G. C. (1975). Site and rate of vanadium assimilation in the tunicate *Ciona intestinalis*. *J. Gen. Physiol.* 65, 315.

Lake, P. S., Swain, R., and Mills, B. (1979). Lethal and Sublethal Effects of Cadmium on Freshwater Crustaceans. Australian Water Resources Technical Paper No. 37. Australian Government Publishing Service, Canberra.

Leckie, J. O. and James, R. O. (1974). "Control Mechanisms for Trace Metals in Natural Waters." In A. J. Rubin (Ed.), *Aqueous–Environmental Chemistry of Metals*. Ann Arbor Science, Ann Arbor, Michigan, p. 1.

Lee, G. F. (1975). "Role of Hydroxous Metal Oxides in the Transport of Heavy Metals in the Environment." In P. A. Krenkel (Ed.), *Heavy Metals in the Aquatic Environment*. Pergamon Press, Oxford, pp. 137–147.

Lu, C. S. J. and Chen, K. Y. (1977). Migration of trace metals in interfaces of seawater and polluted surficial sediments. *Environ. Sci. Technol.* 11, 174.

Mackay, N. J., Kazacos, M. N., Williams, R. J., and Leedow, M. I. (1975). Selenium and heavy metals in black marlin. *Mar. Pollut. Bull.* 6, 57.

Malcolm, C. V. and Stoneman, T. C. (1976). Salt encroachment – the 1974 salt land survey. *J. Agric. W. Aust.* 17 (4th Ser.), No. 2.

Malmquist, P.-A. (1975). Heavy metals in urban storm water. *Abstr. Int. Conf. Heavy Met. Environ.*, Toronto, Canada, C-46/48.

McElroy, A. D., Chiu, S. Y., Nebgen, J. W., Aleti, A., and Vandegrift, E. (1975). Water pollution from non-point sources. *Water Res.* 9, 675.

Ministry of Technology, U.K. (1970). *Water Pollution Research* 1969. H.M.S.O., London, p. 58.

Morel, F. M. M. and Morgan, J. J. (1972). A numerical method for computing equilibria in aqueous chemical systems. *Environ. Sci. Technol.* 6, 58.

Morel, F., McDuff, R. E., and Morgan, J. J. (1973). "Interactions and Chemostasis in Aquatic Chemical Systems: Role of pH, pE, Solubility and Complexation." In P. C. Singer (Ed.), *Trace Metals and Metal-Organic Interactions in Natural Waters*. Ann Arbor Science, Ann Arbor, Michigan, p. 157.

Mueller, J. A., Anderson, A. R., and Jeris, J. S. (1976). Contaminants entering the New York Bight: Sources, mass loads, significance. *Am. Soc. Limnol. Oceanogr.* Spec. Symp. 2, 162.

Mulcahy, M. J. (1978). Salinisation in the southwest of Western Australia. *Search* **9**, 269.

Murphy, Jr., C. B. (1981). Bioaccumulation and toxicity of heavy metals and related trace elements. *J. Water. Pollut. Control Fed.* **53**, 6, 993.

Murphy, P. M. (1979). A manual for toxicity tests with freshwater macroinvertebrates and a review of the effects of specific toxicants. University of Wales Institute of Science and Technology Publication.

Nieboer, E. and Richardson, D. H. S. (1980). The replacement of the nondescript term "heavy metals" by a biologically and chemically significant classification of metal ions. *Environ. Pollut. Ser. B* **1**, 3.

Nissenbaum, A. and Swaine, D. J. (1976). Organic matter — metal interactions in recent sediments: The role of humic substances. *Geochim. Cosmochim. Acta* **40**, 809.

Ochiai, E. (1977). *Bioinorganic Chemistry: An Introduction*. Allyn and Bacon, Boston.

Pearson, R. G. (1968a). Hard and soft acids and bases, HSAB, Part I. Fundamental principles. *J. Chem. Educ.* **45**, 581.

Pearson, R. G. (1968b). Hard and soft acids and bases, HSAB, Part II. Underlying theories. *J. Chem. Educ.* **45**, 643.

Pels, S. (1978). Waterlogging and salinisation in irrigated semiarid regions of New South Wales. *Search* **9**, 273.

Pentreath, R. J. (1973a). The accumulation and retention of ^{65}Zn and ^{54}Mn by the plaice, *Pleuronectes platessa* L. *J. Exp. Mar. Biol. Ecol.* **12**, 1.

Pentreath, R. J. (1973b). The accumulation and retention of ^{59}Fe and ^{58}Co by the plaice, *Pleuronectes platessa* L. *J. Exp. Mar. Biol. Ecol.* **12**, 315.

Pentreath, R. J. (1973c). The accumulation from water of ^{65}Zn, ^{54}Mn, ^{58}Co and ^{59}Fe by the mussel, *Mytilus edulis. J. Mar. Biol. Assoc. U.K.* **53**, 127.

Pillai, T. N. V., Desai, M. V. M., Mathew, E., Ganapathy, S., and Ganguly, A. K. (1971). Organic materials in the marine environment and the associated metallic elements. *Curr. Sci.* **40**, 75.

Phillips, D. J. H. (1977). The use of biological indicator organisms to monitor trace metal pollution in marine and estuarine environments — A review. *Environ. Pollut.* **13**, 281.

Phillips, D. J. H. (1980). *Quantitative Aquatic Biological Indicators*. Applied Science Publishers, London.

Prosi, F. (1979). "Heavy Metals in Aquatic Organisms." In U. Förstner and G. T. W. Wittman (Eds.), *Metal Pollution in the Aquatic Environment*. Springer-Verlag, Berlin, pp. 271–323.

Rashid, M. A. (1971). Role of humic acids of marine origin and their different molecular weight fractions in complexing di- and tri-valent metals. *Soil Sci.* **111**, 298.

Rashid, M. A. and Leonard, J. L. (1973). Modifications in the solubility and precipitation behavior of various metals as a result of their interaction with sedimentary humic acid. *Chem. Geol.* **11**, 89.

Ratkowsky, D. A., Dix, T. G., and Wilson, K. C. (1975). Mercury in fish in the Derwent Estuary, Tasmania, and its relation to the position of the fish in the food chain. *Aust. J. Mar. Freshwater Res.* **26**, 223.

Reisch, D., Rossi, S. S., Mearns, A. J., Oshida, P. S., and Wilkes, F. G. (1979). Marine and estuarine pollution. *J. Water Pollut. Control. Fed.* **51**, 6, 1477.

Renfro, W. C., Fowler, S. W., Heyraud, M., and La Rosa, J. (1975). Relative importance of food and water in long term [65]Zn accumulation by marine biota. *J. Fish. Res. Board Can.* **32**, 1339–1345.

Sholkovitz, E. R. (1976). Flocculation of dissolved organic and inorganic matter during the mixing of river water and seawater. *Geochim. Cosmochim. Acta* **40**, 831.

Sibley, T. H. and Morgan, J. J. (1977). "Equilibrium Speciation of Trace Metals in Fresh Water–Seawater Mixtures," T. C. Hutchinson (Ed.). In *Proceedings of the International Conference on Heavy Metals in the Environment*, Toronto, 1975, Vol. I, p. 319.

Sillén, L. G. (1961). "The Physical Chemistry of Seawater." In M. Sears (Ed.), *Oceanography*. American Association for Advancement in Science, Washington, D.C., p. 549.

Spehar, R. L., Lemke, A. E., Pickering, Q. H., Roush, T. H., Russo, R. C., and Yount, J. D. (1981). Effects of pollution on freshwater fish. *J. Water Pollut. Control. Fed.* **53**, 1028.

Stumm, W. and Belinski, H. (1972). "Trace Metals in Natural Waters: Difficulties in Interpretation Arising from Our Ignorance on their Speciation." In S. H. Jenkins (Ed.), *Advances on Water Pollution Research*, Proceedings of the 6th International Conference, Jerusalem. Pergamon Press, New York, pp. 39–52.

Stumm, W. and Brauner, P. A. (1975). "Chemical Speciation." In J. P. Riley and G. Skirrow (Eds.), *Chemical Oceanography*. Academic Press, New York, p. 173.

Stumm, W. and Morgan, J. J. (1970). *Aquatic Chemistry*. John Wiley & Sons, New York.

Suarez, D. and Langmuir, D. (1976). Heavy metal relationships in a Pennsylvania soil. *Geochim. Cosmochim. Acta* **40**, 589.

Talsma, T. and Philp, J. R. (1971). *Salinity and Water Use*. Macmillan Press, London.

Thorp, V. J. and Lake, P. S. (1974). Toxicity bioassays of cadmium on selected freshwater invertebrates and the interaction of cadmium and zinc on the freshwater shrimp, *Paratya tasmaniensis* Rick. *Aust. J. Mar. Freshwater Res.* **25**, 97.

Tinsley, I. J. (1979). *Chemical Concepts in Pollutant Behavior*. John Wiley & Sons, New York.

Turner, J. C., Solly, S. R. B., Mol-Krijnen, J. C. M., and Shanks, V. (1978). Organochlorine, fluorine and heavy metal levels in some birds from New Zealand estuaries. *N.Z. J. Sci.* **21**, 99.

Venugopal, B. and Luckey, T. P. (1975). "Toxicology of Nonradioactive Heavy Metals and Their Salts." In T. D. Luckey, B. Venugopal, and D. Hutchinson (Eds.), *Heavy Metal Toxicity, Safety and Hormology*, 4-73 Supplement, Vol. 1, F. Coulston and F. Korte (Eds.), *Environmental Quality and Safety*. George Thieme, Stuttgart.

Walker, T. I. (1976). Effects of species, sex, length and locality on the mercury content of school shark, *Galeorhinus australis* (Macleay) and gummy shark

Mustelus antarcticus (Guenther) from south-eastern Australian waters. *Aust. J. Mar. Freshwater Res.* **27**, 603.

Ward, T. J. and Young, P. C. (1982). Effects of sediment trace metals and particle size on the community structure of epibenthic seagrass fauna near a lead smelter, South Australia. *Mar. Ecol. Prog. Ser.* **9**, 137.

Weatherley, A. H., Beevers, J. R., and Lake, P. S. (1967). "The Ecology of a Zinc Polluted River." In A. H. Weatherby (Ed.), *Australian Inland Waters and Their Fauna*. Australian National University Press, Canberra.

Weatherley, A. H., Dawson, P., and Peuridge, L. (1975). Assessment and eradication of heavy metal pollution in a planned urban environment. *Verh. int. Verein. theor. angew. Limnol.* **19**, 2112.

Whipple, Jr., W. and Hunter, J. V. (1977). Nonpoint sources and planning for water pollution control. *J. Water. Pollut. Control. Fed.* **49**, 15.

Williams, S. L., Aulenbach, D. B., and Clesceri, N. L. (1974). "Sources and Distribution of Trace Metals in Aquatic Environments." In A. J. Rubin (Ed.), *Aqueous–Environmental Chemistry of Metals*. Ann Arbor Science, Ann Arbor, Michigan, p. 77.

Wilson, R. C. H. (1972). Prediction of copper toxicity in receiving waters. *J. Fish. Res. Board Can.* **29**, 1500.

Wittmann, G. T. W. (1979). "Toxic Metals." In U. Förstner and G. T. W. Wittmann, (Eds.), *Metal Pollution in the Aquatic Environment*. Springer-Verlag, Berlin, p. 3.

Wood, J. M. (1974). Biological cycles for toxic elements in the environment. *Science* **183**, 1049.

Wright, D. A. (1978). Heavy metal accumulation by aquatic invertebrates. *Appl. Biol.* **3**, 331.

Young, D. R., Mearns, A. J., Tsu-Kai, Jan, Heisen, T. C., Moore, M. D., Eganhouse, R. P., Hershelman, G. P., and Gossett, R. W. (1980). Trophic structure and pollutant concentrations in marine ecosystems of southern California. CalCOF1 Rep. Vol. XXI, 197–206.

11

ATMOSPHERIC POLLUTANTS

The atmospheric environment consists of a mixture of gases extending approximately 10–16 km from the earth's surface. It consists of oxygen (21%), nitrogen (78%), carbon dioxide (about 0.03%), argon (less than 1%), and traces of other gases plus varying amounts of water vapor. This composition has evolved since the origin of life on earth, before which the amount of carbon dioxide greatly exceeded the oxygen content. With the evolution of green plants, carbon dioxide was converted by photosynthesis into atmospheric oxygen and the carbon deposited in sedimentary layers.

A heterogeneous mixture of potentially harmful substances, such as dust, salt, and various gases, enter the atmosphere from natural and anthropogenic sources. Important anthropogenic additions have resulted from the use of fossil fuels, particularly in the internal combustion engine, electricity generation, and the smelting of mineral ores. An outline of the various pollutants is shown in Table 11.1 where the major pollutants are identified as photochemical oxidants, sulfur oxides, and fluorides. In addition, a slow but steady increase in the carbon dioxide concentration in the atmosphere has been noted in recent years. In the long term, the significance of carbon dioxide in the atmosphere may be of major importance to global ecosystems. In contrast to aquatic pollutants, atmospheric pollutants are often dispersed worldwide and have global significance.

SULFUR DIOXIDE AND ACID RAIN

In natural systems decomposing animal and vegetable matter, volcanic action, and wind erosion can result in the release of several gases. These gases usually contain carbon, sulfur, and nitrogen, which are needed in the photosynthetic process for the production of proteins, nucleic acids, and other substances in plants and animals. In addition to obtaining nutrients from the atmosphere, plants can derive a portion of their requirements dissolved in precipitation as well as from soils (Gorham, 1976; Cowling, 1980). Although the molecular species in the atmosphere and dissolved in precipitation are beneficial at low concentrations, at higher concentrations they may be injurious to plants, animals, or microorganisms (see Table 11.1).

Combustion of fossil fuels has provided a new source of airborne substances. Tables 11.2 and 11.3 compare biogenic sulfur and nitrogen emissions with those

Table 11.1.   Major and Minor Air Pollutants and Pollutant Mixtures that Affect Plant Ecosystems

Pollutant	Common Source	Phytotoxic concentration (ppm hr^{-1})
	Major Pollutants	
Photochemical Oxidants		
Ozone	Internal combustion engine	0.04–0.7
Nitrates	Internal combustion engine	0.004–0.01
Oxides and nitrogen	Combustion furnace and internal combustion engine	0.21–100
Sulfur Oxides	Combustion furnace (coal or oil) Other commercial processes	0.1–0.5
Fluorides	Production of aluminum, phosphates, brick-making, sintering	0.0001
	Minor Pollutants	
Ammonia	Various industrial processes — spills	
Boron	Same	
Chlorine	Same	
Ethylene and propylene	Internal combustion engine	0.0005–10
Hydrogen chloride and hydrochloric acid	Various industrial processes	
Particulates and heavy metals	Combustion furnace and various industrial processes	Varies
Sodium sulfate	Combustion furnace	
	Pollutant Combinations	
Ozone and sulfur dioxide	All of above sources	
Ozone and peroxyacetyl nitrate		
Sulfur dioxide and nitrogen dioxide		

Source: Manning and Feder (1980). Reprinted with permission from Applied Science Publishers Ltd.

produced from fossil fuels and other sources (Gorham, 1976). Thus significant additions have been made to atmospheric sulfur and nitrogen from fossil fuel combustion. Direct toxicity and toxic effects, produced by solubilization of sulfur and nitrogen gases in precipitation, have had a deleterious effect on natural ecosystems, particularly in western Europe and the northeastern United States.

Table 11.2. Estimated Rates of Sulfur Mobilization from Biogenic Sulfur Emissions, Fossil Fuel Combustion, Industrial Sources, and Volcanoes

Source	Sulfur (10^6 t yr^{-1})	
	Estimate I	Estimate II
Natural		
Sea spray[a]	44	43
Biogenic (sea)[a]	48	89
Biogenic (land)	58	
Volcanoes	2	0.7
Total	152	133
Anthropogenic		
Fossil fuel combustion	51	50
Non fuel sources	14	
Total	65	
Examples of Individual Anthropogenic Sources		
Sudbury metal smelters, 1972	> 1.8	
Thirty-five gas processing plants, Alberta	0.17	
Coal-fired power plant, 600 MW, burning coal with 3% S	0.14	

Source: Gorham (1976). Reprinted with permission from D. Reidel Publishing Company.

[a] Most marine emissions return directly to the sea.

Table 11.3. Worldwide Emissions of Nitrogen Oxides by Combustion Processes and Natural Sources

Source	NO_x Emissions as N (10^6 t yr^{-1})	
Coal combustion		8.2
Power generation	3.7	
Industrial	4.2	
Domestic/commercial	0.3	
Petroleum combustion		6.6
Residual oil	2.7	
Gasoline	2.3	
Fuel oil	1.1	
Other	0.5	
Natural gas combustion		0.6
Forest fires		< 0.2
Natural sources		150

Source: Gorham (1976). Reprinted with permission from D. Reidel Publishing Company.

Chemistry of Acid Precipitation

The atmosphere includes 0.03% of carbon dioxide which, in equilibrium with water as precipitation, produces a pH of about 5.7 (Gorham, 1976). In rainwater affected by atmospheric pollutants, additional acidity is usually due to three mineral acids: sulfuric, nitric, and hydrochloric acids. Generally, sulfate ions are dominant with lesser proportions of nitrate ions and comparatively low proportions of chloride ions.

Sulfur occurs in coal in proportions of 1–3% and in petroleum products the proportions can be somewhat higher. Other important sources of sulfur are sulfide ore smelters and volcanoes (Gorham, 1976). Combustion of the fossil fuel produces sulfur dioxide which can be then oxidized and converted to sulfuric acid:

$$2SO_2 + H_2O + O_2 \longrightarrow H_2SO_4$$

The oxidation of sulfur dioxide in an effluent gas is strongly influenced by the relative humidity (Butler, 1979). Little oxidation occurs at relative humidities below 70% but at higher humidities there is relatively rapid oxidation and conversion to sulfuric acid. Manganese and iron salts derived from flyash catalyze these reactions.

The oxides of nitrogen and nitric acid are naturally produced in the atmosphere by the energy discharged in lightning flashes (Butler, 1979). However, the major pollution source of these substances is the internal combustion process (see Figure 11.1). Acid precipitation can be partially neutralized by the occurrence of bases in the atmosphere such as ammonia and seaspray (Gorham, 1976).

Mechanisms for the Interaction of Acid Rain with Biota

Sulfur dioxide in the atmosphere can interact with organisms in a number of ways. It can be absorbed into plants directly in the molecular form or it may be adsorbed onto the moist surfaces of plants, soils, aquatic systems, and so on (Reuss, 1977). Alternatively, it can be converted into sulfuric acid and remain in the atmosphere as aerosol droplets to be removed by precipitation. Precipitation accounts for 40–80% of the sulfur deposited on land with the remainder being directly adsorbed onto surfaces (Nisbet, 1975). In addition, the impact of aerosols and acid forming soot particles onto surfaces may be another important mechanism of transfer of sulfur-containing materials from the atmosphere to the earth's surface (Gorham, 1976). The use of tall smokestacks together with suitable atmospheric and meteorological conditions can lead to transfer and deposition of sulfates thousands of kilometers from the emission source.

Significantly, an indirect effect of acid rain and the lowering of the pH of aquatic areas is the release of toxic metals adsorbed on bottom sediments. Thus, effects in natural aquatic ecosystems may be due to these metals as well as low pH.

Reactions in the Combustion Chamber

1. *Generation of atomic oxygen*

$$O_2 \longrightarrow O + O$$

Step (i) $CO_2 + OH \longrightarrow CO_2 + H$

Step (ii) $H + O_2 \longrightarrow OH + O$

2. *Formation of nitric oxide utilizing atomic oxygen and atmospheric nitrogen*

Step (i) $O + N_2 \longrightarrow NO + N$

Step (ii) $N + O_2 \longrightarrow NO + O$

Reactions in the Atmosphere

1. *Formation of nitrogen dioxide and nitrogen trioxide*

$$2NO + O_2 \longrightarrow 2NO_2$$

$$O_3 + NO \longrightarrow NO_2 + O_2$$

$$NO_2 + O_3 \longrightarrow NO_3 + O_2$$

2. *Formation of N_2O_5 and reactions of nitrogen trioxide*

$$NO_3 + NO_2 \longrightarrow N_2O_5$$

$$NO_3 + NO \longrightarrow 2NO_2$$

3. *Formation of nitrous and nitric acids in the presence of moisture*

$$N_2O_5 + H_2O \longrightarrow 2HNO_3$$

$$NO_2 + NO + H_2O \longrightarrow 2HNO_2$$

Figure 11.1.   Reaction sequence resulting in the formation of nitrous and nitric acids from the combustion of fossil fuels. [Adapted from Butler (1979).]

Physiological Responses to Acid Precipitation and Related Atmospheric Pollutants

Sulfur dioxide and acid rain have a variety of interactions with the physiology and biochemistry of plants (Varshney and Garg, 1979). Sulfurous acid may remove magnesium ions from the tetrapyrole ring of the chlorophyll molecule thus converting chlorophyll to phaeophytin, a photosynthetically inactive pigment. In addition, chlorophyll may be oxidized to other photosynthetically inactive substances. Protein molecules may be detrimentally altered by oxidation of the disulfide links by sulfurous acid. These interactions can result in lethal and other damaging effects as summarized by Tamm and Cowling (1977) in Table 11.4.

A modest amount of data is available on the effects of pH on aquatic invertebrates (see Giddings and Galloway, 1976). Generally, pH values below 5 are seriously detrimental to aquatic invertebrates. A comparatively large body of information is available on the effects of pH on fish (see Table 11.5).

Table 11.4.   Potential Effects of Acidic Precipitation on Vegetation

Direct Effects

1. *Damage to protective surface structures such as cuticle.*   Damage to surface
    structures may occur due to accelerated erosion of the cuticular layer that pro-
    tects most foliar organs. It also could result from direct injury to surface cells
    by high concentrations of sulfuric acid and other harmful substances that are
    concentrated by evaporation or adherence of soot particles on plant surfaces.
2. *Interference with normal functioning of guard cells.*   Malfunction of guard
    cells will lead to loss of control of stomata and thus altered rates of transpi-
    ration and gas-exchange processes and possibly increased susceptibility to
    penetration by epiphytic plant pathogens.
3. *Poisoning of plant cells after diffusion of acidic substances through stomata or
    cuticle.*   This could lead to development of deep necrotic or senescent spots
    on foliar organs including leaves, flowers, twigs, and branches.
4. *Disturbance of normal metabolism or growth processes without necrosis of
    plant cells.*   Such disturbances may lead to decreased photosynthetic effi-
    ciency, altered intermediary metabolism, as well as abnormal development or
    premature senescence of leaves or other organs.
5. *Alteration of leaf- and root-exudation processes.*   Such alterations may lead
    to changes in populations of phyllosphere and rhizosphere microflora and
    microfauna, including N-fixing organisms.
6. *Interference with reproduction processes.*   Such interference may be achieved
    by decreasing the viability of pollen, interference with fertilization, decreased
    fruit or seed production, decreased germinability of seeds, etc.
7. *Synergistic interaction with other environmental stress factors.*   Such rein-
    forcing interactions may occur with gaseous SO_2, O_3, fluoride, soot particles,
    and other air pollutants as well as with drought, flooding, etc.

Indirect Effects

1. *Accelerated leaching of substances from foliar organs.*   Damage to cuticle and
    surface cells may lead to accelerated leaching of mineral elements and organic
    substances from leaves, twigs, branches, and stems.
2. *Increased susceptibility to drought and other environmental stress factors.*
    Erosion of cuticle, interference with normal functioning of guard cells, and
    direct injury to surface cells may lead to increased evapotranspiration from
    foliar organs and vulnerability to drought, air pollutants, and other environ-
    mental stress factors.
3. *Alteration of symbiotic associations.*   Changes in leaf- and root-exudation
    processes and accelerated leaching of organic and inorganic substances from
    plants may affect the formation, development, balance, and function of
    symbiotic associations such as mycorrhizae, N-fixing organisms, lichens, etc.
4. *Alteration of host–parasite interactions.*   Resistance and/or susceptibility to
    biotic pathogens and parasites may be altered by predisposing plants to in-
    creased susceptibility, altering host capacity to tolerate disease, altering patho-
    gen virulence, etc. The effects of acidic precipitation may vary with: the nature
    of the pathogen involved (whether a fungus, bacterium, mycoplasma, virus,
    nematode, parasitic seed plant, or multiple pathogen complex), the species,
    age, and physiological status of the host, and the stage in the disease cycle in
    which the acidic stress is applied — for example, acidic rain might decrease the
    infective capacity of bacteria before infection and increase the susceptibility of
    the host to disease development after infection.

Source: Tamm and Cowling (1977). Reprinted with permission from D.
Reidel Publishing Company.

Table 11.5.   Summary of Effects of pH Changes on Fish

pH	Effects
3.0–3.5	Toxic to most fish; some plants and invertebrates survive.
3.5–4.0	Lethal to salmonids. Roach, tench, perch, and pike survive.
4.0–4.5	Harmful to salmonids, tench, bream, roach, goldfish, and common carp; resistance increases with age. Pike can breed, but perch, bream, and roach cannot.
4.5–5.0	Harmful to salmonid eggs and fry; harmful to common carp.
5.0–6.0	Not harmful unless > 20 ppm CO_2, or high concentrations of iron hydroxides present.
6.0–6.5	Not harmful unless > 100 ppm CO_2.
6.5–9.0	Harmless to most fish.
9.0–9.5	Harmful to salmonids and perch if persistent.
9.5–10.0	Slowly lethal to salmonids.
10.0–10.5	Roach and salmonids survive short periods, but lethal if prolonged.
10.5–11.0	Lethal to salmonids; lethal to carp, tench, goldfish, and pike if prolonged.
11.0–11.5	Lethal to all fish.

Source: European Inland Fisheries Advisory Committee (1969). Reprinted with permission from Pergamon Press Ltd.

Effects on Terrestrial Ecosystems

In terrestrial ecosystems plants are principally affected. The differences in susceptibility of different plant species have been well documented (Tamm and Cowling, 1977; Holdgate, 1979). This is consistent with the absence of various plant species from city centers and industrial areas while the same species are present in closely adjacent areas. Susceptibility usually reflects differences in genetic factors, age, or physiological condition. Not only are there differences between species but there are often variations between plant genotypes (Hawkesworth and Rose, 1976). In a number of cases genetic selection has occurred within some natural plant communities for resistance to atmospheric pollution (e.g., Bell and Mudd, 1976). For example, a species of rye grass exposed over a considerable period of time to coal smoke pollution had built up resistance to this form of atmospheric pollution which was not present in the same species from unpolluted areas.

In natural systems it would be expected that relationships between species present and susceptibility to atmospheric pollution would occur. The lichens have been the subject of continuing investigations over about the last 100 years. The relationship between sulfur dioxide levels in the atmosphere and presence of lichen species is shown in Table 11.6. This illustrates how the number and diversity of epiphytic flora increases from zero at mean winter sulfur dioxide levels of greater than 170 μg m^{-3} to a wide diversity of sensitive species with no

Table 11.6.   Relation Between Winter Sulfur Dioxide Concentrations and Corticolous Lichen Flora of Tree Trunks with Moderately Acid Bark[a] (with updated nomenclature)

Zone	Epiphyte Flora	Mean Winter SO_2 (μg m^{-3})
0	None	?
1	*Desmococcus viridis* s.l. confined to the base	> 170
2	*Desmococcus viridis* s.l. extends up the trunk: *Lecanora conizaeoides* confined to the bases	about 150
3	*Lecanora conizaeoides* extends up the trunk; *Lepraria incana* becomes frequent on the bases	about 125
4	*Hypogymnia physodes* and/or *Parmelia saxatilis* or *P. sulcata* appear on bases but do not extend up trunk. *Hypocenomyce scalaris, Lecanora expallens,* and *Chaenotheca ferruginea* often present	about 70
5	*Hypogymnia physodes* or *P. saxatilis* extends up the trunk to 2.5 m or more. *P. glabratula, P. subrudecta, Parmeliopsis ambigua,* and *Lecanora chlarotera* appear. *Calicium viride, Chrysothrix candelaris,* and *Pertusaria amara* may occur. *Ramalina farinacea* and *Evernia prunastri,* if present, confined to the bases: *Platismatia glauca* may be present on horizontal branches	about 60
6	*P. caperata* present at least on base: rich in *Pertusaria* species (e.g., *P. albescens, P. hymenea*) and *Parmelia* (e.g., *P. revoluta,* except in NE, *P. tiliacea, P. exasperatula* (in N). *Graphis elegans* appearing; *Pseudevernia furfuracea* and *Bryoria fuscescens* present in upland areas	about 50
7	*Parmelia caperata, P. revoluta* (except in NE), and *P. tiliacea, P. exasperatula* (in N) extend up trunk: *Usnea subfloridana, Pertusaria hemisphaerica, Rinodina roboris* (in S), and *Arthonia impolita* (in E) appear	about 40
8	*Usnea ceratina, Parmelia perlata,* or *P. reticulata* (S and W) appear: *Rinodina roboris* extends up trunk (in S); *Normandina pulchella* and *U. rubicundo* (in S) usually present	about 35
9	*Lobaria pulmonaria, L. amplissima, Pachyphiale cornae, Dimerella lutea,* or *Usnea florida* present: if these are absent, crustose flora are well developed, often with more than 25 species on larger, well lit trees	under 30
10	*L. amplissima, L. scrobiculata, Sticta limbata, Pannaria* spp., *Usnea articulata, U. filipendula,* or *Teloschistes fiavicans* present to locally abundant	"pure"

Source: Hawksworth and Rose (1976). Reprinted with permission from Macmillan Journals Ltd.

[a] A different series of indicators have been defined for trees with basic or nutrient-enriched bark (see Hawksworth and Rose, 1976).

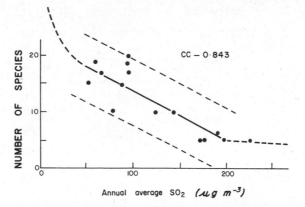

Figure 11.2.   A comparison between levels of air pollution and the diversity of lichens and bryophytes growing on asbestos roofs. Correlation coefficients and 95% confidence limits have been calculated. [From Gilbert (1970a). Reprinted with permission from the Trustees of the New Phytologist.]

sulfur dioxide present. Also, Gilbert (1970a,b) has found a regular relationship between the average concentration of sulfur dioxide and the number of species of lichens and bryophytes present (see Figure 11.2). These relationships have been extensively used for the biomonitoring of atmospheric pollutants (e.g., Manning and Feder, 1980). Thus species diversity of plants would generally be expected to decrease with increasing concentration of atmospheric pollutants. Kozlowski (1980) has pointed out that agricultural ecosystems consist of monospecific sets of plants and therefore may be subject to severe disturbance because of their lack of diversity in susceptibility to atmospheric pollution.

The effects of sulfur dioxide and acid precipitation are often most visible and dramatic in forest ecosystems adjacent to smelters or some other concentrated source of pollutant. A forest ecosystem in North America adjacent to a coal burning power plant has been investigated by Rosenberg et al. (1979). Consistent with other investigations, lichen species richness and diversity increase with increasing distance from the power plant. In addition, this relationship was extended over a greater distance downwind from the plant as compared with the upwind side. Particular species of trees, the Sweet Birch and White Pine, are found to be most susceptible to atmospheric pollution. Similar relationships have been found by Gordon and Gorham (1963) in investigating the effects on vegetation produced by sulfur dioxide pollution from an iron sintering plant in Ontario (see Figure 11.3). Some species, which were the most tolerant of air pollution, were infrequent in the normal forest vegetation.

Effects on Aquatic Ecosystems

Effects of acid precipitation have had their longest and most intensive study in the Scandinavian countries of northern Europe. Highly polluted precipitation falls

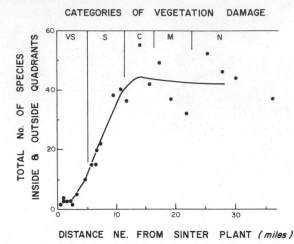

Figure 11.3. The relation between floristic variety of the ground flora, aerial estimates of damage to vegetation, and distance from pollution source (VS = very severe; S = severe; C = considerable; M = moderate; N = not obvious). [From Gordon and Gorham (1963). Reprinted with permission from the National Research Council of Canada.]

over large areas of Scandinavia and it has been suggested that the pollutants causing this originate in other countries of western Europe. In addition, the intensity and geographical extent of acid precipitation may be increasing (Wright et al., 1977). As early as the 1920s, there were Norwegian reports that fish populations were disappearing due to acid precipitation. The National Academy of Sciences (1978) has listed three conditions that strongly influence the extent and severity of acid precipitation in aquatic areas. These are: (1) location with particular emphasis on those adjacent to pollution sources; (2) the nature of the base rocks in the area; and (3) low watershed to surface ratios. The National Academy of Sciences (1978) was also able to summarize the effects of pH on fish (see Table 11.5). The implications of these data are revealed by the survey carried out by Likens (1976) on the status

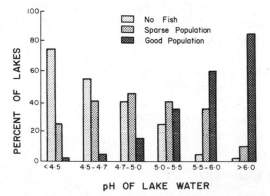

Figure 11.4. Status of fish populations in Norweigian lakes in relation to the pH of the water. [From Leivestad et al. (1976). Reprinted with permission from SNSF Project, Research Report FR6/76.]

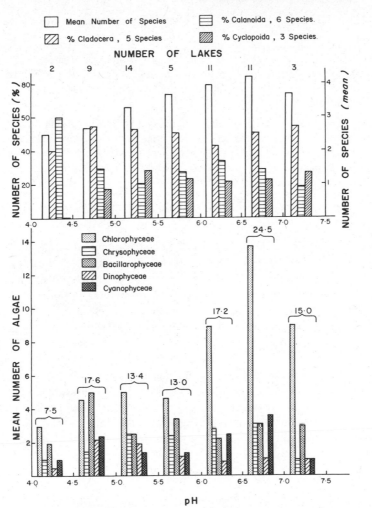

Figure 11.5. Mean number of algal species and percent contributions from three classes of zooplankton to total species number and mean number of zooplankton species against pH in samples from 57 lakes of Norway. [Adapted from Wright et al. (1976). Reprinted with permission from D. Reidel Publishing Company.]

of fish populations in Norwegian lakes in relation to pH of water (see Figure 11.4). This shows clearly the deterioration of fish populations in relation to pH. In Norway acid precipitation has also had an effect on commercial fisheries. Wright et al. (1977) have reported a decrease in the salmon catch over the last 100 years in rivers which have been subject to steady decreases in pH.

The data obtained on algal and zooplankton species from a survey of 57 lakes in Norway are shown in Figure 11.5 (Wright et al., 1977). Both the algae and zooplankton show a maximum number of species at pH values between 6.5 and 7.0, the normal pH which would be expected in unaffected aquatic areas. But pH

Table 11.7.   Generalized Changes in Aquatic Areas Resulting from Acid Precipitation[a]

Chemical Changes: Loss of buffering capacity; occurrence of short-term decreases in pH related to meltwater from snow or precipitation; reduced availability of nutrients due to decreased remineralization of detritus; possible elevated heavy metal concentrations.

Biological Changes: Less activity by decomposer organisms giving increased accumulations of organic matter; at pH < 6 considerable decrease in phytoplankton, zooplankton, bottom fauna, and invertebrates; shift from populations of higher plants to mosses; at pH < 5.5 serious depletion of fish populations with particular damage to sensitive newly hatched larvae.

---

[a] Compiled from Cowling (1980) and National Academy of Sciences (1978).

values above and below this range show decreases in number of species in relation to the difference from the normal pH. Also the composition of both the algal and zooplankton communities is changed, with the more sensitive species removed from the communities and the more resistant species becoming more important in the community structure (see Figure 11.5). An additional effect is a decrease in the production of algal cells with deviation from the 6.5–7.0 pH range. Similar general effects to those mentioned above have been noted in eastern North America (Gorham, 1976).

With decreasing pH a series of chemical changes occur which result in a reduction in the rate of nutrient cycling within the aquatic system. Thus there is a decreased amount of organic matter in an area and a shift toward oligotrophic status with lakes. The ecological changes follow the generalized effects of toxic agents on ecosystems (see Chapter 4). Overall, the generalized effects of acid precipitation on aquatic areas are shown in Table 11.7.

PEROXYACETYL NITRATE (PAN), OZONE, AND OTHER OXIDANTS

Petroleum hydrocarbons (PHCs) are discharged into the atmosphere from the incomplete combustion of fossil fuels and evaporation (see Chapter 8). Substantial amounts are involved and estimates made in 1969 indicate that 88×10^6 tonnes were discharged annually (Manahan, 1975). The major source of PHCs in the atmosphere is the internal combustion motor used in motor vehicles. Natural sources of hydrocarbons are mainly vegetation which releases large quantities of terpenoids into the atmosphere. But in addition methane is generated in substantial quantities by the anaerobic decay of vegetation.

PHCs are not significant primary pollutants since they do not have a strong physiological impact on animals or plants. However, the secondary reactions of partly combusted PHCs with oxides of nitrogen result in the production of important atmospheric pollutants. A simplified sequence of reactions leading to the production of ozone and PAN is shown in Figure 11.6. The primary pollutants

$$N_2 + O_2 \xrightarrow[\substack{\text{internal combustion} \\ \text{motor}}]{\text{complex reactions}} NO + NO_2$$

where $NO \gg NO_2$.

But NO_2 triggers the following reaction sequence:

$$NO_2 + h\nu \longrightarrow NO + O$$

$$O + O_2 + M \longrightarrow O_3 + M$$

$$O_3 + NO \longrightarrow NO_2 + O_2$$

where M is any third body.

Further reactions now occur with olefins also derived from the internal combustion motor:

$$O_3 + R\text{--}CH\text{=}CH\text{--}R \longrightarrow R\text{--}\underset{H}{\overset{O}{C}} + RO\cdot + H\dot{C}O$$

$$R\text{--}\underset{H}{\overset{O}{C}} + O + h\nu \longrightarrow R\text{--}\overset{O}{C}\cdot + \cdot OH$$

$$R\text{--}\overset{O}{C}\cdot + O_2 \longrightarrow R\text{--}\overset{O}{C}\text{--}O\text{--}O\cdot$$

$$R\text{--}\overset{O}{C}\text{--}O\text{--}O\cdot + NO_2 \longrightarrow R\text{--}\overset{O}{C}\text{--}O\text{--}O\text{--}NO_2$$

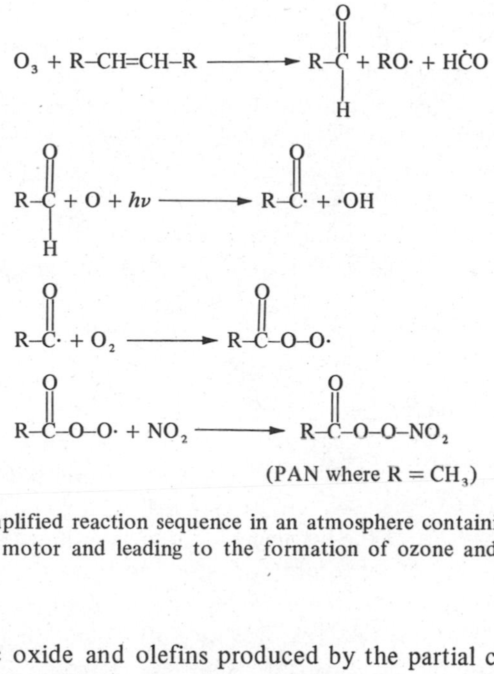

(PAN where $R = CH_3$)

Figure 11.6. A simplified reaction sequence in an atmosphere containing products from the internal combustion motor and leading to the formation of ozone and peroxyacetyl nitrate (PAN).

involved are nitric oxide and olefins produced by the partial combustion of PHC fuels. As atmospheric pollutants, ozone and PAN can be considered together since they are derived from the same primary pollutants and have similar environmental properties, primarily related to their oxidizing capacity. Ozone and PAN are the major atmospheric pollutants produced by the process described in Figure 11.6, but a great number of other substances with pollutant properties are produced also. Of particular importance are PAN homologs and aldehydes.

The mixture of ozone, PAN, and the other substances is generally described as "photochemical smog." Photochemical smog is particularly important in urban areas with a large number of cars and also a high incidence of solar radiation. Los Angeles is considered to be a classic example of the development of photochemical smog. Glass (1979) has reported that it has been a chronic problem in this area for more than 30 years, causing losses to agricultural crops, ornamental plants, and native vegetation valued at more than $55 million dollars by the 1970s.

In addition, various areas have been reported to be unsuitable for certain types of crops. Bell (1976) has produced estimates of a $500 million dollar loss to U.S. agriculture up to 1968. Agricultural problems cited were damage to tobacco crops, beans, grapes, and potatoes in the eastern states of the United States.

Until recent years, it was considered that the incidence of solar radiation was not sufficient to result in damaging concentrations of photochemical oxidants in northern Europe. However, Bell (1976) reports a number of cases of atmospheric oxidant damage to crops in European countries.

The effects of photochemical oxidants have been most studied with plants. Bell (1976) reports that there is no well documented evidence that free ranging mammals, birds, and reptiles are directly affected by photochemical oxidants, although he points out the habitat damage may result in an impact on these populations.

Photochemical oxidants are highly reactive and are removed from the atmosphere comparatively rapidly by reactions with vegetation and soil. However, ozone shows low water solubility so rain-out is not an important factor with this compound. These substances show a classic diurnal cycle of concentration of the various pollutants as related to the incidence of solar radiation and use of motor vehicles (Manahan, 1975). Clear unpolluted air generally has a concentration of ozone ranging between 0.01 and 0.02 ppm. However, slightly smoggy Los Angeles atmosphere generally ranges from 0.1 to 0.2 ppm with the highest concentrations detected ranging from 0.5 to 0.8 ppm. Usually, the concentration of ozone is 10–20 times greater than the concentration of PAN (Mudd, 1975).

Physiological and Metabolic Effects

Photochemical smog enters leaves through the stomata and can inhibit and prevent various biochemical processes (Bell, 1976). Generally, PAN on contact with cellular material causes a lowering of the number of sulfydryl groups (Mudd, 1975). Thus PAN and ozone affect enzymes according to their sulfhydryl content (Mudd, 1975; Heath, 1975). Heath (1975) has reported that ozone will oxidize sulfhydryl groups to $-S-S-$, $-SO_2H$, and $-SO_3H$. The first two of these reactions are reversible, whereas the formation of $-SO_3H$ is irreversible and causes permanent alteration to the metabolism of the plant.

Carbohydrate metabolism, particularly the synthesis of cellulose, is inhibited by contact with PAN (Mudd, 1975). It acts by inhibiting the enzyme phosphorylase and possibly the enzyme phosphoglucomatase. In addition, Mudd (1975) has noted that hormonal regulation can be affected by the oxidation of indoleacetic acid giving lowered growth rates. Mudd (1975) has also reported that the lowering of numbers of sulfhydryl groups in enzymes results in the inhibition of important processes involved in photosynthesis. Thus the rate of photosynthesis is lowered by contact with PAN.

As well as the processes described above, PAN and ozone react with a variety of substances, such as olefins and amides, which occur in cells (Mudd, 1975). Unsaturated fatty acids have been suggested as particularly important sites for reactions of this kind.

Table 11.8.   Classification of Plants According to Production
of Minimal Injury After 1 hr Exposure

Sensitive (0.1 ppm O_3)	Intermediate (0.2 ppm O_3)	Resistant (0.35 ppm O_3)
Spinach	Begonia	Zinnia
Radish	Onion	Beet
Muskmelon	Chrysanthemum	Radish
Oat	Dogwood	Poinsettia
Pinto bean	Sweet corn	Black walnut
White pine	Wheat	Strawberry
Potato	Lima bean	Carrot
Tomato		

Source: Heck (1968). Reprinted with permission from Fluor-
moy Publishers, Inc.

In plants, ozone causes the palisade cells to collapse (Bell, 1976). This leads to
visible necrotic spots on the leaf surface. With PAN, the process leads to silvering,
bronzing, or glazing of the lower leaf surface. Additional evidence suggests that
PAN alters cell permeability and the extent of PAN injury is related to the periods
of light exposure (Bell, 1976).

Phytotoxicity of Ozone and PAN

Plant species show differing susceptibility to damage by photochemical oxidants
(see Table 11.8). An important area of interest has been the loss of production in
crops. Many varieties of crops have been shown to suffer a substantial loss of yield
due to exposure to atmospheric pollutants (see Table 11.9).

Effects on Ecosystems and Communities

Woodwell (1970) has summarized the effects of atmospheric pollutants on ecosystems as: (1) elimination of sensitive species; (2) reduction in species diversity
and numbers; (3) removal of overstorey plants and favoring of small plants; (4)
reduction in standing crop of organic matter leading to a reduction of nutrient
elements in the system; and (5) increase in insect pests and some diseases.

Treshow and Stewart (1973) carried out an extensive study of the sensitivity
of natural communities to ozone (see Table 11.10). These data indicate significant
differences in the sensitivity of different groups and species within those groups.
This suggested that an alteration in the structure of natural communities would
occur as a result of exposure to ozone with an initial loss of the most sensitive
species. In accord with this, the Sub-Committee on Ozone and Other Photochemical
Oxidants (1976) has reported investigations carried out in southern California

Table 11.9. Effects of Ambient Oxidants on Growth and Yield in Selected Plants[a]

Species	Variety	Facility[b]	Oxidant Concentration (ppm)	Exposure Time[c]	Plant Yield Percent Reduction from Control
Lemon	–	Field, CT	> 0.10	Growing season	32
Oranges	–	Field, CT	> 0.10	Growing season	{52 {54
Grape	Zinfandel	Field, CT	> 0.25	Growing season (1st year) (2nd year) (3rd year)	12 61 47
Cotton	Acala	Field, CT	Ambient[d]	Growing season (2 yr)	,5–29
Potato	2-Sensitive	Greenhouse	≥ 0.05	326–533 hr (2 yr)	34–50
Tobacco	Bel W$_3$	Field, OT	> 0.05	Growing season	22
Soybean	4	Field, OT	≥ 0.05	Growing season (3 yr)	20[e]

Source: Heck et al. (1977). Reprinted with permission from the Soil Conservation Society.

[a] Summary data for several studies of ambient oxidants.
[b] CT = closed top chamber; OT = open top chamber.
[c] The concentrations shown are hourly averages that occur often during the growing season.
[d] Dose not reported.
[e] Reduction shown is the average for four varieties over 3 yr.

Table 11.10.   Injury Threshold for Exposure to Ozone (2 hr) for Plants in Natural Communities in Salt Lake Valley, Utah, U.S.[a]

Type	Injury Threshold (pphm O_3)
Grassland-Oak Communities	
Trees and shrubs	25 to > 40
Perennial forbs	15 to > 40
Grasses	15–30
Aspen and Conifer Communities	
Trees and shrubs	15 to > 30
Perennial forbs	15 to < 40
Annual forbs	25 to > 30
Grasses	25

[a] Compiled from Treshow and Stewart (1973).

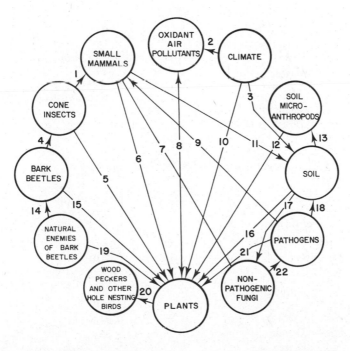

Figure 11.7. Community-level interactions in a mixed-conifer forest ecosystem. Types of interaction are indicated by the numbered arrows. [From Taylor et al., (1974). Reprinted with permission from the Statewide Air Pollution Research Centre, University of California.]

where it was noted that sensitive pine species in natural forests were exposed to photochemical oxidants and were suffering injury. Overall, this organization concluded that: (1) ozone injury limits biomass production by the primary producers and their capacity to reproduce; (2) the decrease of biomass or energy flow to consumers and decomposers in the ecosystem affects the population of these organisms; (3) essential recycling processes may be interrupted, further limiting primary production; and (4) stand structure is altered rapidly in some areas by salvage logging of high risk trees; as a result species composition is changed and wildlife habitat altered. The Sub-Committee on Ozone and Other Photochemical Oxidants (1976) has also reported on modeling investigations of the effects of atmospheric oxidants on ecosystems (see Figure 11.7).

FLUORIDES

Fluorine is chemically combined as fluoride in the earth's lithosphere to the extent of 0.06–0.09% of the total weight (Committee on Biological Effects of Atmospheric Pollutants, 1971). The major fluoride minerals are fluorospar (CaF_2), cyrolite (Na_3AlF_6), and fluorapatite ($Ca_{10}F_2$)(PO_4)$_6$. Fluoride is ubiquitous throughout the environment occurring in all major segments as indicated by the fluoride content of a variety of environmental samples (see Table 11.11).

Table 11.11.   Fluoride Content of Various Environmental Samples[a]

Sample	Concentration (ppm)
Soils	
U.S. surface soils (0–3 in.)	20–500 (av. 190)
U.S. surface soils (0–12 in).	20–1620 (av. 292)
USSR	30–320 (av. 200)
New Zealand	68–540 (av. 200)
Natural Waters	
North America	0–16
Europe	0–10
Australia	0–13.5
Seawater	
Atlantic Ocean	0.344–12
Pacific Ocean	1.16–1.3
Indian Ocean	0.8 (one sample)

[a] Compiled from Smith and Hodge (1979), and Committee on Biological Effects of Atmospheric Pollutants (1971).

Table 11.12.   Estimated Total Fluoride Emissions from Major Industrial Sources in the United States in 1968[a]

Source	Atmospheric Emissions (tons/yr)
Manufacture of normal superphosphate fertilizer	9,700
Manufacture of wet-process phosphoric acid	3,000
Manufacture of triple superphosphate fertilizer	300
Manufacture of diammonium phosphate fertilizer	100
Manufacture of elemental phosphorus	5,500
Manufacture of phosphate animal feed	100
Manufacture of aluminum	16,000
Manufacture of steel (open-hearth furnace)	16,800
Manufacture of steel (basic-oxygen furnace)	8,400
Manufacture of steel (electric furnace)	14,900
Welding operations	2,700
Nonferrous-metal foundries	4,000
Manufacture of brick and tile products	18,500
Manufacture of glass and frit	2,700
Combustion of coal	16,000
Total	118,700

Source: Committee on the Biologic Effects of Atmospheric Pollutants (1971).
[a] Data from U.S. Department of Health, Education and Welfare.

Fluorides are discharged into the ecosphere by a variety of different processes. Gaseous fluoride, in the form of hydrogen fluoride, and other products originate from volcanoes and fumaroles. In addition to these natural discharges there are many discharges made as a result of industrial activity (see Table 11.12).

With fertilizers and phosphoric acid manufacture, fluoride-containing phosphate rock, used in the process, undergoes a number of reactions to form hydrofluoric acid (HF) and silicon tetrafluoride (SiF_4) which are gases and discharged to the atmosphere (Committee on Biological Effects of Atmospheric Pollutants, 1971). In aluminum manufacture the process involves the use of calcium fluoride during electrolysis and the same gases (HF and SiF_4) are released. Fluoride-containing minerals are often used in steel, brick, tile, pottery, cement, glass, and enamel manufacture and this can lead to the release of fluoride-containing dusts and gases. In addition, coal usually contains 0.001–0.048% fluoride, which leads to the formation of hydrofluoric acid and silicon tetrafluoride in the combustion process. These are discharged to the atmosphere together with fluoride-containing particulate matter. Fluoride discharges can originate from a number of other sources as well, for example, uranium enrichment plants and fluoride compounds used in insecticides.

In the environment the silicon tetrafluoride reacts with water to yield fluorosilic acid (H_2SiF_6). Both fluorosilic acid and hydrofluoric acid, the principal gaseous forms of fluorine, are readily absorbed by animals and plants (Smith and Hodge,

1979). Common particulates (cryolite, sodium fluorosilicate, aluminum fluoride, sodium fluoride) are emitted to the atmosphere and have a water solubility ranging from 0.04 to 4.0 g per 100 mL at 100°C (Committee on the Biological Effects of Atmospheric Pollutants, 1971). With these substances, rainfall and other climatic conditions have a strong influence on the toxic effects exhibited.

Physiological Effects on Individuals

There is no convincing evidence that fluorine or fluoride are essential trace substances for the growth of animals or plants. However, there is evidence to show that it does have an influence on bone metabolism in all organisms and dental health in humans (Committee on the Biological Effects of Atmospheric Pollutants, 1971).

Fluoride inhibits many enzymes. In most cases, the enzymes affected contain a metal ion with which the fluoride combines to form a metal fluoride complex. This gives a general inhibiting effect on overall metabolism and the growth of cells (Committee on the Biological Effects of Atmospheric Pollutants, 1971).

Phytotoxicity is usually due to the gaseous fluorides. At the cellular level there is a collapse of mesophyll causing a collapse in epidermal cells. In addition, a reduction in photosynthetic activity has also been observed with a resultant adverse effect on growth. The symptoms of injury are most evident with leaves which show necrosis on the leaf tips and margins (Committee on the Biological Effects of Atmospheric Pollutants, 1971). In assessing the impact of fluoride compounds on plants, the dose–rate relationship is often used:

$$F = KCT$$

where F is the response attributed to atmospheric fluoride; K a constant depending on environmental conditions, plant type, and so on; C the concentration of fluoride in the atmosphere; and T the period of exposure.

This relationship has limitations, but it takes into account the two critical factors regulating the response effects of fluoride, that is, concentration and period of exposure. Quite a number of studies have revealed statistically significant relationships of this form, as well as between fluoride concentration alone and plant response (see Table 11.13).

Possible air quality criteria in terms of concentration and period of exposure have been developed and are shown diagrammatically in Figure 11.8. It should be remembered that toxicity is influenced by a wide variety of biological, environmental, and chemical factors which may lead to marked differences between expected and actual toxicity.

There have been many studies on the effects of fluoride on animals and humans, particularly in relationship to dental care in humans. With animals, most studies have been concerned with domestic animals and aspects of economic importance.

There are limited data available on the effects of dissolved fluoride on aquatic animals, some of which are contained in Table 11.14. Groth (1975a,b) concluded that short-term effects were evident with aquatic organisms at concentrations as

Table 11.13.   Regression Equations Relating Fluoride Concentrations to Plant Response

Plant Studied	Range of Fluoride Concentrations	Regression Equation[a]
Citrus	0.14–0.45 ppb	$Y(\%) = 99.7 - 176F$
Citrus	Not stated	$Y = 381.91 - 1.3132C$
		$Y = 417.25 - 0.8797C$
Pine	Not stated	$BI = 0.06 + 0.1607F$
		$BI = 0.93 + 0.0027F$
Bean	2.2–13.9 $\mu g\ m^{-3}$	$C = 14 + 102F$
		$Y(\%) = 102.2 - 3.45F$
Orchard grass	Up to 11 $\mu g\ m^{-3}$	$C = 1.13FT - 1.17$
Alfalfa	Up to 11 $\mu g\ m^{-3}$	$C = 1.89FT + 0.74$
Vegetation	20–128 $\mu g\ m^{-3}$	$C = 8.50 + 0.314F$
Strawberry	0.55–10.4 $\mu g\ m^{-3}$	$Y(\%) = 99.5 - 5.1F$
Timothy and red	2.3 $\mu g\ m^{-3}$	$C = 2.555 + 4.120FT$
clover mix	5.0 $\mu g\ m^{-3}$	$C = 30.288 + 3.820FT$

Source: Rose and Marier (1977). Reprinted with permission from the National Research Council, Canada.

[a] Y = yield; BI = tip burn index; F = airborne fluoride concentration; C = concentration of fluoride in foliage; T = time; all expressed in units used by the original authors except where (%) indicates our calculations as a percentage of the control value.

low as 3 ppm. Water temperature, hardness, chlorinity, and other environmental factors influenced the exhibited effects.

There is an accumulation of fluorine in the skeleton of all animals and Groth (1975a,b) has suggested that this is one of the most notable characteristics of fluoride as a pollutant. Thus, serious adverse effects are possible even at very low

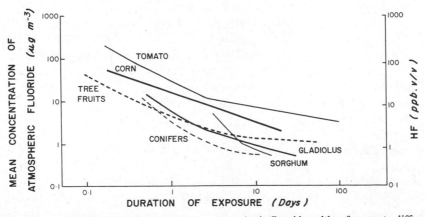

Figure 11.8.   Possible air-quality criteria for atmospheric fluoride, with reference to different plant species. [From: McCune (1969). Reprinted with permission from the American Petroleum Institute.]

Table 11.14.   Effects of Fluoride on Animals[a]

Animal	Concentration in Water[b]/Body[c] Tissues (ppm)	Effect
Aquatic Animals		
Blue crab (*Callineatus sapidus*)	20	4.5% growth increment reduction per moult
Brown trout fry	5 ppm (200 hr)	Increased mortality
Brown mussels (*Perma perma*)	1.4–7.2 (15 days)	Mortality
Oysters	32 or 128 (60 days)	100% Mortality
Sand shrimp	52 (72 days)	23–45% Mortality
Rainbow trout	2.7–4.7 (48–240 hr)	LC_{50}
Terrestrial Animals		
Honeybees	130–170	Death
Bumblebees	406	Endangered
Sphinx moth	394	Endangered
Fruitflies	5.5 (gaseous) (4 days)	Death

[a] Compiled from Rose and Marier (1977) and Groth (1975b).
[b] Aquatic organisms.
[c] Body concentrations with terrestrial organisms.

levels of exposure if this persists for lengthy periods of time. Even some plant varieties have been found to accumulate fluoride in their tissues to levels approximately 1 million times that of the surrounding air.

Kay et al. (1976) have found that fluoride accumulation is related to age in animals older than 6 yr. Rose and Marier (1977) have described several factors which suggest that wild, as compared to domestic, animals may be more susceptible to the accumulation of high levels of fluorine in skeletal tissues leading to fluorosis and lameness. In addition, fluorides are known to cause chromosome damage and mutation in animals and plants (Rose and Marier, 1977).

Effects on Communities and Populations

Plant species exhibit a wide range of susceptibility (see Table 11.15). Thus for a given exposure a plant community would be expected to exhibit a decrease in species diversity and a change in the species composition. In accord with this, Treshow (1971) found a dominance of resistant species in some affected areas. In proximity to an aluminum smelter a reduction in species number from 26 to 1 was

Table 11.15.   Summation of Susceptibility of Plants to Fluoride[a]

Plant Class	Susceptibility
Pines (Eastern white, Lodge pole, Scotch, Mugo, and Ponderosa)	Susceptible
Broad–leaved deciduous trees of importance in forestry	Intermediate susceptibility – relatively tolerant
Common vegetable crops and field crops	Tolerant to moderately susceptible
Important fruit and berry crops	Moderately susceptible

[a] Compiled from Committee on Biological Effects of Atmospheric Pollutants (1971).

found, and at the same time there was a shift in the community composition whereby the dominant Scots Pines were replaced by the more resistant *Alnus glutinosa, Carpinus betulus, Fagus sylvatica*, and *Quercus petrea.* Similar alterations in species composition have been observed elsewhere.

Groth (1975a,b) has suggested that reduced photosynthesis of certain algal species in aquatic areas would lead to lower primary productivity and resultant ecological effects. In addition, the accumulation of fluoride in skeletal tissue may lead to a decrease in the activity of vertebrates resulting in possible ecological alterations.

METALS

A background level of metals in atmospheric particulates has always existed. However, aerial pollution due to metals can be considered to have begun with the smelting of metallic ores. In fact, analyses for lead in peat deposits in Europe have shown that smelting activities resulted in the atmospheric deposition of lead (e.g., Lee and Tallis, 1973). While smelting remains an important source of metal-containing particulates in the atmosphere, recent sources also include the combustion of fossil fuels and the addition of metals to fuels as anti-knock substances and their use in the chemical industry. Generally, precipitation of particulates occurs within a comparatively short distance of the source although in some cases residence times in the atmosphere can be up to a month (Bowen, 1975).

The major metals having an ecological impact are generally referred to as the "heavy metals." But there is considerable confusion surrounding the definition of the term heavy metals which is discussed in Chapter 10. Most of the metals discussed in this section would generally be considered heavy metals.

Atmospheric Forms of Metals

Metals are rarely discharged into the atmosphere in the elemental form, with the notable exception of mercury. Significant quantities of mercury are discharged into the atmosphere in the elemental form (Manahan, 1975) but other metals can take a variety of physical forms from molecules of organometallic compounds to particulates of various types. However, the most important forms are particulates in which metals can occur adsorbed onto the surface or in chemical combination forming the particulate itself.

The chemical form of metals in the atmosphere is dependent on the origin and history of the material present. The natural background particulates generally consist of materials originating from soil and minerals and reflect this composition (Kirchoff, 1977). Particulates produced by the combustion of coal contain a high proportion of metal oxides (Wawerka et al., 1976). However, the presence of variable quantities of nitrogen and sulfur in the fuel causes differences in the chemical combinations in which the metals occur.

The chemical form of metals on deposition is particularly important as it has a powerful influence on the consequent mobility of the metal. In addition, the size and physical nature of particles have also been found to affect the adherence of particles, and thus metals, to surfaces (Chamberlain, 1967).

The principal form of lead discharged from vehicles using leaded fuel is lead chlorobromide (Smith, 1976; Wesolowski et al., 1973). After discharge into the atmosphere a rapid exchange of the bromide for chloride occurs, particularly in coastal areas. Figure 11.9 shows the patterns of distribution of lead after leaving a motor vehicle. Siccama and Smith (1978) have found that, in a forest watershed, lead is strongly retained in the forest humus which, in many situations, is rapidly increasing in lead content. A similar situation has been found by Van Hook et al. (1977) with cadmium and zinc as well as with lead. In urban areas in excess of 90% of lead originating from motor vehicles remains in the catchment soils (Rolfe and Haney, 1975; Bogges and Wixson, 1977). As a general rule, heavy metals are strongly retained in organic matter and soil and have turnover rates of the order of thousands of years (Hughes et al., 1980).

Behavior of Metals in Terrestrial Ecosystems

There are two main routes of entry into plants: (1) through the plant surface above the ground and (2) through the root system. Properties regulating the access of particles to plant and soil surfaces have been summarized by Hughes et al. (1980) as (1) particle size, (2) morphology of the deposition surface, (3) aerosol age, and (4) wind speed. Once deposited, the chemical properties of the particular metal form present are important. The solubilization of metallic constituents on the leaf's surface allows entry of these materials into the plant while those entering through the root system must first pass through the soil. The interaction of metals with soil is an extremely complex process depending on the organic content of the soil as well as other important constituents such as clay and hydrous metal oxides. A number of different theoretical models have been developed to explain metal activity in soils (Hughes et al., 1980).

After entry, metals can be distributed by complex processes to other parts of the

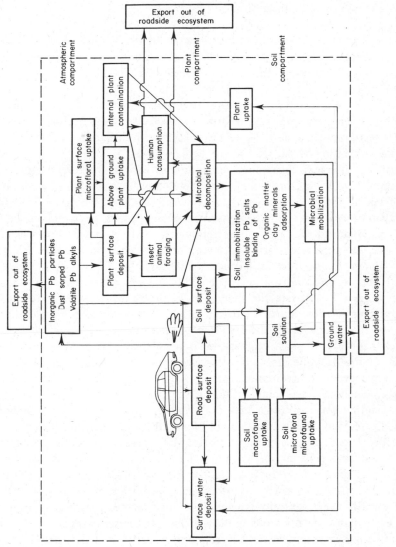

Figure 11.9. Potential lead distribution and transfer routes in the roadside ecosystem. [From Smith (1976). Reprinted with permission from the Air Pollution Control Association.]

plant. But they may also become immobilized by adsorption onto surfaces or detoxified by chelation with compounds present in the plant (Malone et al., 1974). However, in general terms metal accumulation occurs in the surfaces of leaves, stems, and roots.

Food Chain Transfer and Biomagnification

With terrestrial animals the major sources of metals are in food and water, although significant adsorption can occur through the lungs in some situations. In terrestrial ecosystems the major source is through food organisms. Therefore, the transfer of metals in food from food to consumer is an important process influencing the final concentration of metal in the consumer organism. Hughes et al. (1980) have pointed out that there is a limited amount of information available on the transfer of metals through terrestrial food chains. However, sufficient information is available to examine the process in general.

The transfer of metals from food to consumer can be considered using the concentration ratio (CR), the ratio of the metal concentration in the consumer to that present in the food. In assessing CRs in different food chains the following situations can be envisaged: (1) The consumer retains none of the metal present in the food material and thus the concentration ratio will approach zero. (2) The consumer has the ability to retain the metal but not to a concentration greater than that present in the food organisms. In this case the CR will be between 0 and 1. (3) The organism not only has the ability to accept metals from the food organism but can concentrate the metals to levels higher than that present in the food organism. Thus the CR is greater than 1. If this occurs in a sequence of organisms within a food chain there will be an increasing concentration of metal with increasing trophic level and biomagnification will occur.

The use of this simplified technique may result in errors since it does not take into account concentration variations within a population due to feeding habits, geographical location, time period utilizing the diet to generate the metal concentration, and a number of other factors. Nevertheless, it gives a general guide to the possibility of food chain transfer and biomagnification. Results of a limited number of investigations have been summarized by Hughes et al. (1980). With lead the CRs found were generally between 0 and 1, although in a limited number of cases they were greater than 1. This suggests that while lead can be transferred through members of a food chain it does not concentrate or biomagnify. With cadmium the CRs were generally greater than unity with earthworms, an isopod, herbivorous molluscs, detritivorous molluscs, and, to a lesser extent, other herbivores, carnivorous invertebrates, and insectivorous small mammals. Table 11.16 is an illustrative set of results. Martin et al. (1976) have suggested that cadmium can be readily transferred across the gut wall. In addition, there is some evidence to suggest that cadmium accumulation in some mammals is age dependent (Schlesinger and Potter, 1974; Munshower, 1977). Zinc exhibited a wide variety of different CRs and organisms appear to be able to bioregulate this element. In general, it

Table 11.16.   Elemental Concentrations of Cadmium in Consumers and Their Food at Different Sites in a Region Contaminated by Re-entrained Mine Waste[a]

Animal Consumers	Consumer Concentration (ppm dry wet)	Food Type	Food Concentration (ppm dry wet)	Concentration Ratio
Herbivorous	1.2 ± 0.3	Vegetation	0.7 ± 0.2	1.71
invertebrates	4.3 ± 1.2		2.2 ± 0.5	1.95
	11.1 ± 2.4		3.8 ± 1.0	2.92
Herbivorous	1.2 ± 0.3	Litter	1.9 ± 0.3	0.63
invertebrates	4.3 ± 1.2		5.4 ± 0.9	0.80
	11.1 ± 2.4		10.3 ± 1.8	1.08
Carnivorous	5.7 ± 2.0	Herbivorous	1.2 ± 0.3	4.75
invertebrates	15.9 ± 4.3	invertebrates	4.3 ± 1.2	3.70
	34.3 ± 4.6		11.1 ± 2.4	3.09

[a] Adapted from Hughes et al. (1980).

was concluded that zinc can be transferred within a food chain but not biomagnified. Very little information is available on the concentration of copper in food chains. Copper is an essential trace element and used in the formation of the respiratory pigment hemocyanin.

Physiological Effects

Plants

Excessive concentrations of most heavy metals cause reductions in growth and plant productivity and in some cases death. Table 11.17 indicates some concentrations of heavy metals at which toxicity is exhibited. The reductions in growth and productivity lead in many cases to stunting and chlorosis (Hughes et al., 1980). However, the interaction of metals with plants is not well understood and is often complicated by the presence of other atmospheric pollutants such as acid rain, sulfur dioxide, and particulates. In many cases reduced productivity has been caused by iron deficiency. In these cases the metals — cadmium, copper, nickel, and zinc — have been found to interact with iron and prevent the entry of iron into the plant's metabolic processes (Hughes et al., 1980). Barber (1974) has found with the root system that the presence of heavy metals may result in limiting levels of phosphorus, potassium, and iron being present in the root tissue.

With copper, reductions in growth and primary productivity are also observed and have been attributed to chlorophyll damage (Rao et al., 1977). With algae, copper porphyrin is formed by the incorporation of copper rather than magnesium into the porphyrin nucleus utilized in the formation of chlorophyll. Copper porphyrin has little photosynthetic activity and thus primary productivity and growth drop in proportion to the amount of copper porphyrin formed. Robitaille et al. (1976) have found that lead, copper, zinc, cadmium, and arsenic all cause decreases in the chlorophyll content of lichens.

Table 11.17. Foliar Heavy Metal Concentrations which Exhibit Threshold Toxicity to Plants

Plant	Metal	Threshold Toxicity[a] (ppm)	Reference
Higher plants	Lead	1000	Lag et al. (1969)
	Zinc	100	Jordan (1975)
	Cadmium	10	Jordan (1975)
Hypogymnia physodes (lichen)	Zinc	101	Seaward (1974)
	Iron	3150	
	Lead	58	
Hypogymnia physodes (lichen)	Lead	718–918	Rao et al. (1977)
	Copper	210–405	
	Zinc	168–233	
	Cadmium	5	
Parmelia squarrosa	Lead	175–575	Rao et al. (1977)
	Copper	255–283	
	Zinc	200–260	
	Cadmium	3.5–4.7	

[a] Foliar levels.

The toxicity of lead has been shown to be due to a somewhat different mechanism but one that also involves chlorophyll. Lead released into the cytoplasm inhibits two enzymes — γ-aminolevulinic acid dehydratase and prophobilinogenase — involved in the biogenesis of chlorophyll (Hampp and Lendzian, 1974).

The development of metal-tolerant populations of plants occurs where exposure to heavy metals has occurred and a resistant population has built up by a selection process over a period of time (e.g., Antonovics et al., 1971). Ernst (1976) has summarized the mechanisms which plants have developed to combat heavy metal toxicity as: (1) the exclusion of metals from metal-sensitive sites; (2) formation of specific metal-resistant enzymes; and (3) the alteration of metabolic pathways. Plants sensitive to copper, zinc, lead, nickel, cobalt, cadmium, and manganese contain enzymes with sulfhydryl and carboxylic groups which react with the metals in different ways. Tolerance results from the presence of enzymes that do not exhibit a similar affinity for the heavy metals.

Animals

Lead. Lead is a well-known toxic agent and has been studied for an extensive period. It affects the formation of blood cells in the bone marrow and inhibits the synthesis of hemoglobin. It is interesting to note that the chemical mechanism for the inhibition of the synthesis of hemoglobin is very similar to the chemical mechanism described for the inhibition of chlorophyll formation in plants. In fact, similar enzymes in a similar pathway are involved. The pathway for the formation of hemoglobin and other hemopigments is shown in Figure 11.10. The inhibition of the two enzymes δ-aminolevulinic acid dehydratase (alad) and prophobilinogenase

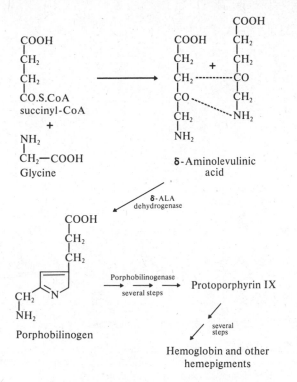

Figure 11.10.   Biosynthesis of hemoglobin by animals. (Adapted from: Ottaway, 1980).

(pgba) leads to the excretion of δ-aminolevulinic acid and prophobilinogen and the inhibition of the formation of heme and related pigments.

Mercury.   Inorganic mercury damages the liver and kidneys and interferes with the brain. Organic mercury, usually in the form of methyl or dimethyl mercury, is soluble in body tissues, circulates freely in the bloodstream, and causes damage to the brain (Waldbott, 1978).

Cadmium.   This metal and its compounds are absorbed readily into mammalian tissue. Cadmium is detrimental to many different functions of mammals (Waldbott, 1978). Ottaway (1980) has noted that mammals can sequester the protein metallothionein which contains approximately one-third of its amino acid as cystein and thus has a high proportion of – SH groups. It is likely that the sulfhydryl groups may form coordination complexes with heavy metals and thus the toxicity can be modified.

Ecological Impact

Atmospheric particulates containing heavy metals are usually deposited within a short distance of the source. Common sources include areas of dense motor vehicle traffic (e.g., lead particulates) and metal smelters. Atmospheric fallout results in an

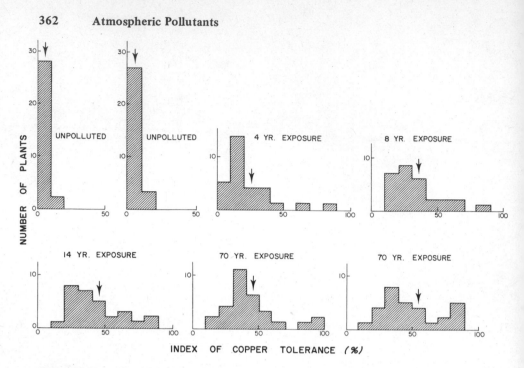

Figure 11.11.   The distribution of copper tolerance in populations of different ages around a copper refinery near Liverpool. Arrows indicate mean values. [Adapted from Wu et al. (1975). Reprinted with permission from the Longman Group Ltd.]

accumulation of metals in the litter and upper soil horizons of natural systems (Smith, 1976; Tyler, 1975a). The decomposition of organic matter, nitrogen and phosphorus mineralization rates, soil urease, and acid phosphatase activity are negatively related to copper plus zinc concentration, with copper concentrations in the range of 30–200 ppm (Tyler, 1975b). This effect may be complicated by the presence of high calcium levels in the soil (Ebregt and Boldewijn, 1977). Various other workers have shown somewhat similar effects with copper, nickel, lead, and zinc. Overall, it would be expected that these effects would produce a slower rate of remobilization of nutrients which would have an effect on the availability of nutrients to living organisms.

A study of the community structure of soil invertebrates indicates no observable effect on number or distribution of species up to 399 ppm of lead (Williamson and Evans, 1973). Strojan (1978) and Watson et al. (1976) observed a depression of litter arthropod populations due to heavy metals in soils in the vicinity of mining and smelter activities. Soil arthropods play a major role in the remobilization of nutrients and as a food source for other members of the ecosystem.

Natural populations usually contain a wide range of genotypes exhibiting variable resistance to heavy metals. The introduction of metals into these populations results in a strong selection pressure for metal tolerance. Resistant populations can develop comparatively rapidly with plants having short life spans. Such plants are grasses and some dicotyledons, many of which have successfully developed tolerant populations, but long-lived species such as trees and shrubs either cannot develop tolerant populations or take long periods. Figure 11.11 illustrates the development of a copper tolerant population of lawn grass over different periods of time.

GLOBAL EFFECTS OF ATMOSPHERIC POLLUTANTS

Carbon Dioxide

Carbon dioxide occurs in the earth's atmosphere in concentrations of about 0.03%. Although it is present in low concentrations it plays an important role in the earth's climate. Incoming solar radiation contains a range of different wavelengths but on striking the earth's surface most of the energy is ultimately converted into infrared radiation. Carbon dioxide is a strong infrared absorber and this property helps prevent infrared radiation from leaving the earth, thus playing a major role in the regulation of the earth's surface temperature. This "greenhouse" effect is influenced by the proportion of carbon dioxide in the earth's atmosphere. Rainfall, seasonal temperature variations, and sea level are all affected by changes in the earth's temperature and climate.

Since 1958, accurate observations on the carbon dioxide content in the atmosphere have been made (see Figure 11.12). Apart from these results it has been estimated that there has been a steady increase in the carbon dioxide content of the atmosphere over the last 100 years (Woodwell, 1978). Figure 11.12 shows an annual cycle on the upward trending graph with a maximum in the late winter in the northern hemisphere when photosynthesis would be expected to be at a minimum.

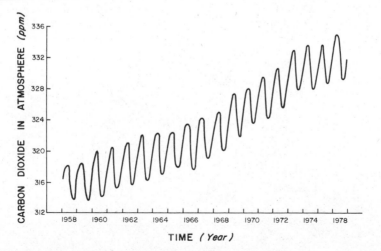

Figure 11.12.   Changes in monthly average atmospheric concentrations of carbon dioxide with time. Oscillations are caused by seasonal growth patterns. [From Woodwell (1978). Reprinted with permission from Scientific American, Inc.]

Fleagel (1979) has reported that the rate of atmospheric increase is about one-half the rate of production of carbon dioxide from the use of fossil fuels and that the remainder enters the ocean and vegetation. But the role of vegetation and the oceans in the carbon dioxide cycle is not clear at the present time. In addition, Woodwell (1978) has suggested that the extensive clearing of forests during the last 100 years has also contributed to the rise in carbon dioxide content in the atmosphere.

Climatic Effects

Many changes in the earth's climate have occurred over the geologic time scale. Over the last 50 million years global temperatures have generally declined (Fleagel, 1979). In the last 2 million years temperature cycles on time scales of about 100, 40, and 20 thousand years with an amplitude of 6–10°C have been observed. There have been variations in temperature of 1–2°C during the last 100 thousand years. The changes in mean temperature for the Northern Hemisphere since 1850 are shown in Figure 11.13 (Mitchell, 1977).

Numerical modeling by Manabe and Wetherald (1975) has indicated that the recent rise in temperature can be attributed to the buildup in carbon dioxide concentrations. Woodwell (1978) has produced curves for the relationship between fossil fuel consumption and carbon dioxide concentration in the atmosphere (see Figure 11.14). But there is uncertainty in such predictions due to our lack of knowledge of all the factors affecting atmospheric carbon dioxide content. A general idea of the significance of the upper and lower estimates in Figure 11.14 can be obained from the model of Manabe and Wetherald (1975). For example, if carbon dioxide concentrations double, average global surface temperatures are predicted to increase 2.9°C. However, in Figure 11.13 a fall in surface temperatures is indicated in the period from 1940 to the present during which carbon dioxide concentrations have continued to increase. A number of authors have suggested that this is due to the increase in particulate pollution in the atmosphere reducing incoming radiation.

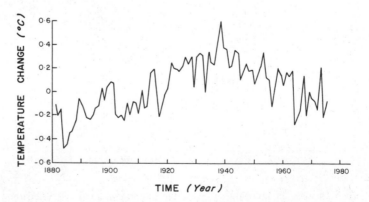

Figure 11.13. Recorded changes of annual mean temperatures of the Northern Hemisphere. [From Mitchell (1977). Reprinted with permission from the National Academy of Sciences.]

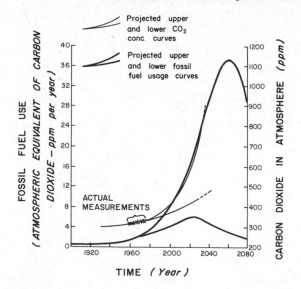

Figure 11.14. Projections of fossil fuel consumption and atmospheric carbon dioxide concentrations. [From Woodwell (1978). Reprinted with permission from Scientific American, Inc.]

Biological and Ecological Effects

It is extremely difficult to predict the effects of increased carbon dioxide levels on biological and ecological systems but an attempt has been made by Fleagel (1979). It would be expected that there would be increased primary production, in general, at the earth's surface since carbon dioxide is presently less than the optimum concentration for plant growth. This would result in a reduction in atmospheric carbon dioxide but balancing this there would be an increase in respiration due to the increased temperatures. The precipitation and evaporation regime, as well as temperature distribution over the various land masses, would be expected to be altered. Thus, agriculture and vegetation in general would be affected. In some regions this effect could be favorable, but in others adverse effects could be expected.

Trace Gases and Stratospheric Ozone

In recent years, there has been concern that trace substances introduced into the stratosphere may react with the ozone present. The substances involved include halogenated hydrocarbons, used as propellants in commercial products, and nitrogen oxide produced during combustion processes in planes or rockets (Fleagel, 1979). A possible reaction sequence, shown in Figure 11.15, may result in the removal of ozone from the upper atmosphere but the amount involved is subject to speculation. Ozone is a strong absorbent of short-wavelength ultraviolet solar

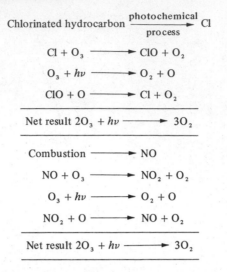

Figure 11.15. Postulated reaction sequences involving chlorinated hydrocarbons and NO which could cause loss of O_3 from the stratosphere.

radiation, preventing its entry to the earth's surface. It has been suggested that increased amounts of short-wavelength ultraviolet radiation may result in damage to biota and human health.

REFERENCES

Antonovics, J., Bradshaw, A. D., and Turner, R. G. (1971). "Heavy Metal Tolerance in Plants." In J. B. Cragg (Ed.), *Advances in Ecological Research*, Vol. 7. Academic Press, New York and London, p. 1.

Barber, S. A. (1974). "Influence of the Plant Root on Iron-Movement in the Soil." In E. W. Carson (Ed.), *The Plant Root and its Environment*. Charlottesville University Press, Charlottesville, Virginia, p. 525.

Bell, J. N. B. (1976). Effects of Ozone and Peroxyacetyl Nitrate on Plants, Some Gaseous Pollutants in the Environment, Report of a Seminar held at the Royal Society, 1974, London, p. 17.

Bell, J. N. B. and Mudd, C. H. (1976). "Sulphur Dioxide Resistance in Plants: A Case Study of *Lobium perenne*." In T. A. Mansfield (Ed.), *Effects of Air Pollution on Plants*, Society for Environmental Biology Seminar Series No. 1. Cambridge University Press, Cambridge.

Bogges, W. R. and Wixson, B. G. (1977). Lead in the Environment, Report PB-278-278, National Technical Information Service, Springfield, Virginia, p. 272.

Bowen, H. J. M. (1975). "Residence Times of Heavy Metals in the Environment." In *Proceedings of the International Conference on Heavy Metals in the Environment*, Vol. I. Institute for Environmental Studies, Toronto, p. 1.

Butler, J. D. (1979). *Air Pollution Chemistry*. Academic Press, London.

Chamberlain, A. C. (1967). Deposition of Particles to Natural Surfaces, Airborne Microbes, *Symposium of the Society of General Microbiology*, Vol. XVIII. Cambridge University Press, Cambridge, p. 138.

Committee on Biological Effects of Atmospheric Pollutants (1971). *Fluorides.* National Academy of Sciences, Washington, D.C., p. 295.

Cowling, E. B. (1980). Acid precipitation and its effects on terrestrial and aquatic ecosystems. *Ann N.Y. Acad. Sci.* **338**, 533.

Ebregt, A. and Boldewijn, J. M. A. M. (1977). Influence of heavy metals in spruce forest soil on amylase activity, carbon dioxide evolution from starch and soil respiration. *Plant Soil* **47**, 137.

Ernst, W. (1976). "Physiology and Biochemistry of Metal Tolerance." In T. A. Mansfield (Ed.), *Effect of Air Pollutants on Plants.* Cambridge University Press, Cambridge, p. 115.

European Inland Fisheries Advisory Committee (1969). Water quality criteria for European freshwater fish, Report on extreme pH values and inland fisheries. *Water Res.* **3**, 593.

Fleagel, R. G. (1979). The impact of man on climate, *Water Air Soil Pollut.* **12**, 9.

Giddings, J. and Galloway, J. N. (1976). The Effects of Acid Precipitation on Aquatic and Terrestrial Ecosystems, Lit. Rev. on Acid Precipitation, EEP − 2. Cornell University, Ithaca, New York.

Gilbert, O. L. (1970a). Further studies on the effect of sulphur dioxide on lichens and bryophytes. *New Phytol.* **69**, 605.

Gilbert, O. L. (1970b). A biological scale for the estimation of sulphur dioxide pollution. *New Phytol.* **69**, 629.

Glass, N. R. (1979). Environmental effects of increased coal utilization: Ecological effects of gaseous emissions from coal combustion. *Environ. Health Perspect.* **33**, 249.

Gordon, A. G. and Gorham, E. (1963). Ecological aspects of air pollution from an iron sintering plant at Waa Waa, Ontario. *Can. J. Bot.* **41**, 1063.

Gorham, E. (1976). Acid precipitation and its influence upon aquatic ecosystems − an overview. *Water Air Soil Pollut.* **6**, 457.

Groth, E. (1975a). Fluoride pollution. *Environment* **17**, 29.

Groth, E. (1975b). An evaluation of the potential for ecological damage by chronic low-level environmental pollution by fluoride. *Fluoride* **8**, 224.

Hampp, R. and Lendzian, K. (1974). Effect of lead irons on chlorophyll synthesis. *Naturwissenschaften* **61**, 218.

Hawkesworth, D. L. and Rose, F. (1976). *Lichens as Pollution Monitors*, Institute of Biology Studies in Biology No. 66. Edward Arnold, London.

Heath, R. L. (1975). "Ozone." In J. B. Mudd and T. T. Kozlowski (Eds.), *Responses of Plants to Air Pollution.* Academic Press, New York, p. 23.

Heck, W. W. (1968). Comments made after Taylor's paper. *J. Occup. Med.* **10**, 496.

Heck, W. W., Heagle, A. S., and Cowling, E. B. (1977). "Air Pollution: Impact on Plants." Proceedings, Soil Conservation Society of America, 32nd Meeting, p. 193.

Holdgate, M. W. (1979). Targets of pollutants in the atmosphere. *Phil. Trans. R. Soc. Lond. Ser. A* **290**, 591.

Hughes, M. K., Lepp, N. W., and Phipps, D. A. (1980). "Aerial Heavy Metal Pollution and Terrestrial Ecosystems." In A. MacFadyen (Ed.), *Advances in Ecological Research*, Vol. 11. Academic Press, London, p. 217.

Jordan, M. J. (1975). Effects of zinc smelter emissions and fire on a chesnut-oak woodland. *Ecology* **56**, 78.

Kay, C. E., Tourangeau, P. C., and Gordon, C. C. (1976). Population variation of fluoride parameters in wild ungulates from the western United States. *Fluoride* **9**, 73.

Kirchoff, W. N. (Ed.) (1977). Proceedings of the 8th Materials Research Symposium; Methods and Standards for Environmental Research, National Bureau of Standards, Special Publication 464, U.S. Government Printing Office, Washington, D.C.

Kozlowski, T. T. (1980). Impact of air pollution on forest ecosystems. *Bioscience* **30**, 88.

Lag, T. O., Hvatum, O., and Bolvika, B. (1969). An occurrence of naturally lead-poisoned soil at Kastad near Gjovik, Norway. *Nor. Geo. Unders.* **266**, 141.

Lee, J. A. and Tallis, J. H. (1973). Regional and historical aspects of lead pollution in Britain. *Nature* **245**, 216.

Leivestad, H., Hendrey, G., Muniz, P., and Snekvik, E. (1976). "Effects of Acid Precipitation on Freshwater Organisms." In F. H. Braekke (Ed.), *Impact of Acid Precipitation on Forest and Freshwater Ecosystems in Norway*. SNSF Project, Research Report FR 6/76, pp. 87–111.

Likens, G. E. (1976). Acid precipitation. *Chem. Eng. News* **54**, 29.

Malone, C., Coppee, D. E., and Miller, R. J. (1974). Localisation of lead accumulated by corn plants, *Plant Physiol.* **53**, 388.

Manabe, S. and Wetherald, R. T. (1975). The effects of doubling the CO_2 concentration on the climate of a general circulation model. *J. Atmos. Sci.* **32**, 3.

Manahan, S. E. (1975). *Environmental Chemistry*, 2nd ed. Willard Grant Press, Boston, Massachusetts, p. 416.

Manning, W. J. and Feder, W. A. (1980). *Biomonitoring Air Pollutants with Plants*. Applied Science Publishers, London.

Martin, M. H., Coughtrey, P. J., and Young, E. W. (1976). Observations on the availability of lead, zinc, cadmium and copper in woodland litter and the uptake of lead, zinc and cadmium by the woodlouse (*Oniscus asellus*). *Chemosphere* **5**, 313.

McCune, D. C. (1969). *On the Establishment of Air Quality Criteria, with Reference to the Effects of Atmospheric Fluorine on Vegetation*. Air Quality Monograph 69-3. American Petroleum Institute, New York.

Mitchell, J. M. (1977). *The Changing Climate, Energy and Climate*, Geophysics Research Board, National Academy of Science, National Research Council, Washington, D.C., p. 70.

Mudd, J. B. (1975). "Peroxyacylnitrates." In J. B. Mudd and T. T. Kozlowski (Eds.), *Responses of Plants to Air Pollution*. Academic Press, New York, p. 97.

Munshower, F. (1977). Cadmium accumulation in plants and animals of polluted and non-polluted grasslands. *J. Environ. Qual.* **6**, 411.

National Academy of Sciences (1978). *Nitrates, An Environmental Assessment.* National Academy of Sciences, Washington, D.C.

Nisbet, I. (1975). In Air Quality and Stationary Source Emission Control, Committee on Public Works, U.S. Senate, Serial No. 94-4, U.S. Government Publishing Service, Washington, D.C., Chap. 7.

Ottaway, J. H. (1980). *The Biochemistry of Pollution.* The Institute of Biology's Studies in Biology No. 123. Edward Arnold, London.

Rao, D. N., Robitaille, G., and Leblanc, F. (1977). Influence of heavy metal pollution on lichens and bryophytes. *J. Hattori bot. Lab.* **42**, 213.

Reuss, J. O. (1977). Chemical and biological relationships relevant to the effect of acid rainfall on the soil–plant system, *Water Air Soil Pollut.* **7**, 461.

Robitaille, G., Leblanc, F., and Rao, D. N. (1976). Acid rain: A factor contributing to the paucity of epiphytic cryptograms in the proximity of a copper smelter. *Rev. Bryol. Lichenol.* (in press).

Rolfe, G. L. and Haney, A. (1975). An Ecosystem Analysis of Environmental Contamination by Lead. Institute for Environmental Studies, University of Illinois, Urbana-Champaign, p. 133.

Rose, D. and Marier, J. R. (1977). *Environmental Fluoride.* National Research Council, Canada, NRCC No. 16081. Ottawa, p. 151.

Rosenberg, C. R., Hutnik, R. J., and Davis, D. D. (1979). Forest composition at varying distance from a coal-burning power plant. *Environ. Pollut.* **19**, 307.

Schlesinger, W. H. and Potter, G. L. (1974). Lead, copper and cadmium concentrations in small mammals in the Hubbard Brook experimental forest. *Oikos* **25**, 148.

Seaward, M. R. D. (1974). Some observations on heavy metal toxicity and tolerance in lichens. *Lichenologist* **6**, 158.

Siccama, T. J. and Smith, W. H. (1978). Lead accumulation in a northern hardwood forest. *Environ. Sci. Technol.* **12**, 593.

Smith, F. A. and Hodge, H. C. (1979). Airborne fluorides and man: Part 1. *Crit. Rev. Environ. Control* **8**, 293.

Smith, W. H. (1976). Lead contamination of the roadside ecosystem. *J. Air Pollut. Control Assoc.* **26**, 758.

Strojan, C. L. (1978). Forest leaf litter decomposition in the vicinity of a zinc smelter. *Ecologia (Berlin)* **32**, 203.

Sub-Committee on Ozone and Other Photochemical Oxidants (1976). Ozone and other photochemical oxidants, Vol. 2. U.S. Environmental Protection Agency, Office of Research and Development, *Environmental Health Effects Research Series* Pub. No. 600/1-76-0276 National Technical Information Service.

Tamm, C. O. and Cowling, E. B. (1977). Acidic precipitation and forest vegetation. *Water Air Soil Pollut.* **7**, 503.

Taylor, O. C. (1974). Oxidant Air Pollutant Effects on a Western Coniferous Forest Ecosystem, Annual Progress Report 1973–1974, Task D. Statewide Air Pollution Research Center, University of California, Riverside.

Treshow, M. (1971). Fluorides as air pollutants affecting plants. *Annu. Rev. Phyto pathol.* **9**, 22.

Treshow, M. and Stewart, D. (1973). Ozone sensitivity of plants in natural communities. *Biol. Conserv.* **5**, 209.

Tyler, G. (1975a). Heavy metal pollution and mineralisation of nitrogen in forest soils. *Nature (London)* **255**, 701.

Tyler, G. (1975b). "Effects of Heavy Metal Pollution on Decomposition and Mineralisation Rates in Forest Soils." In *Proceedings of the International Conference on Heavy Metals in the Environment*, Vol. II. Institute for Environmental Studies, Toronto, p. 217.

Van Hook, R. I., Harris, W. F., and Henderson, G. S. (1977). Cadmium, lead and zinc distributions and cycling in a mixed deciduous forest. *AMBIO* **6**, 281.

Varshney, C. K. and Garg, J. K. (1979). Plant responses to sulphur dioxide pollution. *Crit. Rev. Environ. Control* **9**, 27.

Waldbott, G. L. (1978). *Health Effects of Environmental Pollutants*, 2nd ed. C. V. Mosby, St. Louis, Missouri.

Watson, A. P., Van Hook, R. I., Jackson, D. R., and Reichle, D. E. (1976). Impact of a Lead Mining–Smelting Complex on the Forest Floor Litter Arthropod Fauna in the New Lead Belt Region of South-west Missouri, ORNL/NSF/EATC-30, Oak Ridge National Laboratory, Oak Ridge, Tennessee.

Wawerka, E. M., Williams, J. M., Wanek, P. L., and Olsen, J. D. (1976). Environmental Contamination from Trace Elements in Coal Preparation Waste. Report of the U.S. Environmental Protection Agency EPA/600/7-76-007, NTIS, Springfield, Illinois, p. 59.

Wesolowski, J. J., John, W., and Kaifer, R. (1973). "Lead Source Identification by Multi-Element Analysis of Diurnal Samples of Ambient Air." In E. L. Kothny (Ed.), *Trace Elements in the Environment*, Advances in Chemistry, Series 123. American Chemical Society, Washington, D.C., p. 1.

Williamson, P. and Evans, P. R. (1973). A preliminary study of the effects of high levels of inorganic lead on soil fauna. *Pedobiologia* **13**, 16.

Woodwell, G. M. (1970). Effects of pollution on the structure and physiology of ecosystems. *Science* **168**, 429.

Woodwell, G. M. (1978). The carbon dioxide question. *Scient. Am.* **238**, 34.

Wright, R. F., Torstein, D., Gjessling, E. T., Hendry, G. R., Henriksen, A., Johannessen, M., and Muniz, I. P. (1977). Impact of acid precipitation on freshwater ecosystems in Norway. *Water Air Soil Pollut.* **6**, 483.

Wu, L., Bradshaw, A. D., and Thurman, D. A. (1975). The potential for evolution of heavy metal tolerance in plants. III. The rapid evolution of copper tolerance in *Agrostis stolonifera. Heredity* **34**, 165.

12

THERMAL POLLUTION

HEATED EFFLUENTS

A large proportion of the heat which is of concern as pollution originates from electricity generation. The production of electricity involves the processes shown diagrammatically in Figure 12.1. The proportion of usable energy produced is governed by the following relationship:

$$E = \frac{\text{usable energy produced}}{\text{energy consumed}} = \left(1 - \frac{T_{\text{sink}}}{T_{\text{source}}}\right) \times 100$$

where T_{sink} is the absolute temperature of the coolant and T_{source} is the temperature of the heat source (boiler or nuclear reactor). This relationship indicates that by decreasing the sink temperature or increasing the source temperature greater efficiency can be achieved (Majewski and Miller, 1979). With flow-through cooling, the water used is derived from rivers, lakes, or reservoirs and so the sink temperature is governed by the prevailing natural conditions. However, closed-cycle cooling can lead to increased sink or intake water temperature which gives rise to reduced efficiency (see Table 12.1). Some cooling systems discharge heat to the atmosphere but little is known of the physiological or ecological effects of this. The rate of transfer of heat from an electricity generating station to water is given by the relationship

$$H = Q\rho C_p \delta_t$$

where H is the rate of waste heat transfer (J sec^{-1} or kcal sec^{-1}); Q the cooling water discharge (m^3 sec^{-1}); ρ the water density (kg m^{-3}); C_p the specific heat capacity of water (J kg^{-1} K^{-1}); and δ_t the temperature rise across condensers (°C).

This relationship indicates that discharge temperatures can be controlled by the rate of water flow. However, most electricity generating stations operate under conditions and water flow rates that produce discharges showing a temperature rise of between 6 and 16°C.

There are two major trends influencing the nature of heated effluent discharges. These are an increasing rate of construction of generating stations and the tendency for electricity generation to be concentrated into larger stations. This has led to an increase in the overall volume of heated effluent and larger point sources of discharges. In addition, there is increasing use of nuclear fuels rather than fossil fuels (Majewski and Miller, 1979). Fossil fuel stations generally exhibit energy efficiencies

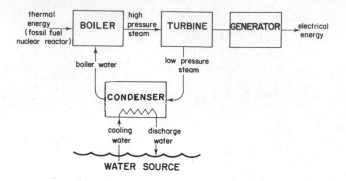

Figure 12.1.   Diagrammatic representation of an electricity generating station.

in the range of 40–42% whereas nuclear power stations are less efficient, having efficiencies in the range of 30–33% (Majewski and Miller, 1979). Thus waste heat production from nuclear power stations is larger than from conventional fossil fuel stations.

Physical and Chemical Aspects

Increased discharge water temperatures have an effect on physical properties, such as density, viscosity, vapor pressure, surface tension, gas solubility, and diffusion, which may be ecologically important. A rise in water temperature during passage through a power station can lead to supersaturation of the dissolved atmospheric gases and thus these gases may bubble out of solution. In addition, increases in the evaporation rate can result in substantial losses of water from the supply. The International Atomic Energy Agency (IAEA, 1974) has reported that 2000 m^3 hr^{-1} are lost to the atmosphere from the wet cooling towers of a nuclear power plant. Heated discharges can lead to the formation of patterns of stratification which differ substantially from those normally observed in aquatic areas.

Previously, it was shown that the efficiency of electricity generation is markedly affected by the temperatures operating in the condenser – in other words, the

Table 12.1.   Efficiency Reduction for a Nuclear Power Plant Using Different Cooling Systems in a Temperate Zone

Cooling Systems	Yearly Average Reduction (%)	Maximum Summer Daily Reduction (%)
Once through	0	0
Cooling pond	2.1	3.4
Wet cooling tower	3.1	6.8
Dry cooling tower	4.9	7.8

Source: Majewski and Miller (1979). Reprinted with permission from the United Nations Educational, Scientific and Cultural Organization.

efficiency of the condensers. Serious loss in efficiency can occur due to organism growth on the condenser surfaces. Therefore, chlorine is commonly added to intake water to inhibit biological fouling. However, chlorine can also have toxic effects on biota in the discharge area.

Atmospheric discharges from cooling towers and ponds may have a significant impact on the atmosphere in the adjacent area. Majewski and Miller (1979) have described these effects as: (1) ground level fog and icing; (2) clouds and precipitation; (3) severe weather effects; (4) plume length and shadowing; and (5) drift.

Effects on Individuals

Entrapment and impingement at cooling water intakes and passage through power plant cooling systems result in physical damage to entrained organisms (USEPA, 1976a; Schubel and Marcy, 1978). Organisms are subject to acceleration forces, abrasion, collision, sudden fluctuations in pressure, velocity sheer forces, and physical buffetting which may result in major kills (Majewski and Miller, 1979).

Small planktonic and weakly swimming pelagic organisms are most affected. Survival of entrained organisms depends on operating conditions, temperature differential between intake water and condenser, and other factors. Larger copepods, fish eggs, and larvae tend to be the most sensitive organisms and losses of these populations are reported as ranging from 70 to 100% at a variety of different power plants (Majewski and Miller, 1979). Schubel et al. (1978) have made a comparison of mortality of entrained organisms due to (1) physical buffeting and related effects and (2) thermal stress. They concluded that thermal stress is the major cause of mortality of entrained organisms.

Water temperature is one of the most important factors influencing the distribution of aquatic organisms because it regulates a wide variety of metabolic and physiological processes. Most aquatic organisms are poikilotherms and therefore have little heat production and little insulation from the environment (Spotila

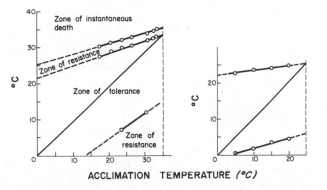

Figure 12.2.   The thermal relations of the roach (left) and the coho salmon (right) relating test and acclimation temperatures. [From Erichsen-Jones (1964). Reprinted with permission from Butterworth and Company (Publishers) Ltd.]

Table 12.2.   Some Thermal Death Points for Aquatic Organisms[a]

Organism	Acclimation Temperature (°C)	Incipient Upper Lethal Temperature[b]
Fish		
Carassius auratus	20	34.8
Cyprinus carpio	20	31–34
Lepomis macrochirus	20	31.7
Salmo trutta	14–18	25
Salmo gairdneri	11	24
Invertebrates		
Gammarus fasciatus	–	31
Ascellus intermedius	–	33
Procladius	–	30
Cryptochironomus	–	32.8
Chironomus (3 sp.)	–	34.5–35.5

[a] Compiled from Spotila et al. (1979).
[b] 1 ULT is defined as the test temperature that killed 50% of the test population over unlimited time.

et al., 1979). Thus body temperatures closely follow those in the external environment.

The effect of thermal stress on aquatic organisms has been the subject of extensive investigation over the last few decades (e.g., Krenkel and Parker, 1969; Coutant, 1970; Clark and Brownell, 1973). Some of the general characteristics of the response of fish are summarized in Figure 12.2. For any particular acclimation temperature a fish has an upper temperature at which there is a 50% lethality with indefinite exposure. This is described as the thermal death point or upper incipient lethal temperature (see Table 12.2), and there is also, at lower temperatures, a cold death point or lower incipient lethal temperature. At higher and lower temperatures than these, the organism can survive but for a limited period and this is the zone of resistance (see Figure 12.2). In between these extreme temperatures is a zone of tolerance where progressive rises in the acclimation temperature lead to corresponding, but smaller, rises in the upper thermal limit. Finally, a point is reached where the acclimation temperature and the thermal death point are the same. These features are a characteristic of each different species of fish (Erichsen-Jones, 1964). Acclimation periods are variable, depending on the size of the change from previous conditions, and can take from hours to many days.

Within the zone of tolerance indicated in Figure 12.2 there are other zones for optimal activity and for spawning (see Figure 12.3). Thus, there is a zone for overall optimal activity which is more limited than the tolerance zone.

The plots shown in Figures 12.2 and 12.3 are temperature tolerance polygons and data derived from them can be used as overall measures of thermal tolerance

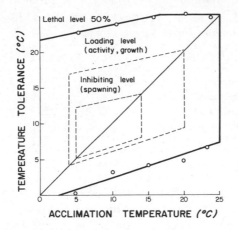

Figure 12.3. Upper and lower lethal temperatures for young sockeye salmon, the thermal zone outside which growth is poor, and the zone outside which temperature is likely to inhibit normal reproduction. [From Brett (1960).]

(e.g., Brett, 1956). The area of the polygon (in °C²) is a measure of the thermal tolerance of the organism. However, tolerance to elevated temperatures is of major interest and is related to the area enclosed by: (1) the line where the test is equal to the acclimation temperature (45° line intersecting the origin); (2) the y axis; and (3) the upper LC_{50} line (see Figure 12.4). Tolerances to elevated temperatures are shown for a variety of fish species in Table 12.3. The upper temperature tolerances of various estuarine invertebrates have also been measured (McErlean et al., 1969).

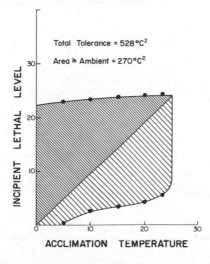

Figure 12.4. Temperature tolerance polygon for the coho salmon, *Onchyohnchus kisutch*. The contribution to the polygon of above ambient testing is shown in small cross lining. [From Brett (1952). Reprinted with permission from the Fisheries Research Board of Canada.]

Table 12.3.   Tolerance to Elevated Temperatures[a] (see Figure 12.4)

Fish	Tolerance ($°C^2$)
Brook trout	306
Mosquito fish	676
Goldfish	595
Chum salmon	260
Sockeye salmon	282

[a] Compiled from McErlean et al. (1969).

Examples of the change in thermal sensitivity of fish at different stages of their life cycle and at different seasons, due to the different natural acclimation temperatures, have been assembled by Brett (1970) (see Figure 12.5). Swimming behavior is affected by environmental temperature with many species exhibiting an optimum swimming temperature (see Figure 12.6). This would be expected to affect the organism's ability to capture prey or conversely avoid predation.

Susceptibility to thermal extremes is affected by other environmental factors (see Figure 12.7). Generally, stress caused by lowered dissolved oxygen concentrations or the presence of pollutants causes a reduction in tolerance to temperature changes (Hutchinson, 1976). Vannote and Sweeney (1980) have conducted a detailed analysis of natural thermal effects on aquatic insects due to seasons, diurnal cycle, and stream order. They suggest that an optimal temperature regime exists where adult size and fecundity are maximized.

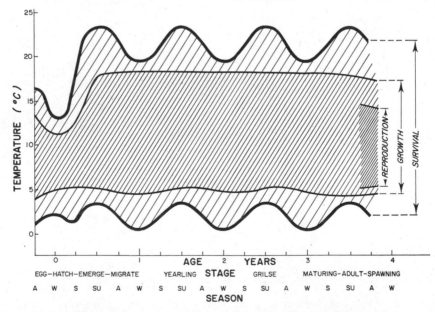

Figure 12.5.   Schematic representation for minima and maxima of different life processes in Pacific salmon. [From Kinne (1963). Reprinted with permission from University Press.]

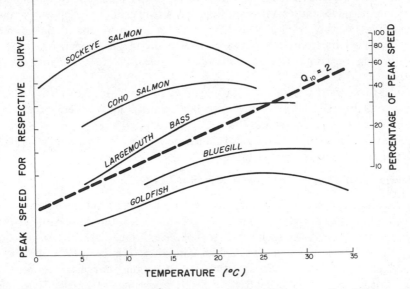

Figure 12.6. Relative cruising speeds of various fish in relation to acclimation temperature. The scale for percentage of peak speed is set to the peak of the topmost curve. [From Fry (1967). Reprinted with permission from Academic Press, Inc. (London) Ltd.]

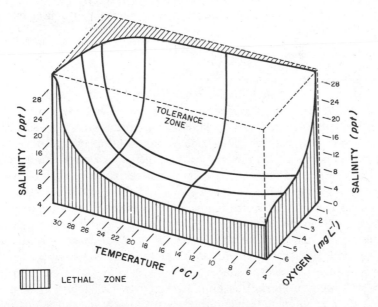

Figure 12.7. Diagram of the boundary of lethal conditions for lobsters for various combinations of temperature, salinity, and oxygen. [From Hutchinson (1976). Reprinted with permission from Academic Press Inc. (London) Ltd.]

Table 12.4 shows a comparison between optimum and sublethal and lethal upper limiting temperatures for semitropical and tropical biota. Ambient summer temperatures in many tropical and subtropical areas can be about the optimum temperature and thus only a few degrees below the upper temperature limit. In this situation, organisms can be particularly susceptible to thermal discharges (Thorhaug et al., 1978).

Similar data to that illustrated in Figure 12.3 would be expected for most aquatic organisms. Thus, if the environmental temperature was below that which was most satisfactory for activity and growth, a rise in temperature would be beneficial for that particular species. This has been observed in a number of cases and reported on by Gibbons et al. (1980).

Mechanism of Thermal Tolerance and Death

Thermal death has been attributed to the coagulation of cell proteins and the inactivation of enzymes at temperatures beyond their optimum level of activity. Organisms that have been acclimated to cold conditions have higher metabolic rates than those acclimated to warmer conditions (Haschemeyer, 1978). To accommodate this, higher concentrations of metabolic enzymes are required, giving activity increases from 16 to 200% (Haschemeyer, 1978). Thus, acclimation to different temperatures requires the adjustment of the organism to appropriate levels of enzyme activity. Temperature reduction causes a decrease in protein synthetic rate which is partially compensated by an elevation in the rate of polypeptide chain elongation. This leads to more efficient enzymic activity and thus acclimation to the lower temperatures (Haschemeyer, 1978). Presumably, somewhat similar but reverse processes operate when the organism experiences elevated temperatures.

Reproduction Growth Development and Metabolism

Brett (1969) has reported that organisms with the greatest capacity to accommodate temperature change are found in temperate environments where the temperature fluctuations (daily and seasonal) are highest. In general, there is a relationship between the lethal temperature of adult aquatic organisms, as measured in the laboratory, and their geographical distribution. Brett (1969) has identified the key environmental factors as: (1) the temperature tolerance of organisms as related to other factors such as salinity and different life stages; (2) reductions of 20% below the optimal sustained swimming speed are considered to be undesirable; (3) the metabolic rate which is related to temperature and salinity with a good supply of oxygen; and (4) growth rate which is related to temperature and other environmental factors. A decrease of more than 20% would be considered undesirable.

The response of organism growth to different water temperatures can be of a positive or negative nature depending on factors such as the organism, state of development, availability of nutrients, salinity, and many other factors (Gibbons, 1978). There is usually an optimum temperature range for growth, and above and below that range decreases are noted. However, the growth of organisms can also be affected by food supply, activity, and many other environmental factors (Cravens, 1981).

Table 12.4.   Comparison Between Optimum and Upper Limiting Temperatures for Semitropical and Tropical Biota

Biotic Group	Optimum Temperature (°C)	Thermal Stress/Limiting Temperature (°C)	
Molluscs	26.7[a]	31.4	50% species exclusion
Echinoderms	27.2[a]	31.8	50% species exclusion
Coelenterates	25.9[a]	29.5	50% species exclusion
Porifera	24.0[a]	31.2	50% species exclusion
Fouling community (larval settlement)	25.4–27.8	28.0	50% reduction
Lytechinus variegatus growth and gonadol development	27.0	29.9	50% reduction gonadol volume
		32.0	86 hr TL$_{50}$
Thallasia testudinum productivity	30.0	31.0	(daily avg.) long-term decreased growth
Survival of larval fish to 12 hr post-hatch	28.3	30.32	tolerance limit

Source: USEPA (1976b).
[a] Temperature for high species diversity.

The final body size is also affected by the prevailing temperature regime and is related to the optimal temperature (Gibbons et al., 1980). Studies of discharges from an electricity generating plant on the Savannah River indicated that with fish, turtles, and some insects, increased temperatures led to increased body size at maturity. With poikilotherms, increases in environmental temperature will lead to corresponding increases in body temperature and thus increases in the metabolic rate. These higher metabolic rates would be expected to cause changes in such factors as body fat content, overall live weight, calorific value, and so on (Gibbons et al., 1980).

Gonad development and spawning are highly temperature dependent and species specific and a substantial amount of information is available on factors affecting these phenomena (Cravens, 1981). In addition, embryonic and larval development within organisms occur within a limited range of temperatures (Cravens, 1981). With eggs, the time for hatching is systematically related to the temperature with progressive increases at lower temperatures (McLaren, 1966). De Sylva (1969) has reported that since eggs and juvenile stages of fish have limited capacity for avoiding unfavorable environments, the temperature requirements are generally much narrower than those for the adults. Thus thermal additions would be expected to have a comparatively large effect on the reproduction of aquatic species. However, Gibbons et al. (1980) have found that areas subject to thermal discharges for periods of 26–28 yr contain organisms that show evidence of genetic adaptation. With fish and insects there is evidence for selection of genotypes with tolerance to higher temperatures. In fact, larvae of the genus *Libellula*, collected from pools along a thermal effluent stream, exhibited a sequence of higher lethal temperatures. In support of this, electrophoretic techniques have provided evidence of these genetic changes, which have been related to the relative preponderance of the malate dehydrogenisms isozymes. Alternatively, some organisms in different thermal regimes showed no evidence of genotypic selection when examined by the electrophoretic technique. Quite a large number of behavioral characteristics relating to feeding, emergence time, movement, burrowing activity, swimming speed, and migration are affected by altered temperatures (see Talmage and Coutant, 1980 and Cravens, 1981).

Effects on Ecosystems and Populations

Dispersal

Changes in water temperature result in short- or long-term compensatory geographical movements with poikilotherms. The optimum temperature range for swimming, reproduction, and a number of other characteristics, is presumably sought out by organisms when the temperature regime becomes unfavorable. Adverse temperature changes can occur in natural situations with stratification, seasonal changes, and numerous other natural effects, as well as from pollution.

Congregations of fish and alligators have been noted in winter at thermal discharges with some fish (e.g., the large mouth bass) being attracted to the warm

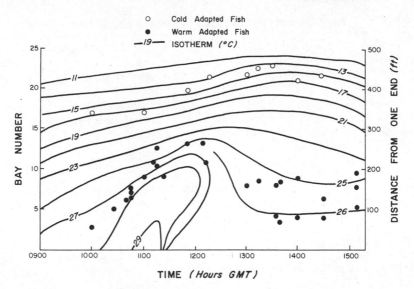

Figure 12.8.   Positions of different thermally adapted bream shoals and distribution of temperatures in a winding channel. [From Alabaster (1969). Reprinted with permission from Vanderbilt University Press.]

effluent in the winter and repelled in summer (Gibbons et al., 1980). Alabaster (1963) has experimentally demonstrated that fish move to the most suitable temperature conditions according to their acclimation history (see Figure 12.8). Other results by Alabaster (1963) have indicated that natural populations of fish move in response to water temperatures in a similar manner. In addition, Gibbons et al. (1980) have noted that water temperature changes, as a result of thermal discharges, affect water fowl populations.

Primary Producers

Aquatic areas contain a wide range of aquatic algal species each with its own temperature tolerance and optimum range (Patrick, 1969). These characteristics are an important factor governing the occurrence of algae in natural aquatic areas (see Table 12.5). Thus there will be an ecological response to changes in water temperature in terms of algal community composition. Figure 12.9 shows the change of dominance in a mixed algal community with changes in temperature.

Attached aquatic plants lack the capacity to move and thus can be more vulnerable to thermal changes. These plants can be of considerable ecological importance in the tropics and subtropics where estuarine and coastal ecosystems are often heavily dependent on detritus-based food chains (Blake et al., 1976). Thorhaug et al. (1978) have shown that tropical and subtropical communities of seagrass may be very sensitive to elevated temperatures. With *Thalassia* seagrass communities, it was found that the minimum summer water temperature is 30°C and temperature elevations of 4°C cause damage. This occurs at all latitudes, with less damage as a result of lower temperature rises, and in communities more remote from the tropics.

Table 12.5.   Algal Occurrence as Related to Environmental Temperature

Achnanthes marginulata	26–41.5°C	*Phormidium valderianum*	48°C
Cocconeis schlettum	36–34°C	*P. corium*	48°C
Diploneis interrupta	48°C	*P. angustissima*	47°C
D. oculata	20–30°C	*Lyngbya sp.*	48.2°C
Eunotia tenelia	25°C	*Oscillatoria chalybaea*	47°C
Mastogloia smithii	20–25°C	*O. proboscidea* var. wests	41.5°C
Navicula variostriata	11–14°C	*O. tenuis*	44°C
Nitzschia filiformis	31–35°C	*O. sancta*	41.5°C
Nitzschia tryblionella	10–25°C	*Scytonema varium*	47.7°C

Source: Schwabe (1936). Reprinted with permission from E. Schweizbartsche Verlagsbuchhandlung.

Secondary Production

An increase in temperature introduces additional energy to an aquatic system but in a form that is not readily utilized by organisms. However, if the existing temperature regime is below optimum, increased productivity can result from temperature elevation. This can occur with primary producers, and the flow-on of energy gives increased secondary production. Gibbons (1976) has outlined a number of examples where the productivity of aquatic ecosystems has been increased by increases in temperature. In many of these cases an extension of the favorable summer seasonal conditions has been caused by the addition of heated water to the system.

Changes in the thermal regime may alter interspecific relationships in a number of ways. Elevated temperatures can eliminate a predator species, a prey species, or even a parasite, resulting in significant alterations to other species. Although these interactions are believed to occur, there are very few cases where they have been accurately and quantitatively identified.

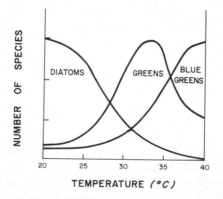

Figure 12.9.   Response of a mixed algal community to a gradual increase in temperature. [From Cairns (1969). Reprinted with permission from the National Resources Institute of the University of Maryland.]

Table 12.6.   Variations in Aquatic Communities Due to a Thermal Discharge at Yallourn into the Latrobe River, Victoria

Sampling zone (in Relation to the Discharge Point)	Productivity (kg ha^{-1})	Fish Population (Percentage Composition by Weight)			
		Brown Trout	Blackfish	Carp	Smelt
73 km upstream	56.8	83	17	0	0
37 km upstream	18.2	34	65	0	1
15 km upstream	34.3	47	52	0	1
1 km downstream	2.1	0	0	100	0
17 km downstream	6.1	0	0	100	0
80 km downstream	45.2	0	0	98.2	1.5

Source: Connell (1981). Reprinted with permission from the University of Queensland Press.

Community Structure, Species Composition, and Diversity

Thermal effluents cause alterations in the species composition of aquatic areas by the elimination or reduction of some species sensitive to temperature changes. On the other hand, some species may increase due to their improved survival and growth at the elevated temperatures. A good example of tolerance to elevated temperature regimes is indicated by the prolific growth of blue-green algal mats (see Figure 12.9) which occur where green algae, macrophytes, and other plants cannot survive (Gibbons et al., 1980). In addition, Sharitz et al. (1974) have found that trees comprise 40% of the species present in an undisturbed swamp habitat but less than 10% in areas subjected to thermal stress. Gibbons et al. (1980) have reported that this change in tree density has resulted in changes in the dependent bird population.

Gibbons et al. (1980) have also reported that the number of species of aquatic insects in thermally stressed areas is lower than comparable unstressed areas. Connell (1981) has reported a number of variations in aquatic communities due to thermal discharge into a river in Australia (see Table 12.6). There is a loss in productivity in the stream around the thermal discharge and also a loss in species sensitive to elevated temperatures. At the same time there has been an increase in the number of fish tolerant of elevated temperatures.

Continuous exposure of a community to higher than normal temperatures usually results in a reduction in organism numbers as well as species diversity and evenness. Synergistic effects may occur where pollutants, salinity, and dissolved oxygen, as well as temperature, may interact together. Generally, organisms suffering temperature stress are less tolerant of the presence of pollutants in aquatic systems.

DISCHARGES FROM DAMS AND RESERVOIRS

Outflow water is usually taken from the lower levels of dams and reservoirs and discharged downstream of the dam wall. This water is often cooler (up to 10–15°C)

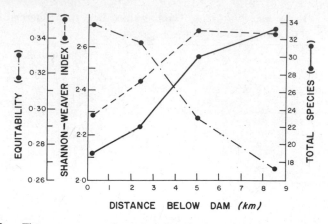

Figure 12.10.   Three measures of species diversity as a function of the distance below Cheesman Dam, Colorado, where unpolluted streams have Shannon–Weaver diversity indices between 3 and 4. [From Ward (1976).]

than inflow water due to stratification. Such discharges can cause low-temperature thermal stress on stream communities but usually the lower lethal limit is not reached. Ward (1976) has described the sublethal effects on invertebrate community structures as: (1) reduction of niche overlap and a shift toward an equilibrium community as a consequence of reduced environmental fluctuation; (2) more intense competition associated with greater productivity; (3) elimination of major invertebrate predators; and (4) failure of the limited temperature range to provide optimal temperatures for various physiological processes. For example, Connell (1981) has reported that in Australia dam construction has resulted in a reduction in native fish species since the stream water does not reach normal spawning temperature for these fish. Ward (1976) has reported a reduction in species below a dam site in Colorado due to low-temperature stress with recovery not complete 9 km from below the dam (see Figure 12.10).

REFERENCES

Alabaster, J. S. (1963). The effect of heated effluents on fish. *Int. J. Air Water Pollut.* **7**, 541.

Alabaster, J. S. (1969). "Effects of Heated Discharges on Freshwater Fishes in Britain." In P. A. Krenkel and F. L. Parker (Eds.), *Biological Aspects of Thermal Pollution.* Vanderbilt University Press, Nashville, p. 354.

Blake, N. I., Doyle, L. J., and Pyle, T. E. (1976). "The Macro-Benthic Community of a Thermally Altered Area of Tampa Bay Florida." In G. W. Esch and R. W. McFarlane (Eds.), *Thermal Ecology II. Proceedings of a Symposium in Augusta Georgia, April 1975.* U.S. Energy Research and Development Administration, U.S. Government Printing Office, Washington, D.C., p. 296.

Brett, J. R. (1952). Temperature tolerance in young Pacific salmon, genus *Oncorhynchus. J. Fish Res. Board Can.* **9**, 265.

Brett, J. R. (1956). Some principles in the thermal requirements of fishes. *Q. Rev. Biol.* **31**, 75.

Brett, J. R. (1960). "Thermal Requirements of Fish – Three Decades of Study, 1940–70." In C. M. Tarzwell (Ed.), *Biological Problems in Water Pollution, Transactions of the 1959 Seminar.* R. A. Taft, Sanit. Eng. Center Tech. Report W60-3, Cincinnati, p. 110.

Brett, J. R. (1969). Temperature and fish. *Chesapeake Sci.* **10**, 275.

Brett, J. R. (1970). "Temperature – Fishes: Functional Responses." In O. Kinne (Ed.), *Marine Ecology.* Vol. 1, *Environmental Factors.* John Wiley & Sons, New York, p. 515.

Cairns, Jr., J. (1969). The response of freshwater protozoan communities to heated wastewaters. *Chesapeake Sci.* **10**, 177.

Clark, J. and Brownell, W. (1973). Electric Power Plants in the Coastal Zone: Environmental Issues. Special Publication No. 7, American Littoral Society, Highlands, New Jersey.

Connell, D. W. (1981). *Water Pollution – Causes and Effects in Australia and New Zealand.* University of Queensland Press, Brisbane.

Coutant, C. C. (1970). Biological aspects of thermal pollution. I. Entrainment and discharge canal effects. *Crit. Rev. Environ. Control* **1**, 341.

Cravens, J. B. (1981). Thermal effects. *J. Water Pollut. Control Fed.* **53**, 949.

de Sylva, D. P. (1969). "Theoretical Considerations of the Effects of Heated Effluents on Marine Fishes." In P. A. Krenkel and F. L. Parker (Eds.), *Biological Aspects of Thermal Pollution.* Vanderbilt University Press, Nashville, p. 229.

Erichsen-Jones, J. R. (1964). *Fish and River Pollution.* Butterworths, London.

Fry, F. E. J. (1967). "Responses of Vertebrate Poikilotherms to Temperature." In A. H. Rose (Ed.), *Thermobiology.* Academic Press, New York, p. 375.

Gibbons, J. W. (1976). "Thermal Alteration and the Enhancement of Species Populations." In G. W. Esch and R. W. McFarlane (Eds.), *Thermal Ecology II. Proceedings of a Symposium in Augusta, Georgia, April 1975.* U.S. Energy Research and Development Administration, U.S. Government Printing Office, Washington, D.C., p. 27.

Gibbons, J. W. (1978). Effects of thermal effluent on body condition of large-mouth bass. *Nature (London)* **274** (5670), 470.

Gibbons, J. W., Sharitz, R. R., and Brisbin, I. L. (1980). Thermal ecology research at the Savannah River plant: A review. *Nucl. Safety* **21**, 367.

Haschemeyer, A. E. V. (1978). "Protein Metabolism and its Role in Temperature Acclimation." In D. C. Malins and J. R. Sargent (Eds.), *Biochemical and Biophysical Perspectives in Marine Biology*, Vol. 4. Academic Press, London, p. 29.

Hutchinson, V. H. (1976). "Factors Influencing the Tolerances of Individual Organisms." In G. W. Esch and R. W. McFarlane (Eds.), *Thermal Ecology II. Proceedings of a Symposium in Augusta, Georgia, April 1975.* U.S. Energy Research and Development Administration, U.S. Government Printing Office, Washington, D.C., p. 10.

IAEA (1974). Thermal Discharges at Nuclear Power Stations – Their Management and Environmental Impact. IAEA Tech. Reports Serial No. 155. International Atomic Energy Agency, Vienna, p. 155.

Kinne, O. (1963). The effects of temperature and salinity on marine and brackish water animals I. Temperature. *Oceanogr. Mar. Biol. Ann. Rev.* **1**, 309.

Krenkel, P. A. and Parker, S. L. (Eds.) (1969). *Biological Aspects of Thermal Pollution.* Vanderbilt University Press, Nashville.

Majewski, W. and Miller, D. C. (1979). Predicting Effects of Power Plant Once-through Cooling on Aquatic Systems. IHP Working Group 6.2, Report on the Effects of Thermal Discharges, United Nations Educational, Scientific and Cultural Organization, Paris.

McErlean, A. J., Mihursky, J. A., and Brinkley, H. J. (1969). Determination of upper temperature tolerance triangles for aquatic organisms. *Chesapeake Sci.* **10**, 293.

McLaren, I. A. (1966). Predicting development rate of copepod eggs. *Biol. Bull.* **131**, 457.

Patrick, R. (1968). "Some Effects of Temperature on Freshwater Algae." In P. A. Krenkel and F. L. Parker (Eds.), *Biological Aspects of Thermal Pollution.* Vanderbilt University Press, Nashville, p. 161.

Schubel, J. R. and Marcy, B. C. (1978). *Power Plant Entrainment — A Biological Assessment.* Academic Press, New York, p. 271.

Schubel, J. R., Coutant, C. C., and Woodhead, P. M. J. (1978). "Thermal Effects of Entrainment." In J. R. Schubel and B. C. Marcy (Eds.), *Power Plant Entrainment — A Biological Assessment.* Academic Press, New York, p. 19.

Schwabe, G. H. (1936). Contribution to the understanding of the Iceland thermal biotope. *Arch. Hydrobiol. Suppl. Bd.* **6**(2), 161.

Sharitz, R. R., Irwin, J. E., and Christy, E. J. (1974). Vegetation of swamps receiving reactor effluents. *Oikos* **27**, 7.

Spotila, J. R., Terpin, K. M., Koons, R. R., and Bonati, R. L. (1979). Temperature requirements of fishes from Eastern Lake Erie and the Upper Niagara River: A review of the literature. *End. Biol. Fish.* **4**, 281.

Talmage, S. S. and Coutant, C. C. (1980). Thermal effects. *J. Water Pollut. Control Fed.* **52**, 1575.

Thorhaug, A., Blake, N., and Schroeder, P. B. (1978). The effect of heated effluents from power plants on seagrass (*Thalassia*) communities quantitatively comparing estuaries in the sub-tropics to the tropics. *Mar. Pollut. Bull.* **9**, 181.

USEPA (1976a). Development Document for Best Technology Available for the Location, Design, Construction and Capacity of Cooling Water Intake Structures for Minimizing Adverse Environmental Impact. U.S. Environmental Protection Agency, No. EPA 440/1-76/015-a. U.S. Government Printing Office, Washington, D.C.

USEPA (1976b). Quality Criteria for Water. U.S. Environmental Protection Agency, 0-222-094. U.S. Government Printing Office, Washington, D.C., p. 256.

Vannote, R. L. and Sweeney, B. W. (1980). Geographic analysis of thermal equilibria: A conceptual model for evaluating the effect of natural and modified thermal regimes on aquatic insect communities. *Am. Naturalist* **115**, 667.

Ward, J. W. (1976). "Effects of Thermal Constancy and Seasonal Temperature Displacement on Community Structure of Stream Macro-Invertebrates." In

G. W. Esch and R. W. McFarlane (Eds.), *Thermal Ecology II. Proceedings of a Symposium in Augusta, Georgia, April 1975.* U.S. Energy Research and Development Administration, U.S. Government Printing Office, Washington, D.C., p. 302.

13

RADIONUCLIDES

All elements are made up of mixtures of isotopes. Most isotopes are stable but some, because of the nature of the nucleus, are unstable and subject to nuclear disintegration. When these radioactive isotopes — radionuclides — disintegrate, they produce radiation of various kinds as described in Table 13.1. The radiation basically consists of two types. The first type is short-wavelength electromagnetic radiation, which takes the form of γ-rays and X-rays, with shorter wavelengths than ultraviolet radiation and consequently higher energy. The second type of radiation consists of particles and takes three forms: α- and β-rays and neutrons as described in Table 13.1.

The rate of decay or disintegration is different for each radionuclide. However, for a given radionuclide the rate of decay is constant and independent of temperature, pressure, physical form, or chemical combination. In all cases, the rate of decay follows first-order kinetics and thus

$$N = N_0 e^{-\lambda t}$$

where N is the number of atoms at any time t; N_0 the initial number of atoms at time 0; and λ the radioactive decay constant.

The rate of decay of any radionuclide is a characteristic of that element and is commonly measured as the half-life, $t_{1/2}$, which is constant. It can be simply demonstrated that [see Chapter 2, Eq. (5)]

$$t_{1/2} = \frac{0.693}{\lambda}$$

When an isotope disintegrates it gives rise to a "daughter" element. The daughter could be a stable element and the process of radioactive decay would stop at that point, or it could be an unstable element subject to further disintegration. If unstable, radioactive decay of this daughter element would occur and give rise to a stable or unstable element. If it was an unstable element, a radioactive decay chain of elements would occur. Table 13.2 shows the uranium radioactive decay series.

Biological effects of radiation result from the amount of radiation emitted and also the amount of radiation absorbed by an organism. Measures of both of these factors are outlined in Table 13.3. Of particular importance is the fact that different forms of radiation have different biological effects and thus the energy of radiation

Table 13.1. Types of Radiation Emitted by Radioactive Substances

Type	Nature	Energy and Relative Biological Effectiveness (RBE)	Range	Source
α-Rays	$^4\text{He}^{2+}$; charged helium nuclei	High, RBE ≈ 20	Approximately 5 cm in air; stopped by thin sheets of many solids	Disintegration of unstable isotopes of elements with atomic weight > 150
β-Rays	Electrons (or positrons)	Fairly high, RBE ≈ 1	Approximately 3 m in air; stopped by thin sheets of many solids	Light or heavy nuclei
γ-Rays	Electromagnetic radiation of short wavelength (ca. 0.0001–0.01 Å)	Fairly high, RBE ≈ 1	Approximately 4 m in air; and several centimeters in lead	Emitted on nuclear disintegration or when a nucleus captures another particle
Neutrons	Particles of unit mass and no charge	Fairly high, RBE ≈ 5	Comparatively long range in air	Liberated when susceptible elements are bombarded with α- or γ-rays from the heaviest elements. No significant natural emitters.

Table 13.2. The Uranium Radioactive Decay Series

Isotope	Historical Name	Half-Life	Radiation	α Energy (MeV)[b]	Remarks
^{238}U	Uranium I	4.5×10^9 yr	α		β-rays in uranium ore come from these
^{234}Th	Uranium X$_1$	24.1 days	β, γ		
^{234}Pa	Uranium X$_2$	1.18 min	β, γ		
^{234}U	Uranium II	2.50×10^5 yr	α, γ		
^{230}Th	Ionium	7.6×10^4 yr	α		End up in tailings
^{226}Ra	Radium	1.620 yr	α		
^{222}Rn	Radon	3.82 days	α	5.49	Gas collects on dust in mine
^{218}Po	Radium A[a]	3.05 min	α	6.00	β- and γ-rays in ore from these
^{214}Pb	Radium B	26.8 min	β, γ		
^{214}Bi	Radium C	19.7 min	β, γ		
^{214}Po	Radium C'[a]	2.7×10^{-6} min	α	7.69	
^{210}Pb	Radium D	22.0 yr	β, γ		May be used to monitor ^{222}Rn exposure
^{210}Bi	Radium E	5.0 days	β		
^{210}Po	Radium F	138.4 days	α	5.30	
^{206}Pb	Radium G	Stable			

Source: Fry (1975). Reprinted with permission from Atomic Energy in Australia.

[a] Ra-A and Ra-C' constitute the α hazard in exposure to ^{222}Rn and its daughters.

[b] MeV = million electron volts. It is a measure of the energy with which radiation is emitted from radioactive isotopes. A 5-MeV α particle will penetrate some 40 μm of soft tissue.

Table 13.3.   Radiation Units

Type	Units	Measure
Radioactivity	Curie (Ci) (often used as mCi, i.e., 10^{-3} Ci and μCi, i.e., 10^{-6} Ci)	Radiation given off by 1 g radium: 3.7×10^{10} disintegrations per second
Radioactivity	Becquerel (Bq)	1 disintegration per second
Radiation absorbed	rad	100 ergs g^{-1} of irradiated material
Radiation absorbed for human safety purposes	rem (radiation equivalent man) SI unit the sievert (Su) = 100 rem	Dose in rads × quality factor or relative biological effectiveness

emitted is not relevant, in itself, as a measure of biological effects. The *relative biological effectiveness*, as this factor is known in biology, or *quality factor*, as it is called in human health, is used to measure the amount of a particular type of radiation which is necessary to produce a given biological effect. This, combined with the radiation dose, provides a relevant measure of the potential biological effects of radiation absorbed by a living organism.

NUCLEAR ENERGY USAGE

Nuclear energy is used for two principal purposes: weapons and electricity generation. It can be derived from two different types of nuclear processes. Firstly, fission, the splitting of elements with large atoms into smaller fragments with the production of energy, and secondly, fusion, the combination of lighter elements to produce heavier elements also with the production of energy. Figure 13.1 shows a simplified set of reaction sequences for some of the principal fission and fusion processes. (Schultz and Whicker, 1980). These reactions are accompanied by a large release of energy which can be used in various ways.

Uranium fission

$$^{235}_{92}U + ^{1}_{0}n \longrightarrow ^{236}_{92}U \longrightarrow ^{136}_{56}Bd + ^{85}_{36}Kr + 3^{1}_{0}n$$

$$^{238}_{92}U + ^{1}_{0}n \longrightarrow ^{239}_{92}U \longrightarrow ^{239}_{93}Np + ^{0}_{-1}e \longrightarrow ^{239}_{94}Pu + ^{0}_{-1}e$$

Nuclear fusion

$$^{2}_{1}H + ^{2}_{1}H \longrightarrow ^{3}_{2}He + ^{1}_{0}n$$

$$^{1}_{0}n + ^{6}_{3}Li \longrightarrow ^{4}_{2}He + ^{3}_{1}H$$

$$^{3}_{1}H + ^{2}_{1}H \longrightarrow ^{4}_{2}He + ^{1}_{0}n$$

$$^{2}_{1}H + ^{2}_{1}H \longrightarrow ^{3}_{1}H + ^{1}_{1}H$$

$$^{1}_{1}H + ^{7}_{3}Li \longrightarrow ^{4}_{2}He + ^{4}_{2}He$$

Figure 13.1.   Simplified principal reactions used to obtain nuclear energy.

Uranium-235 occurs as 0.7% of natural uranium while the rest of the uranium is principally uranium-238. The uranium-235 is readily fissionable and undergoes the reaction shown in Figure 13.1 on bombardment with neutrons. Uranium-238 is not readily fissionable but can be converted to plutonium by the process shown in Figure 13.1. Plutonium-239 is readily fissionable and serves as an excellent base for weapons and possibly as a fuel for nuclear power generation. With uranium-235 the neutrons produced by the external bombardment produce further neutrons (see Figure 13.1) and these impact other uranium-235 nuclei to continue the process and thus give a chain reaction. This chain reaction can occur very rapidly and be used for nuclear weapons.

The rate of reaction of uranium-235 can be controlled by inserting control rods, or moderators, into the fissionable material. These rods absorb neutrons and thus can control the rate of reaction by controlling the availability of neutrons. The concentration of uranium-235 (0.7%) in natural uranium is too low for fuel purposes and so this material must be enriched with uranium-235 for use as a fuel.

The reaction of uranium-235 occurs with the release of large amounts of radiation and heat. This heat can be used to generate steam and run a turbine to generate electricity. This process is the basis of nuclear electricity generating plants throughout the world.

Currently, nuclear fusion (see Figure 13.1) is only used in weapons. There are no electricity generating plants operating by the fusion process, although research is presently investigating this possibility.

RADIOACTIVE INPUTS INTO THE ENVIRONMENT

Many natural radionuclides are present in all parts of the natural environment and consequently there is a natural background of radioactivity (Haury and Schikarski, 1977). The distribution of radioactivity in the natural environment is reasonably well understood and summarized in Table 13.4. This radioactivity is distributed in a reasonably homogeneous fashion throughout the environment. In this way radioactivity does not usually reach particularly high levels in any specific area. In addition, cosmic radiation can result in the production of radionuclides which add to the natural background activity.

Nuclear reactions from weapons testing and electricity generation and related activities have resulted in a new and significant source of radioactivity in the environment (see Table 13.5). Schultz and Whicker (1980) have reported that weapons testing up to 1973 has resulted in far larger discharges to the environment than nuclear power generation.

Nuclear power generation utilizing nuclear fission reactions based on uranium require a cycle of nuclear fuel activities. Figure 13.2 shows the nuclear fuel cycle from mining and milling through to final storage of spent fuel elements (Schultz and Whicker, 1980). It can be seen that radioactive materials are discharged to the environment at a number of points in this cycle. Mining, milling, refining, and conversion to uranium hexafluoride involve the production of radioactive tailings

Table 13.4.   Average Concentrations and Inventories of Some Natural Radio-nuclides[a]

Nuclide	$t_{1/2}$ (yr)	Decay Type	Average Concentration (Ci g^{-1} or m^3)	Inventory (Ci)
		Lithosphere		
^{14}C	5500	β	4.46×10^{-4}	1.5×10^{8}
^{40}K	1.3×10^{9}	β, γ	1.69×10^{-11}	1.5×10^{15}
^{87}Rb	4.8×10^{10}	β	7.25×10^{-12}	6.5×10^{14}
^{238}U	4.5×10^{9}	α, γ	1.65×10^{-12}	1.5×10^{14}
^{230}Th	8.0×10^{4}	α, γ	3.88×10^{-12}	3.5×10^{14}
^{210}Pb	20	β, γ	3.19×10^{-12}	2.8×10^{14}
		Oceans		
^{3}H	12.26	β	1.0×10^{-12}	1.37×10^{9}
^{87}Pb	4.8×10^{10}	β	2.8×10^{-12}	3.8×10^{9}
^{40}K	1.3×10^{9}	β, γ	3.3×10^{-10}	4.5×10^{11}
^{238}U	4.5×10^{9}	α, γ	1.0×10^{-12}	1.4×10^{9}
		Atmosphere		
^{3}H	12.26	β	5.0×10^{-13}	3.0×10^{6}
^{14}C	5500	β	1.1×10^{-12}	4.0×10^{6}
^{222}Rn	3.825 days	α	1.0×10^{-10}	$\sim 5 \times 10^{7}$

[a] Compiled from Haury and Schikarski (1977). Reprinted with permission from the International Atomic Energy Agency.

and gaseous discharges. Enrichment, conversion, and fuel fabrication result in discharges to air and water of fluorine compounds as well as enriched uranium materials and also the production of solid depleted uranium for storage. Power generation results in further production of solid spent fuel elements for storage as well as atmospheric and water discharges as outlined in Table 13.5.

Probably the most significant substances discharged as a result of this process are tritium (hydrogen-3) and krypton-85. However, over 80 products are generated during the slow fission of uranium-235 and plutonium-239 in nuclear power generators. Plutonium-239 is retained in the spent elements, the storage of which, together with other solid wastes, is the subject of a large body of work and considerable controversy. There are about ten nuclear fuel reprocessing plants throughout the world producing plutonium which can be used in weapons or can be used for nuclear power plant operation. Reprocessing also results in the production of uranium for reuse in electricity generation. Cesium-134 and -137 are also produced as fission products in fuel.

In addition to radioactive materials produced in the fuel itself, neutron activation of other materials external to the fuel can occur. Neutron activation of carbon dioxide, graphite, and air results in the production of carbon-14, oxygen-16, nitrogen-13, nitrogen-16, and argon-41. However, Schultz and Whicker (1980) have reported that the greatest concentration of radioactive materials would be

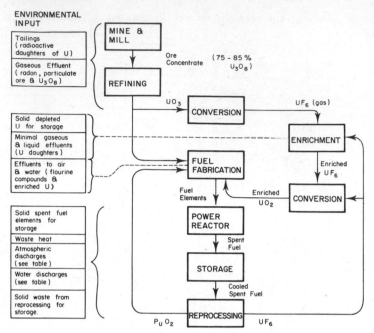

Figure 13.2. Uranium fuel cycle. [Compiled from Schultz and Whicker (1980).

Table 13.5. Estimates of Radioactivity Released to the Environment by Weapons Testing, Nuclear Power Plants, and Fuel Reprocessing Plants

Nuclide	$t_{1/2}$ (yr)	Decay Type	Cumulative Releases (Ci)
Atmosphere – Weapons Tests (to 1973)			
^{3}H	12	β	3.0×10^9
^{85}Kr	11	β	1.3×10^6
^{90}Sr	28	β	2.1×10^7
^{106}Ru	1.0	β, γ	4.2×10^8
^{135}Cs	2.0×10^6		4.7×10^2
Atmosphere – Power and Reprocessing Plants (1975–2000)			
^{3}H	12	β	8.8×10^7
^{85}Kr	11	β	8.0×10^8
^{14}C	5500	β	3.0×10^5
^{129}I	1.6×10^7	β, γ	1.2×10^4
Water – Power and Reprocessing Plants (1975–2000)			
^{3}H	12	β	3.9×10^9
^{60}Co	5.26	β, γ	2.1×10^2
^{134}Cs	2.1	β, γ	2.1×10^2
^{137}Cs	30	β, γ	1.9×10^3

[a] Compiled from Haury and Schikarski (1977).

expected to occur in aquatic systems rather than terrestrial. There are a wide range of radionuclides in liquid effluents from power plants. Many of these occur due to neutron activation of impurities in water passing through the plant although uranium contamination of this water can occur. The most important of these neutron activation products are manganese-56, copper-64, sodium-24, chromium-51, neptunium-239, arsenic-76, silicon-31, cobalt-58, cobalt-60, and zinc-65.

FACTORS AFFECTING DISTRIBUTION OF RADIONUCLIDES IN THE ENVIRONMENT

Radionuclides distributed to the atmosphere and terrestrial systems consist largely of inert gases and are fairly rapidly diluted to low concentrations which do not interact with biota to any significant extent. On the other hand, radionuclides discharged to aquatic environments do not disperse rapidly and are in fact rapidly accumulated by sediments in the immediate vicinity of the discharge. For example, Booth (1976) found that cobalt-60 and cesium-137 discharged from the Yankee Nuclear Power Station exhibited distribution patterns which could be explained, almost completely, by assuming that there was complete accumulation of these substances in the sediments. In marine systems Hetherington (1976), in studies on the Irish Sea, found that 95% of the plutonium introduced into the sea is rapidly lost from the water phase to the sediments. Similar behavior has been observed by Emery and Clopfer (1976) for freshwater systems. In studies of a freshwater pond receiving low levels of plutonium wastes, they found that the sediments, seston, and organic flock covering the sediments were the major sink for plutonium. On examining other parts of the freshwater ecosystem these authors concluded that plutonium and related elements were not mobile in the ecosystem.

Foster and McConnon (1972) conducted extensive studies on the behavior of various radionuclides present in low levels in the Columbia River as a result of discharges from the Hanford reactor. Fish exhibited a distinctive pattern of uptake, with phosphorus-32 and zinc-65 being highly concentrated by fish in muscle tissue. However, the concentration factor was very variable with time and was influenced by a variety of environmental conditions.

Emery and Clopfer (1976) found that algae and macrophytes in a freshwater system were the main biological concentrators of plutonium and americium. However, no living components of the ecological system had higher concentrations than the sediments. Somewhat similar results were obtained by Miettinen (1976) in studies of marine food chains. In this case a reduction in plutonium content was found with increasing trophic level.

PHYSIOLOGICAL EFFECTS OF RADIATION

Nuclear radiation loses energy by passing through matter. Charged nuclear particles displace electrons from atoms by attraction or repulsion depending on the charge.

A.   Interactions of radiation with water

$$H_2O + radiation \longrightarrow H_2O^+ \text{ or } H_2O^-$$

$$H_2O^+ \longrightarrow H^+ + .OH$$

$$H_2O^- \longrightarrow {}^-OH + .H$$

$$OH + OH \longrightarrow H_2O_2$$

$$H + O_2 \longrightarrow HO_2$$

$$HO_2 + e \longrightarrow HO_2^-$$

B.   Interaction of water products with chemical groups in living organisms

$$RH + HO_2^- \longrightarrow R\cdot + H_2O_2$$

$$RH + HO_2^- \longrightarrow RO + H_2O$$

C.   Interaction of neutrons with biologically active elements

$${}^{31}P + n \longrightarrow {}^{28}Al + \alpha + \gamma$$

$${}^{28}Al \longrightarrow {}^{28}Si + \beta$$

Figure 13.3.   Some possible interactions of nuclear radiation with substances in living tissue. [Compiled from Grosch and Hopwood (1979) and Ottaway (1980).]

Also electromagnetic radiation can excite electrons within an atom, resulting in their loss from the atom. Thus the effect of both particulate and electromagnetic radiation is the production of positively charged atoms and free electrons in receiving materials. At high energy, nuclei may absorb γ-radiation to give neutron emission. Neutrons lose energy by collision with the nucleus and this may produce ionization or create an unstable nucleus (see Figure 13.3).

The ionization of atoms and molecules by nuclear irradiation is not selective and any atom or molecule can be affected. However, the major constituent of living matter is water and it is most probably water molecules that interact directly with irradiation. A possible set of reactions are outlined in Figure 13.3. In this set of reactions powerful oxidizing agents are produced, which are capable of significant interactions with sensitive biological tissue. The form of interaction with biological tissue is shown in Figure 13.3B which shows the formation of free radicals or the oxidation of molecules in a living system. In addition, there is the possible interaction of these agents with double bonds, hydrogen bonds, and sulfhydryl groups. Also, neutrons can transmute elements as shown in Figure 13.3C. Such a transformation of phosphorus would be damaging to nucleic acids.

The effects outlined above would be expected to be damaging to biological tissue. Interaction and chemical changes may occur with such substances as enzymes and nucleic acids which may result in death or mutagenic effects.

α- and β-radiation is of limited range (see Table 13.1). Thus this type of radiation would be expected to have its greatest impact when radionuclides are absorbed into living organisms. γ-radiation and neutrons have a greater range and can be expected to be damaging to living organisms in the external environment.

Table 13.6.   Ranges of Acute Lethal Radiation Doses for
Adults of Various Groups

Biota	Radiation (krad)	
Bacteria	4.5–735	(LD_{90})
Blue-green algae	< 400 to > 1200	(LD_{90})
Other algae	3–120	(LD_{50})
Protozoa	?–600	(LD_{50})
Molluscs	20–109	$(LD_{50/30})$
Crustaceans	1.5– 56.6	$(LD_{50/30})$
Fish	1.1– 5.6	$(LD_{50/30})$

Source: Ophel (1976). Reprinted with permission from
the International Atomic Energy Agency.

Ophel (1976) has provided information on the lethal effects of radiation on groups of organisms (see Table 13.6). The LD_{50} is expressed, in a number of cases, as lethality for a 30-day period. Different LD_{50} values are reported for different stages of the life cycle with some fish species. With lower levels of chronic radiation, exposure leads to a competition between injury and repair of the radiation damage. Eggs of freshwater and marine fish species, when exposed to radiation in excess of 10^{-7} Ci L^{-1} of strontium-90, showed reduced numbers on hatching, early mortalities, and increased abnormalities in the progeny (Ophel, 1976).

Exposure can be increased by the preferential absorption of radionuclides from water by organisms. Some radionuclides, for example, phosphorus-32, are incorporated into molecules of DNA which carry genetic information. This can result in cell death or modification of the genetic information leading to mutations.

EFFECTS OF RADIONUCLIDES ON COMMUNITIES, POPULATIONS, AND ECOSYSTEMS

In commercial fisheries it has been suggested that the sensitivity of fish eggs to low levels of irradiation may be a significant problem (see Ophel, 1976). Species with high fecundity may exhibit little effect since other factors may have a greater influence on survival, for example, availability of food. However, species with low fecundity could be affected. In fact, high survival rates have been reported for some species receiving low levels of irradiation. For example, plaice numbers landed from the Irish Sea have increased with discharges of radionuclides from the Windscale plant (Ophel, 1976).

It has also been suggested that fish and other populations may be affected by the mutation rate leading to abnormal organisms. It has been shown that two factors have a major influence on the mutation rate: the amount of radiation received by the organism and the amount of DNA per haploid genome (see Figure 13.4). This effect explains the different mutation rates among different species in a population.

Table 13.7. Radiosensitivity of Plant Components of Terrestrial Ecosystems after 3 Months

Northern Forest of Enterprise, Wisconsin	Exposure (in kR)[a] to Produce Depression of Biomass Production (%)			Other Studies	Exposure (in kR) to Produce Growth Inhibition and Lethality		
	10–25	25–50	35–60		Minor	Intermediate	Severe
Deciduous trees	5–10	10–35	35–60	Coniferous trees	0.1–1	1–2	2–10
				Deciduous trees	1–5	5–10	10–15
				Tropical rain forest	4–8	8–10	10–15
Shrubs	3–5	5–20	20–60	Shrubs	1–5	5–20	20–60
Herbaceous							
Under forest	20–40	40–60	60–160	Herbaceous rock outcrop	8–10	10–40	40
Logging road	40–60	60–160	160–200	Old field annuals	3–10	10–100	100
				Grassland	8–10	10–100	100–200
Lichen	60–100	100–200	200 +	Moss and lichen	10–50	50–500	500–?

Source: Grosch and Hopwood (1979). Reprinted with permission from Academic Press, Inc. Based on data in federal documents on radiation in forests in Enterprise, Wisconsin, Brookhaven, Long Island, and El Verde, Puerto Rico.
[a] Kiloroentgen (kR).

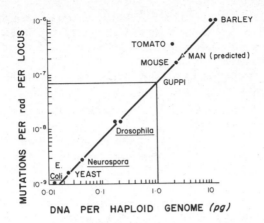

Figure 13.4.   Relation between mutation rate per locus per rad and the DNA content per haploid genome. [From Abrahamson et al. (1973). Reprinted with permission from MacMillan Journals, Ltd.]

The effects of direct exposure on plant communities have been studied in a number of cases (see Grosch and Hopwood, 1979). These investigations have been carried out on a variety of temperate and tropical forest systems. An overall assessment of the results obtained is shown in Table 13.7. It indicates a range of sensitivity from coniferous trees through to moss and lichen with the lowest sensitivity. Changes in the community structure and diversity of forest ecosystems can be related to this general pattern. Primary productivity is not altered in a simple manner. Poor production performance by radiosensitive species can be offset by growth of more-resistant types (Grosch and Hopwood, 1979).

Terrestrial ecosystems have been the subject of little ecological study. But it has been noted that in some cases invasions of insects accompany lethal irradiation of trees resulting in abnormal amounts of dead vegetation. The lizard, *Crotophytus* species, in Nevada exposed to irradiation from cesium-137 was found to exhibit a lack of breeding success 5 yr after the irradiation incident.

REFERENCES

Abrahamson, S., Bender, M. A., Conger, A. D., and Wolff, S. (1973). Uniformity of radiation-induced mutation rates among different species. *Nature* **245**, 460.

Booth, R. S. (1976). "A Systems Analysis Model for Calculating Radionuclide Transport between Receiving Waters and Bottom Sediments." In M. W. Miller and J. N. Stannard (Eds.), *Environmental Toxicity of Aquatic Radionuclides: Models and Mechanisms*. Ann Arbor Science, Ann Arbor, Michigan, p. 133.

Emery, R. N. and Clopfer, D. C. (1976). "The Distribution of Transuranic Elements in a Freshwater Pond Ecosystem." In M. W. Miller and J. N. Stannard (Eds.), *Environmental Toxicity of Aquatic Radionuclides: Models and Mechanisms*. Ann Arbor Science, Ann Arbor, Michigan, p. 269.

Foster, R. F. and McConnon, D. (1972). "Relationship Between the Concentration of Radionuclides in Columbia River Water and Fish." In *Biological Problems in Water Pollution*, Third Seminar. U.S. Department of Health Education and Welfare, Washington, D.C.

Fry, R. M. (1975). Radiation hazards in uranium mining and milling. *At. Energy Australia* **18**, 1.

Grosch, D. S. and Hopwood, L. E. (1979). *Biological Effects of Radiations*, 2nd ed. Academic Press, New York.

Haury, G. and Schikarski, W. (1977). "Radioactive Inputs into the Environment: Comparisons of Natural and Man-made Inventories." In W. Stumm (Ed.), *Global Chemical Cycles and Their Alterations by Man*. Dahlem Konferenzen, Berlin, p. 165.

Hetherington, J. A. (1976). "The Behavior of Plutonium Nuclides in the Irish Sea." In M. W. Miller and J. N. Stannard (Eds.), *Environmental Toxicity of Aquatic Radionuclides: Models and Mechanisms*. Ann Arbor Science, Ann Arbor, Michigan, p. 81.

Miettinen, J. K. (1976). "Plutonium Foodchains." In M. W. Miller and J. N. Stannard (Eds.), *Environmental Toxicity of Aquatic Radionuclides: Models and Mechanisms*. Ann Arbor Science, Ann Arbor, Michigan, p. 29.

Ophel, I. L. (1976). "Effects of Ionising Radiation on Aquatic Organisms." In *Effects of Ionising Radiation on Aquatic Organisms and Ecosystems*, Technical Report Series No. 172. International Atomic Energy Agency, Vienna.

Ottaway, J. H. (1980). *The Biochemistry of Pollution. Studies in Biology*, No. 123. Edward Arnold, London.

Schultz, V. and Whicker, F. W. (1980). Nuclear fuel cycle, ionizing radiation, and effects on biota of the natural environment. *Crit. Rev. Environ. Control* **10**, 225.

14

SUSPENDED SOLIDS AND SILT

Particulate matter is a natural component of aquatic areas and the atmosphere. Aspects of particulates in the atmosphere are considered in Chapter 11 while this chapter is concerned with suspended solids and silt in natural waters.

Silt is a natural component of streams, estuaries, and the oceans. Concentrations vary from thousands of parts per million to very low concentrations, depending on conditions such as rainfall, catchment characteristics, soil types, and bottom sediments. Suspended solids can originate from runoff water or be introduced into the water mass by resuspension of bottom sediments. Particulate pollutants can be derived from activities such as modifications to catchment hydraulics, disposal of dredging spoil, and industrial wastes.

COMPOSITION AND OCCURRENCE

Pollutant particulates vary widely in composition from organic matter derived from plant and animal wastes to inorganic substances such as sand, silt, and clay derived from industrial operations (see Table 14.1).

Table 14.1.   Origin and Composition of Some Common Pollutant Particulates[a]

Origin	Composition and Characteristics
Soil erosion	Soil minerals, sand, clay, silt
Dredging spoil dumping	Sediment minerals, sand, clay, silt, organic detritus
Red mud from alumina manufacture	Iron salts converted to hydrated iron oxides in seawater
Mineral sands refining	As above
Solids from coal washeries	Coal minerals
Clay pits, gravel washing plants, and similar operations	Clay particles

[a] Compiled from Johnson (1976).

Oceanic dredging spoil dumping is carried out in many areas. The New York Bight, for example, receives spoil from the dredging of the Hudson Raritan estuary which is rich in organic matter and industrial pollutants as well as particulates. O'Neal and Sceva (1971) have suggested the following basic characteristics should be measured as a guide to the pollution potential of sediment: total volatile solids, chemical oxygen demand, total Kjeldahl nitrogen, oil and grease, mercury, lead, and zinc.

TRANSFORMATIONS AND ABIOTIC EFFECTS OF SUSPENDED MATTER

The penetration of sunlight in natural waters is substantially decreased by the presence of suspended particulates according to the following relationship:

$$I = I_0 e^{-kCL}$$

where I_0 is the intensity just below the water surface, k the extinction coefficient for the suspended solids, C the concentration of suspended solids, and L the light path length.

Thus sunlight is increasingly absorbed in the surface layers with increasing suspended solids content. By this process, turbid waters may develop thermal stratification more readily than clearer waters. The isolation of bottom waters, in this way, may result in oxygen depletion by respiration processes as described in Chapters 5 and 6.

Suspended matter settles out of the water mass and joins the bottom sediments at a rate dependent on such factors as water turbulence and sediment density. Resuspension may occur through an increase in water turbulence. Suspended particulates have high surface areas, dependent on particle size, and exhibit a variety of surface effects. Cations, anions, and organic compounds may be surface adsorbed onto particulates and strongly held. Pesticides, PCBs, and heavy metals exhibit this property and so their biological effects are modified. Turbid waters thus have a comparatively large capacity to assimilate and deactivate toxic substances.

DELETERIOUS EFFECTS OF SUSPENDED MATTER ON INDIVIDUALS

The effects of suspended solids vary depending on the chemical nature of the suspended matter, particularly the presence of toxic substances. For substances without significant toxic components, such as clays, the settling out of suspended matter and blanketing of benthic plants and invertebrates result in high mortalities (USEPA, 1972). Plants suffer abrasion and mechanical damage (see Table 14.2), smaller invertebrates are smothered, and large invertebrates have gills and ocular and other surfaces clogged. The harmful effects with fish, zooplankton, and other organisms are principally due to clogging of the gills with particles resulting in asphyxiation. Concentrations causing these effects are shown in Tables 14.2 and 14.3.

Table 14.2. Some Observed Effects of Suspended Matter on Aquatic Plants and Invertebrates[a]

Organism	Type of Suspended Matter	Concentration (mg L^{-1})	Time (days)	Effects
Aquatic moss (*Eurhynchium ripariodus*)	Coal dust	{500 {100	{7 {21	Mechanic damage and abrasion
Cladocera and copepoda	Clay	300–500	–	Harmful
Daphnia magna	Kaolinite Montmorillonite Charcoal	392 102 80	–	Harmful

[a] Compiled from Alabaster and Lloyd (1980).

Table 14.3. Some Observed Effects of Suspended Matter on Fish[a]

Organism	Type of Suspended Matter	Concentration (mg L^{-1})	Time (days)	Effects
Goldfish and common carp	Montmorillonite clay	100,000	7	None observed
Rainbow trout	Silt from gravel washing	{160,000 80,000	1 1	Lethal None observed
Harlequin fish (*Rasbora heteromorpha*)	Bentonite clay	40,000	1	Lethal
Rainbow trout	Coal washings	200	280	None observed
Salmon	Suspended solids in streams	137–395	–	Fish select cleaner waters
Cutthroat trout	River turbidity	35	–	Fish sought cover and stopped feeding

[a] Compiled from Alabaster and Lloyd (1980).

404

Eggs from aquatic organisms deposited on the bottom sediments suffer high mortalities from settling of suspended particles. Also, suspended particulates cause mortalities to nonbenthic eggs by adsorption onto the egg surface. Both effects result in a reduction of water flow and dissolved oxygen to the egg (Alabaster and Lloyd, 1980).

Effects on fish behavior have also been noted. The most common is avoidance of turbid waters but inhibition of feeding and an increase in seeking shelter have also been noted (Alabaster and Lloyd, 1980). Turbidity reduces activity and may interfere with migratory paths.

ECOLOGICAL EFFECTS

The energy input to aquatic ecosystems through aquatic plants is restricted due to limited light penetration and other deleterious effects. Thus biomass and number of organisms in the system would be correspondingly reduced. The pattern found in streams receiving mostly suspended solid discharges has been a general reduction in number of species and number of organisms with a recovery downstream in relation to the loss of suspended solid load (Alabaster and Lloyd, 1980).

If the suspended material is rich in organic matter a combination of the effects described above with those due to deoxygenation (see Chapter 5) would be expected. Similarly if the suspended material contained toxic substances a combination of effects would occur.

REFERENCES

Alabaster, J. S. and Lloyd, R. (1980). "Finely Divided Solids." In *Water Quality Criteria for Freshwater Fish*. Butterworths and Food and Agriculture Organisation of the United Nations, London, pp. 1–20.

Johnson, R. (1976). "Mechanisms and Problems of Marine Pollution in Relation to Commercial Fisheries." In R. Johnson (Ed.), *Marine Pollution*. Academic Press, London, pp. 81–101.

O'Neal, G. and Sceva, J. (1971). The Effects of Dredging on Water Quality in the Northwest, July 1971. EPA Office of Water Programs, Region X, Seattle, Washington.

USEPA (1972). Water Quality Criteria, 1972. Environmental Protection Agency, Publication No. EPA-R3-73-033, Washington, D.C.

PART THREE

APPLICATIONS IN ENVIRONMENTAL MANAGEMENT

15

POLLUTION MONITORING

OBJECTIVES OF MONITORING

Previous sections of this book have been concerned with the ecological relationships between pollutants and the environment. This chapter is concerned with environmental management, particularly the management of environmental pollution. The occurrence of pollution implies a detrimental effect of some kind and environmental management is concerned with the control and management of detrimental effects. The effective management of pollution requires knowledge of the detrimental effects that occur in the natural environment and the pollution factors causing these effects. Holdgate (1979) has suggested that a knowledge of the following factors is needed for an effective monitoring program:

1. The pollutants entering the environment, and their quantities, sources, and distribution.
2. The effects of those substances.
3. Trends in concentration and effect and the causes of these changes.
4. How far these inputs, concentrations, effects, and trends can be modified, and by what means and at what cost.

These form part of a wider environmental management program, not only involving monitoring and surveillance but assessment of risk and the need for any action to be taken. In addition, there are legal, administrative, social, economic, engineering and other aspects to be taken into account in the total program. If a discharge is made to the environment but no detrimental effects are observed, this would not be considered pollution and would not require any remedial action.

Thus, in this aspect of environmental management, environmental quality is of primary concern. The evaluation of quality requires that a subjective judgment of some aspect of the characteristics of the environment be made. This judgment usually involves the use to which the particular segment of the involved environment is put. The uses of different sectors of the environment are manifold. For example, water can be used for domestic consumption, industrial purposes, stock consumption, argicultural use, maintenance of fisheries, recreation, aesthetic appreciation, and conservation of natural ecosystems. The water quality requirements for these different uses are quite different and the judgment of environmental quality will vary accordingly.

In the design of a monitoring program, the environmental quality aspects of interest need to be clearly defined. For example, if domestic water quality was required, aspects of interest in monitoring would include the occurrence of harmful microorganisms. On the other hand, if conservation of aquatic ecosystems is the requirement it could be any environmental pollution factor which causes a detrimental change in the system.

MONITORING STRATEGIES

Within the environment itself, the objectives of environmental management, in relationship to monitoring, can be met by focusing on two aspects:

1. Monitoring the pollutant in different parts of the environment, generally described as *factor monitoring*.
2. Monitoring the effects of the pollutant on the natural ecosystem and associated biota, generally described as *target monitoring*.

Monitoring the pollutant usually involves chemical or physical measurements in various locations or situations, for example, production processes, emissions to the environment, occurrence within the environment, at the surface of a target, within an organism, and so on (e.g., Goldberg, 1976). By measurements of this kind an assessment of the exposure of different biota in the environment to the pollutant can be developed. The extent of detrimental effects can be estimated from a comparison of the occurrence of the pollutant with standards which are set for environmental exposure.

There are many successful examples of the application of physicochemical monitoring methods, as described above, in the assessment and management of pollution, for example, water quality management for domestic and industrial purposes, stock consumption, and agricultural use. This is due to the fact that in these cases, there is a limited number of key parameters to be considered (e.g., salinity, fecal coliforms, nitrate and metal concentration), and effects on only a limited range of organisms (e.g., humans, domestic animals, horticultural crops) are considered important. In addition, these organisms are relatively common throughout the world and, because of their importance, have been intensively studied.

As a general rule physicochemical criteria have probably been used with the least success in the long-term management and conservation of natural ecosystems. Here the measurement of changes in the physical and chemical characteristics arising from environmental pollution of an area is of little practical value if the ecological impact of the changes is unknown, particularly over extended periods of time. Many of the environmental quality criteria and standards have been developed from data on the short-term environmental requirements and toxic responses of a fairly limited range of biota. The advantages of physicochemical and related microbiological techniques can be summarized as follows:

1. Rapid assessments can be made.
2. The quantitative nature of these assessments enables comparison with set standards and a rapid evaluation of the pollution level in an area.
3. The physicochemical evaluations usually reveal the nature of the pollutant.
4. The distribution of the pollutant can be related to introduced control measures.

The disadvantages of these methods can be summarized as follows:

1. There is a lack of accurate data on the response of organisms to pollution factors for many areas.
2. The data available on organism response are related to short exposure times and very little data related to continuous long-term exposure of organisms are available.
3. Very limited data are available on sublethal effects such as effects on reproduction and so on.
4. Very limited data are available on the combined effects of pollution mixtures.
5. Often there is limited information available on the chemical form of the toxic agents present. This applies particularly with metals.

The other aspect of monitoring mentioned above is concerned with the response of the natural environment to the pollutant. Table 15.1 shows the targets that may be monitored and the measurements that can be made on them. The biological systems shown in Table 15.1 are of prime management concern since these are the sectors of the environment which are usually subject to detrimental impact. However, the relationships between the occurrence of an environmental pollutant and a change in the biota are often not readily established. The natural environment and biota may be affected by a wide range of natural factors as well as pollution. There are a set of advantages and disadvantages associated with evaluations of biological systems as a monitoring strategy. Biological evaluations can provide the following:

1. An evaluation of the biological consequences of pollution which are the area of principal concern in environmental management.
2. A time-averaged measure of pollution effects.
3. An overall evaluation of the effects of pollutant mixtures.
4. In some circumstances, the technique is a sensitive indicator of pollution.

However, the disadvantages associated with these evaluations are set out below:

1. They can be expensive and time consuming.
2. There is a lack of quantitative criteria.
3. Variations in the environmental conditions, such as season, substrate type, water depth, shading, and so on, may cause large natural variations in communities and populations of biota, irrespective of pollution.

Table 15.1.   Target monitoring

Target	Measurement
Physical Systems	
Atmospheric systems	Circulation, composition
Estuarine and marine systems	Circulation, composition
Freshwater systems	Circulation, composition
Land systems	Pattern, composition
Structures and materials	Corrosion, wear, soiling
Biological Systems	
Ecosystems	Distribution, composition, overall performance (major fluxes)
Species and populations	
Humans	Distribution, performance (includes epidemiological data)
Crops (including garden and forest crops) and livestock	Performance and distribution
Wildlife	Performance and distribution
"Indicator species"	Performance and distribution
Individuals of target or indicator species	Physiological and biochemical parameters, indicators of performance, residues as aids in analysis

Source: Holdgate (1979). Reprinted with permission from Cambridge University Press.

4.   There is often a lack of taxonomic knowledge available to classify the organisms involved to the degree required.

5.   They do not indicate the nature of the substances causing the pollution and so possible sources cannot be identified.

6.   Their use in assessing water quality for domestic and industrial use as well as stock consumption and agriculture is limited.

It can be seen that to a very large extent the physicochemical and biological evaluations are complimentary. Both methods form an important basis for setting criteria and water quality evaluation. In any investigation the techniques that should be used are those which are most appropriate to the resources and objectives of the study. Cairns and co-workers have reviewed the various applications of biological monitoring in water quality management (see Cairns, 1981).

STANDARDS, CRITERIA, AND INDICES

Environmental standards and criteria give an indication of the acceptable levels of occurrence of a pollutant for the maintenance of environmental quality. In the recent past, acceptable water quality was related to use in domestic supply, industry, and agriculture. Deleterious effects were concerned with short-term toxic effects on humans and stock or the loss of agricultural production. Cause and effect relationships have been reasonably established for a variety of common physicochemical pollutants and microorganisms. Consequently, quite well established standards and criteria have been developed (USEPA, 1976).

In recent years the objectives of environmental management and of water quality management, in particular, have expanded. Water quality management is now also concerned with recreation, aesthetic appeal, and conservation of natural systems. With these new aspects of water quality it is more difficult to define criteria and standards. Subjective judgment is often involved and a wide range of organisms need to be considered rather than particular selected organisms. These latter objectives are also concerned with the long-term effects rather than uses over a limited time period.

These newer aspects of water quality management are those in which physicochemical criteria have had least success and monitoring is probably best carried out by monitoring the natural ecosystem, in other words, target monitoring. A variety of indices can be used to characterize biological changes in natural systems and several of these have been described in Chapter 4. Although this form of environmental monitoring has the disadvantages outlined previously, nevertheless, these procedures offer great potential for monitoring long-term "health" of the natural environment.

Both factor and target monitoring generate enormous amounts of physical, chemical, and biological data. In fact, it is difficult to assimilate, interpret, and understand the meaning of much of this information. But there is a need to understand the data in order to effectively institute environmental management. In many cases, in order to evaluate these data, they have been compressed into simplified measurements which can be used for management purposes. These measures are usually referred to as *indices* which can be calculated by combining a variety of environmental measurements in a number of alternative ways.

Table 15.2 contains a summary of water quality indices (Ott, 1978). Some of these indices have been developed for specific purposes by combining the water quality factors in such a way that the most important aspects are emphasized. Other indices are of a general nature and provide an overall assessment of water quality. In other cases, indices have been developed for use in specific areas, and certain critical water quality factors are weighted to take into account the particular problems that exist in the area. Table 15.3 contains a set of indices used with air quality evaluation. A nationally uniform pollutant standards index (PSI) has been developed for the United States (Ott, 1978).

MONITORING METHODS

In previous chapters, various physicochemical factors and associated biological and ecological effects were described. Many of these have been used in environmental

Table 15.2.   Summary of Water Quality Indices

Index Name[a]	No. of Variables	Scale	Range
Planning Indices			
Prevalence Duration Intensity (PDI) Index	b	Increasing	0 to 1
National Planning Priorities Index (NPPI)	b	Increasing	0 to 1
Priority Action Index (PAI)	b	Increasing	0 to 1
Environmental Evaluation System (EES)	78[c]	Decreasing	0 to 1000
Canadian National Index	b	Increasing	0 to 1
Potential Pollution Index (PPI)	3	Increasing	0 to 1000 +
Pollution Index (PI)	b	Increasing	0 to 100 +
Statistical Approaches			
Composite Pollution Index (CPI)	18	Increasing	− 2 to 2
Index of Partial Nutrients	5	Decreasing	0 to 100
Index of Total Nutrients	5	Decreasing	0 to 100
Principal Component Analysis	b	N.A.[d]	N.A.
Harkins' Index (Kendall ranking)	b	Increasing	0 to 100 +
Beta Function Index	b	Increasing	0 to 1
General Water Quality Indices			
Quality Index (QI)	10	Decreasing	0 to 100
Water Quality Index (NSF WQI)	9	Decreasing	0 to 100
Implicit Index of Pollution	13	Increasing	0 to 15 +
River Pollution Index (RPI)	8	Increasing	0 to 1000 +
Social Accounting System	11	Decreasing	0 to 100
Specific-Use Water Quality Indices			
Fish and Wildlife (FAWL) Index	9	Decreasing	0 to 100
Public Water Supply (PWS) Index	13	Decreasing	0 to 100
Index for Public Water Supply	11/13	Decreasing	0 to 100
Index for Recreation	12	Decreasing	0 to 1
Index for Dual Water Uses	31	Decreasing	− 100 to 100[e]
Index for Three Water Uses	14	Increasing	0 to 1 +

Source: Ott (1978). Reprinted with permission from Ann Arbor Science Publishers, Inc.

[a] When the proper name for an index is unavailable, the index characteristic is listed.

[b] Any number of variables can be included.

[c] Water quality variables account for 14 of the 78 variables used in this system.

[d] N.A. = not applicable.

[e] Index can be less than − 100 and can become a large negative number.

monitoring. In the aquatic environment, common physicochemical factors used in monitoring are temperature, salinity, turbidity, chlorophyll concentration, dissolved oxygen concentration, nutrient concentrations, and selected toxicant concentrations. These have been measured in a wide range of biotic and abiotic components of the environment. Many different biological monitoring methods have been used, including diversity and other biotic indices, primary productivity, and biomass. Bioassay methods involving evaluation of the response of organisms to environmental pollutants or samples are also used. Several compilations of

Table 15.3.   Mathematical Characteristics of Various Air Pollution Indices

Index	Pollutant Variable	Subindices	Aggregation Function
Green's Index	COH, SO_2	Power	Arithmetic mean (weighted sum)
CPI	Fuel burned, ventilation	Estimated	Ratio
MURC	COH	Power	None
AQI	CO, TSP, SO_2	Estimated	Linear sum
Ontario API	COH, SO_2	Linear	Nonlinear sum
PINDEX	CO, NO_2, HC,[a] TSP, SO_2, solar radiation	Linear	Linear sum
ORAQI	CO, NO_2, OX,[b] TSP, SO_2	Linear	Nonlinear sum
MAQI	CO, NO_2, OX, TSP, SO_2	Linear	Root sum square
EVI	CO, OX, TSP, SO_2	Linear	Root sum square
STARAQS	CO, NO_2, OX, TSP, COH, SO_2	Linear	Maximum operator
EQI (air)	CO, NO_2, OX, TSP, COH, SO_2, visibility, population, emissions	Linear	Root mean square

Source: Ott (1978). Reprinted with permission from Ann Arbor Science Publishers, Inc.

[a] HC = hydrocarbons.
[b] OX = photochemical oxidants.

monitoring methods and details of the procedures for their use are available (e.g., Greenberg et al., 1980).

Similarly, with the atmospheric environment a wide variety of physicochemical factors are often monitored, including sulfur dioxide, oxidants, and so on. Manning and Feder (1980) have provided a detailed description of the use of plants to evaluate the occurrence of chemical pollutants, such as oxidants, sulfur dioxide, hydrogen fluoride, heavy metals, ethylene, and dust. The pollutant can be measured directly from the amount accumulated in the plant or evaluated from a physiological or biological response, for example, injury. Community structure is also used for monitoring (see Feder, 1978; Manning and Feder, 1980).

Increasing use is being made of suitable individuals or natural ecosystems, developed in an unpolluted situation and shifted to a test location where changes can be used as a measure of ambient quality. This is perhaps more commonly used with plants and atmospheric pollution. However, Burks and Willhm (1977) describe a method for using samples, or an artificial substrate, colonized in a natural ecosystem and moved to a waste-water situation. Diversity indices, for example, can then be applied to changes in the test ecosystem to assess the quality of the waste water. Artificial substrates can also be valuable in monitoring natural ecosystems at different locations. The insertion of artificial substrates eliminates variability due to substrate characteristics at different locations. Somewhat similarly, caged fish can be placed in different situations in aquatic areas and used to assess the quality of water (e.g., Price, 1978).

More sophisticated monitoring systems are being developed in which the response of caged fish are monitored electronically. This system is usually used in a laboratory situation to monitor the quality of discharges from industry or other sources. The fish species involved are usually those which occur in the local environment (see Cairns and van der Schalie, 1980; van der Schalie et al., 1979; Price, 1978; Morgan, 1977).

A variant of this system of using fish to monitor water quality in a laboratory involves the removal of organisms adapted to the pollution situation at different sites and measurement of their physiological responses. These responses reflect the pollution conditions which exist in the environment. Widdows et al. (1981) have applied this system to mussels and measured factors such as scope for growth, oxygen to nitrogen ratio, and growth efficiency which have allowed the calculation of physiological conditions related to the pollution conditions in the environment.

REFERENCES

Burks, S. L. and Willhm, J. L. (1977). "Bioassays with a Natural Assemblage of Benthic Macroinvertebrates." In F. L. Mayer and J. L. Hamelink (Ed.), *Aquatic Toxicology and Hazard Evaluation*, ASTM Special Technical Publication 634. American Society for Testing and Materials, Philadelphia, pp. 127–136.

Cairns, Jr., J. and van der Schalie, W. H. (1980). Biological monitoring. Part I. Early warning systems. *Water Res.* **14**. 1179.

Cairns, J. (1981). Biological monitoring. Part VI. Future needs. *Water Res.* **15**, 941.

Feder, W. A. (1978). Plants as bioassay systems for monitoring atmospheric pollutants. *Environ. Health Perspect.* **27**, 139.

Goldberg, E. D. (1976). *Strategies for Marine Pollution Monitoring*. John Wiley & Sons, New York.

Greenberg, A. E., Connors, J. J., and Jenkins, D. (1980). *Standard Methods for the Examination of Water and Wastewater*, 15th ed. American Public Health Association, American Water Works Association, and Water Pollution Control Federation, Washington, D.C., p. 1134.

Holdgate, M. W. (1979). *A Perspective of Environmental Pollution*. Cambridge University Press, Cambridge.

Manning, W. J. and Feder, W. A. (1980). *Biomonitoring Air Pollutants with Plants*. Applied Science Publishers, London, p. 142.

Morgan, W. S. G. (1977). "An Electronic System to Monitor the Effects of Changes in Water Quality on Fish Opercular Rhythms." In J. Cairns, Jr., K. L. Dickson, and G. J. Westlake (Eds.), *Biological Monitoring of Water and Effluent Quality*, STP607. American Society for Testing and Materials, Philadelphia.

Ott, W. R. (1978). *Environmental Indices: Theory and Practice*. Ann Arbor Science, Ann Arbor, Michigan.

Price, D. R. H. (1978). Fish as indicators of water quality. *Water Pollut. Control* **77**, 285.

van der Schalie, W. H., Dickson, K. L., Westlake, G. F., and Cairns, Jr., J. (1979). Fish bioassay monitoring of waste effluents. *Environ. Manage.* **3**, 217.

Widdows, J., Phelps, D. K., and Galloway, W. (1981). Measurement of physiological conditions of mussels transplanted along a pollution gradient in Narraganett Bay. *Mar. Environ. Res.* **4**, 181–194.

USEPA (1976). Quality criteria for water, Report No. EPA-440/9-76-023, United States Environment Protection Agency, Washington, D.C.

16

ECOTOXICOLOGICAL ASSESSMENT OF CHEMICALS

The chemical industry has grown enormously in recent decades and has now reached the stage where it is vital to modern society since it provides petroleum fuels, antibiotics and other drugs, plastics, pesticides, food preservatives, and so on. At present there are about 4 million known chemicals with new ones being produced every day. About 100,000 are estimated to be in daily use, and of these approximately 7000 are produced commercially in comparatively large quantities (Freed, 1980). Many of these substances have little or no adverse environmental effects, However, some produce wastes during use or manufacture which may be harmful, for example, petroleum fuels, whereas others, due to their patterns of use, may cause direct adverse environmental effects.

Various chemicals have been described in previous chapters which have deleterious environmental effects. Usually, these effects have become apparent only after wide scale use. Action has been taken in most of these cases to control the offending substances. Clearly, there must be an effective testing and evaluation program to determine chemicals with potential environmental impact before use. If use is appropriate, safeguards may be necessary to minimize potential adverse effects. Thus the ecotoxicology of chemicals is becoming an important aspect of environmental management and the management of pollution. In fact, the ecotoxicological evaluation of commercial chemicals must now be seen as forming part of the feasibility evaluation of new chemicals, together with economic factors, distribution, production, and other aspects usually used by manufacturers (e.g., Kimerele et al., 1977). However, the final decision regarding the use of new chemicals must rest with government agencies responsible for environmental management.

Many national governments have taken action to control the use and dispersal of chemicals in the environment. This usually requires the preparation of specified data involving a series of laboratory and other tests. Perhaps the United States is most advanced in this national approach (see USEPA, 1979a, b). Nevertheless, this is a worldwide environmental management problem since chemicals are international commodities with manufacture being undertaken in one country to supply many others. Without coordination, a complex system of different national requirements in different countries could develop. This would lead to limitations on the effectiveness of moves to control chemicals due to the high costs which would be involved.

The OECD countries have developed a coordinated approach to the ecotoxico-logical and hazard evaluation of chemicals with standard tests and procedures being devised (OECD, 1981). However, this control system for chemicals also requires appropriate complimentary sets of legislation, administration, and monitoring to accompany the testing program in each country.

Clearly, this procedure cannot apply immediately to all chemicals in use. Such action would require immediate cessation of use of a host of essential chemicals and bring the OECD economies to a halt while testing proceeded. Thus, newly developed chemicals will be affected by these procedures. However, with those chemicals already in use a knowledge of the deleterious effects has been learned by direct environmental experience. Thus most OECD countries are providing for the reevalu-ation of chemicals in use that are suspected of being environmentally hazardous.

BASIC DATA REQUIREMENTS

The ecotoxicological assessment of chemicals requires data on the properties of the substance so that predictions can be made as to how the substance will behave in the natural environment and its possible effects on biota including humans. Some of these data are gathered as part of the normal process of developing a chemical to enable its effective commercial use or synthesis. For example, the chemical struc-ture, water solubility, boiling point, and melting point are usually derived for all chemicals to enable their effective use or manufacture. Ecotoxicological assessment requires additional data specific to the behavior of the substance in the natural or human environment. It should be noted that assessment procedures for evaluating the ecotoxicology of chemicals are in their infancy. It can be expected that these procedures will be developed and refined resulting in improvements in future years. For example, some existing procedures may be modified, or eliminated completely, and new ones introduced. Some of the basic ecotoxicological data requirements for the evaluation of chemicals are set out below.

1. *Patterns of Usage, Disposal, and Release.* To evaluate the extent of possible distribution of a chemical in the environment, data are needed on possible production quantities and location, intended use, transport methods, and final disposal of any wastes involved. From this, particular sectors of the environment which may be affected can be identified. Hazards can then be evaluated using other data set out below.

2. *Physicochemical Factors.* Some of the data needs in relation to physico-chemical factors are shown in Table 16.1 (e.g., USEPA, 1979a; OECD, 1981). The structural formula can be used to place the substance in a related class of com-pounds and so prediction of the possible environmental properties from other members in the class can be made. Purity data are required since there are many situations in which impurities in the chemical of prime use have caused the major deleterious environmental effects. The potential movement patterns of the chemical in the environment can be identified from the vapor pressure curve, boiling point,

Table 16.1.   Some Physicochemical Data Needs for New
Chemicals

Structural formula	Vapor pressure curve
Degree of purity	Water solubility
Known impurities	Adsorption/desorption
Additives	Dissociation constant
Spectra	Particle size
Melting point	Octanol/water partition coefficient
Boiling point	Complex formation ability in water
Density	Thermal stability and stability in air
Viscosity	Fat solubility
Surface tension	

melting point, water solubility, density, partition coefficient, lipid solubility, and adsorption/desorption properties. The chemical form the substance is likely to adopt in different sectors of the environment can be evaluated from its dissociation constant, complex formation, particulate size, and viscosity. Spectral data are helpful in identifying the substance in environmental samples and in evaluating the possibility of photoinduced reactions.

3.   *Degradation and Bioaccumulation.*   The persistance of a substance in the environment gives an indication of the distance over which the substance can be dispersed and how long it will persist in organisms. OECD (1981) have described a variety of tests which give a measure of the biodegradation of substances in air, water, and soil. Thermal stability and stability in air provide an indication of the stability of a compound to abiotic factors such as atmospheric oxygen and heat.

The bioaccumulation potential of a substance in organisms can be estimated from its *n*-octanol to water partition coefficient. Also, OECD (1981) describe a number of tests with fish whereby a direct measurement of the bioaccumulation of a substance by organisms can be obtained.

4.   *Biological Factors.*   A wide variety of tests of the reactions of single species to chemicals have been described (USEPA, 1979b; OECD, 1981). The reactions of organisms to the test substances range from acute toxicity to inhibition of growth. Some of these tests are indicated in Table 16.2. Fish have been the most common test organisms but there has been increasing use of other biota to give more relevant ecotoxicological data. Buikema et al. (1982) have outlined factors to be taken into account in consideration of the test species and the interpretation of results.

Table 16.2.   Some Biological Tests for New Chemicals

Algal growth inhibition
Acute toxicity to fish
Daphnia sp. reproduction test
Acute toxicity to aquatic invertebrates
Fish embryo test
Quail dietary test

Table 16.3.   OECD Minimum Premarket Set of Data (MPD)

Chemical Identification Data

Names (IUPAC and trade)
Structural formulas
Spectra
Purity
Known impurities

Production/Use/Disposal Data

Production
Manufacturer
Manufacturing process
Uses
Transportation
Disposal
Precautions and emergency measures

Analytical Methods
Physical/Chemical Data

Melting point
Boiling point
Density
Vapor pressure
Water solubility
Partition coefficient
Hydrolysis
Adsorption/desorption
Dissociation constant
Particle size

Human Toxicity Data

Mutagenicity Data

Ecotoxicity Data

Fish LC_{50}
Daphnia sp. reproduction
Algal growth inhibition

Degradation/Accumulation Data

Biodegradation
Bioaccumulation

Cairns (1981) points out that tests on single or a limited number of species provides useful information on growth rate, fecundity, lethality, and so on but usually do not take into account some important environmental variables. For example, organisms usually show different susceptibilities to toxicants at different life history stages. In addition, ecological factors such as predator to prey relationships, nutrient spiraling, mineralization rates, and so on are not accounted for by these tests. Cairns (1981) advocates the use of microcosms (see, e.g., Gillett, 1980 and Draggan and Van Voris, 1979) and other methods more closely related to natural ecosystems. Nevertheless, he points out that microcosms also have a number of disadvantages due to the simplification and minaturization of the ecosystem which make it difficult to extrapolate results to natural ecosystems.

Another method is to test the effects of pollutants on an actual water body. In between these two techniques there lies an experimental design in which medium-scale systems are set up which are larger than laboratory scale experiments but smaller than complete water enclosures. They usually consist of enclosures of various kinds positioned in natural ecosystems with facility to control many of the physicochemical and other factors. Grice and Reeve (1982) have produced a recent publication which describes in detail the different mesocosms which are used in the marine environment. These systems present difficulties in obtaining reproducible conditions for subsequent replication of tests.

5. *Human Health.*   The OECD testing program (OECD, 1981) takes into account human health factors and tests for exposure of humans to chemicals. This is a very important aspect of ecotoxicology but it is not considered here since this book is concerned with natural ecosystems.

OECD Minimum Premarket Set of Data (MPD)

The OECD has developed a set of data needs which are the minimal information needed to assess the effects of a chemical in the environment. These minimum premarket data (MPD) are set down in Table 16.3. Descriptions of the testing procedures to give the data required are set out in the guidelines for testing chemicals (OECD, 1981). The tests are flexible so that operators can omit or substitute tests if this can be justified.

DATA INTEGRATION AND HAZARD ASSESSMENT

Patterns of Dispersal and Environmental Concentrations

The assessment of hazard to the natural environment by a chemical must be preceded by an evaluation of the fate of the chemical after discharge to the environment. This evaluation should indicate the sector of the environment which will be affected, for example, bottom sediments, water column, atmosphere, and so on, and from this the organisms likely to be affected can be identified. The degradation and persistence of the substance in that sector can be evaluated from the results of standard tests.

Mackay and Patterson (1981) have developed a quantitative approach to assessing the distribution of chemicals discharged to the environment using fugacity (see Chapter 2). Fugacity is usually linearly related to concentration at low

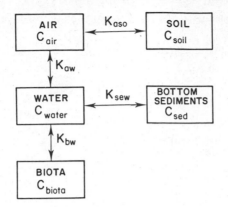

Figure 16.1. Diagrammatic illustration of the interrelationships in the environment for a chemical.

concentrations such as are encountered with environmental contaminants. Thus,

$$C = fZ$$

where C is concentration, f is fugacity, and Z is the fugacity capacity or constant. If there is equilibrium between two phases then the fugacities are also equal and

$$f_1 = f_2$$

Thus,

$$\frac{C_1}{Z_1} = \frac{C_2}{Z_2} \quad \text{or} \quad \frac{C_1}{C_2} = \frac{Z_1}{Z_2} = K_{12}$$

where K_{12} is the partition coefficient between the phases. The interrelationships which occur when a trace toxicant is discharged to the environment are shown in Figure 16.1. For the atmosphere, water, sediments, and biota, it can be shown that the fugacity constants are

$$Z_{air} = \frac{1}{RT}$$

$$Z_{water} = \frac{1}{H}$$

$$Z_{sed} = K_{sew} \frac{\delta}{H}$$

$$Z_{fish} = \frac{K_{bw}}{H}$$

where R is the universal gas constant; H Henry's Law constant; δ sediment density; T absolute temperature; K_{sew} equilibrium constant for sediment to water; K_{bw} equilibrium constant for biota to water.

At equilibrium, fugacities are equal, so

$$f_a = f_w = f_s = f_{fish}$$

Table 16.4.   Dimensions of the Environment for Calculations[a]

Phase	Dimensions
Area	1 km^2
Atmosphere	10 km height
Soil	3 cm depth × 30% area
Water	10 m depth
Bottom sediments	3 cm depth in water area
Water composition–suspended solids	5 ppm
Water composition–biota	0.5 ppm
Water and biota density	1000 kg m^3
Solid phase density	1500 kg m^3
Air density	1.19 kg m^3
Soil composition–organic carbon	2%
Sediment composition–organic carbon	4%
Temperature	$25°C$
Organic carbon sorption coefficient	0.6
Bioconcentration factor	Log BCF $= 0.85 \log P_{ow} - 0.70$

[a] Compiled from Mackay and Patterson (1981).

Thus the concentration ratios for any two compounds can be calculated from the constants listed above:

$$\frac{C_{air}}{C_{water}} = \frac{H}{RT}$$

$$\frac{C_{sed}}{C_{water}} = K_{sew}\delta$$

$$\frac{C_{biota}}{C_{water}} = K_{bw}$$

The biota/water equilibrium can be calculated from the octanol to water partition coefficient (P_{ow}). The sediment to water equilibrium constant can be calculated from MPD data and the organic carbon content of the sediment. Much of the data to make these calculations is in the MPD set or can be derived from it.

Mackay and Patterson (1981) point out that to calculate the amounts in each phase the volume of the phases and other data are also needed. Suitable data and volumes for these calculations are shown in Table 16.4. Using these data, the distributions shown in Table 16.5 have been calculated for a variety of different compounds.

While physicochemical processes affect the distribution of chemicals in the environment the potential concentrations are also influenced by the following factors: geographical patterns of use and discharge: biodegradation processes; hydrolysis; photolysis; and photooxidation. Some of the data relating to these factors is provided in the MPD set.

Table 16.5. Calculated Percentage Mass Distributions for Selected Substances

	Naphthalene	Benzene	Butyl Benzyl Phthalate	Methylene Chloride	Tetrachloro PCB	p-Cresol	DDT
Air	91.5	99.7	0.71	99.4	94.0	30.5	2.35
Soil	0.35	0.0010	15.5	0.0002	0.98	1.45	17.1
Water	6.49	0.31	11.4	0.56	0.42	61.2	0.77
Biota	0.0005	0.000002	0.013	0.000001	0.00075	0.0022	0.0093
Suspended solids	0.003	0.000008	0.12	0.000002	0.0076	0.011	0.13
Sediment	1.64	0.005	72.3	0.001	4.56	6.77	79.7
Molecular weight	128	78	276	84.9	290	108	354.5
Solubility (g m^{-3})	34.4	1780	2.9	13200	0.04	1800	0.0017
Vapor pressure (Pa)	6.56	12670	0.00114	48150	0.053	14.4	0.000025
Log K_{ow}	3.37	2.13	4.77	1.25	5.0	3.01	5.98

Source: Mackay and Patterson (1981). Reprinted with permission from the American Chemical Society.

Evaluation of Ecotoxicological Significance of Environmental Concentrations

The ecotoxicological assessment depends on two basic items of information. Firstly, the environmental concentrations of the substance which occur as a result of discharge and distribution and secondly, the toxicological properties of the chemical at that concentration and in that location. Environmental concentrations and toxicological properties should also include the concentrations of by-products or transformation products as well.

Different methods and procedures can be used to draw all the available ecotoxicological information together to arrive at an assessment. Cairns (1980) has advocated a sequential procedure outlined in Figure 16.2 and other procedures are described in Dickson et al. (1979). These evaluation procedures are now in a process of rapid development.

FUTURE ECOTOXICOLOGICAL MANAGEMENT OF CHEMICALS

The management of new chemicals with the potential to cause environmental damage is only a recent objective of environmental management. In the last 10 years there has been a rapid development of knowledge on the ecotoxicological behavior of chemicals. However, there is a great deal further to be done, particularly with regard to the effects on ecosystems. In evaluating the current information available it seems there is no existing or new chemical for which the full range of significant effects on ecosystems can be predicted at sublethal concentrations over extended periods of time in the natural environment.

Even at the completion of all testing required by present assessment programs, there is still little knowledge of the effects of a chemical at different life stages, sublethal effects over extended time periods, synergistic effects with other pollutants, effects of elevated temperatures, and so on. The sublethal interactions and effects on such factors as nutrient cycling, energy flows, and so on are almost completely unknown for all chemicals.

Thus, although environmental management organizations may be using the best procedures available to protect the environment from deleterious effects of new chemicals by evaluating the ecotoxicological data, this can only be regarded as the beginning. The types of effects mentioned previously such as long-term effects and overall impacts on ecosystems will be learned by actual use of chemicals. Of course, the more obvious problems based on previous experience and testing procedures devised should be avoided by the procedures now being implemented. But, more subtle, nevertheless significant effects, which are difficult to predict from the present testing procedures, will only become apparent when the chemical is used in the natural environment. Thus, the monitoring of chemicals and their impact on natural systems should form an important part of the environmental management program for new chemicals, in addition to the testing and evaluation procedures.

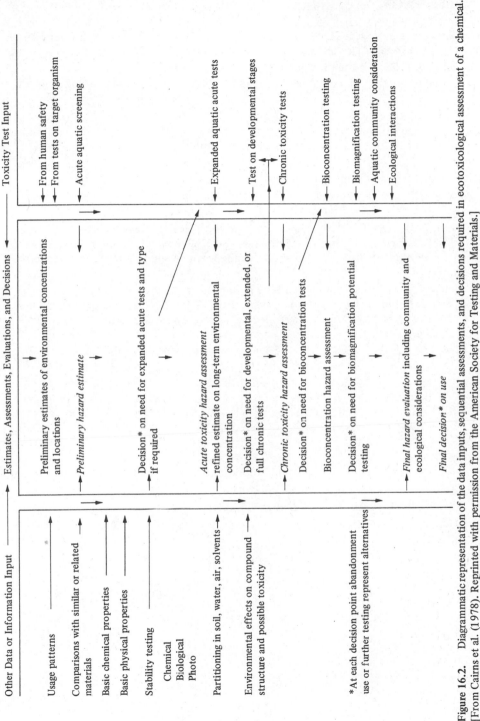

Figure 16.2. Diagrammatic representation of the data inputs, sequential assessments, and decisions required in ecotoxicological assessment of a chemical. [From Cairns et al. (1978). Reprinted with permission from the American Society for Testing and Materials.]

427

REFERENCES

Buikema, A. L., Niederlehner, B. R., and Cairns, J. (1982). Biological monitoring. Part IV. Toxicity testing. *Water Res.* **16**, 239.

Cairns, J. (1980). Estimating hazards. *Bioscience* **30**, 101.

Cairns, J. (1981). Biological monitoring. Part VI. Future needs. *Water Res.* **15**, 941.

Cairns, J., Dickson, K. L., and Mackay, A. W. (1978). Estimating the Hazard of Chemical Substances to Aquatic Life, ASTM Publ No. STP657. American Society for Testing and Materials, Philadelphia, p. 278.

Dickson, K. L., Mackie, A. W., and Cairns, J. (1979). *Analyzing the Hazard Evaluation Process*. American Fisheries Society, Bethesda.

Draggan, S. and Van Voris, P. (1979). The role of microcosms in ecological research. *Int. J. Environ. Studies* **13**, 83.

Freed, V. H. (1980). "The University Chemist – Role in Transport and Fate of Chemical Problems." In R. Haque (Ed.), *Dynamics, Exposure and Hazard Assessment of Toxic Chemicals*. Ann Arbor Science, Ann Arbor, Michigan, p. 27.

Gillett, J. W. (1980). "Terrestrial Microcosm Technology in Assessing Fate, Transport and Effects of Toxic Chemicals." In R. Haque (Ed.), *Dynamics, Exposure and Hazard Assessment of Toxic Chemicals*. Ann Arbor Science, Ann Arbor, Michigan, p. 231.

Grice, G. D. and Reeve, M. R. (Ed.) (1982). *Marine Mesocosms – Biological and Chemical Research in Experimental Ecosystems*. Springer-Verlag, New York.

Kimerele, R. A., Levinskas, G. J., Metcalf, J. S., and Scharpf, L. G. (1977). "An Industrial Approach to Evaluating Environmental Safety of New Products. In F. L. Mayer and J. L. Hamelink (Eds.), *Aquatic Toxicology and Hazard Evaluation*, ASTM STP 634. American Society for Testing and Materials, Philadelphia, p. 36.

OECD (1981). *Guidelines for Testing of Chemicals*. Organisation for Economic Cooperation and Development, Publications Office, Paris.

USEPA (1979a). *Fed. Reg.* **44**, March, p. 16240.

USEPA (1979b). *Fed. Reg.* **44**, May, p. 27334.

Mackay, D. and Patterson, S. (1981). Calculating fugacity. *Environ. Sci. Technol.* **15**, 1006.

INDEX